W9-CMO-891

A Bat Man in the Tropics

ORGANISMS AND ENVIRONMENTS

Harry W. Greene, Consulting Editor

1. The View from Bald Hill: Thirty Years in an Arizona Grassland, *by Carl E. Bock and Jane H. Bock*
2. Tupai: A Field Study of Bornean Treeshrews, *by Louise H. Emmons*
3. Singing the Turtles to Sea: The Comcáac (Seri) Art and Science of Reptiles, *by Gary Paul Nabhan*
4. Amphibians and Reptiles of Baja California, Including Its Pacific Islands and the Islands in the Sea of Cortés, *by L. Lee Grismer*
5. Lizards: Windows to the Evolution of Diversity, *by Eric R. Pianka and Laurie J. Vitt*
6. American Bison: A Natural History, *by Dale F. Lott*
7. A Bat Man in the Tropics: Chasing El Duende, *by Theodore H. Fleming*
8. Twilight of the Mammoths: The Last Entire Earth, *by Paul S. Martin*

A Bat Man in the Tropics

Chasing El Duende

THEODORE H. FLEMING

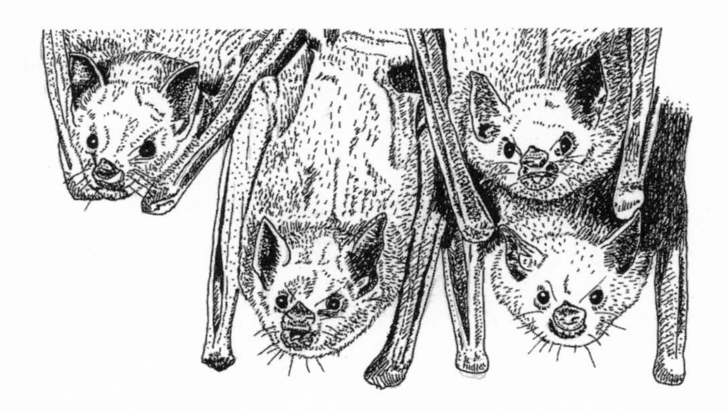

University of California Press
BERKELEY LOS ANGELES LONDON

Passage from Wallace Stegner, *Crossing to Safety*, reprinted by permission of Random House, Inc.

Frontispiece: Four lesser long-nosed bats. Redrawn by Ted Fleming, with permission, from a photo by Merlin D. Tuttle, Bat Conservation International.

University of California Press
Berkeley and Los Angeles, California

University of California Press, Ltd.
London, England

Library of Congress Cataloging-in-Publication Data

Fleming, Theodore H.
A bat man in the tropics : chasing El Duende / Theodore H. Fleming.
p. cm. (Organisms and environments ; 7)
Includes bibliographical references and index (p.).
ISBN 0-520-23606-8 (cloth : alk. paper).
1. Bats—Tropics. I. Title. II. Series.
QL737.C5 F58 2003
599.4—dc21 2002152226

Manufactured in the United States of America
11 10 09 08 07 06 05 04 03
10 9 8 7 6 5 4 3 2 1

The paper used in this publication meets the minimum requirements of ANSI/NISO Z39.48–1992 (R 1997) (*Permanence of Paper*).⊗

To my mentors—Clara Dixon, Emmett Hooper, and Charles Handley—with gratitude, and to Marcia with love

Talent, I tell him, believing what I say, is at least half luck. It isn't as if our baby lips were touched with a live coal, and thereafter we lisp in numbers or talk in tongues. We are lucky in our parents, teachers, experience, circumstances, friends, times, physical and mental endowment, or we are not.

WALLACE STEGNER

Contents

Illustrations

Foreword

A Bat Man in the Tropics: Chasing El Duende is the seventh volume in the University of California Press's series on organisms and environments. Our main themes are the diversity of plants and animals, the ways in which they interact with one another and with their surroundings, and the broader implications of those relationships for science and society. We seek books that promote unusual, even unexpected, connections among seemingly disparate topics; we want to encourage writing that is special by virtue of the unique perspectives and talents of the author.

There are more than nine hundred species of bats, formally known to taxonomists as the Chiroptera; bats constitute about 20 percent of all mammals, and as one moves from temperate regions toward the equator, much of the increase in total species of furry, milk-producing vertebrates is represented by the amazing diversity of tropical chiropterans. One recently discovered species is among the smallest endothermic, or warm-blooded, animals; in contrast, some Old World flying "foxes" achieve the size of small dogs. There are bats with enormous ears, bats with bizarre leaf-shaped noses, and bats with beautiful, white-spotted color patterns. In terms of dietary adaptations, there are bats that can snatch fish from the surface of a pond, a species that eats venomous centipedes and scorpions, one that uses frog songs to locate its preferred amphibian prey, and many kinds that feed entirely on either fruits or small insects. There are bats that build daytime roosting tents out of large understory leaves, bats that pinpoint their erratically flying prey by echolocation, and bats that migrate seasonally over substantial distances.

For at least the past several thousand years, the only mammals capable of powered flight—some of them in total darkness—have also terrified humans and inspired our art; even now, vampires and their legendary kin populate our myths and dreams. Real bats, however, are wonderfully interesting and almost never dangerous to people. As is the case with snakes, spiders, and other widely despised animals, bats need a favorable introduction before we can fully appreciate them.

As a graduate student, Ted Fleming studied tropical rodents and opossums, but soon thereafter switched to the bats that have captivated him for more than three decades. Much of his field research has focused on the ecology of the wet and dry forests of Costa Rica, but in recent years he has shifted emphasis to the essentially northward extension of tropical plants and animals into western North American desert regions. Blessed with the admirable combination of a naturalist's curiosity, a tenacious capacity for descriptive and experimental fieldwork, and a rich understanding of evolutionary and behavioral ecology, Fleming sees bats and asks, Why are they doing that? What circumstances make it possible for them to act in this manner, and what are the consequences of the biology of bats in the larger ecological communities in which they often are major players? Can we generalize about factors that influence organisms as different as nectar-feeding moths and nectar-feeding bats?

Broad conceptual issues and a number of more specific questions about bats have been at the center of Ted Fleming's distinguished career. *A Bat Man in the Tropics* is, accordingly, part scientific natural history; but it also is part personal memoir. Spanning as it does the last four decades of the twentieth century, this book is an account of some of the most important people and places significant to the rise of modern tropical biology. Ted lucidly explains the scientific issues that have inspired his labors, then takes readers through the nitty gritty details of the work: laying out and checking mist nets; measuring and banding bats; climbing ladders to examine the flowers of columnar cacti; and interacting with students, conservationists, and educators. Along the way, he candidly tells us about the shenanigans of fellow biologists, about simple but memorably delicious meals after a night of exhausting work, and about how his adventurous, resilient family accommodated the daily life of an ambitious teacher and scientist.

Books like this are at the heart of modern biology and conservation. Ted

Fleming's research clearly demonstrates that although hypothesis testing is the essence of science, discoveries of new organisms, new things about organisms, and previously unseen patterns in nature repeatedly revise the questions that fascinate biologists. By summarizing much of what we now know about a widespread, diverse, and ecologically important group of vertebrates, *A Bat Man in the Tropics* will promote still more research. In the course of providing readers with an accurate, modern view of these increasingly popular organisms, Fleming also illuminates the human side of science, the transformation of his childhood curiosity into the passion and scholarship of an accomplished professional. And perhaps most important, with its details of natural history and spectacular examples of interactions among organisms, this volume will make it possible for anyone who cares about the natural world to better understand, appreciate, and live with bats.

Harry W. Greene

Preface

My Spanish-English dictionary defines the word *duende* as "hobgoblin" or "ghost." *Duende* can also be defined as "will o' the wisp"—anything that deludes or misleads by luring on. In this book, I use the word *duende* in both senses. The first sense (hobgoblin or ghost) is meant to refer to bats, the principal subject of this book. To most people, bats are mysterious, ghost-like creatures, primarily because of their ability to navigate on the wing at night, often in total darkness. Many myths and legends have arisen about bats because of their seemingly mysterious behavior. In most of these legends, bats are frightening, loathsome creatures. One of my aims in this book is to help dispel this view by portraying bats as the fascinating and ecologically important creatures they really are.

I use *duende* in the second sense as a metaphor for the process of scientific discovery. As a professor of biology, I have been studying bats (and other organisms) in the field for over thirty years. My overall aim in many of these studies has been to document the lives and ecological importance of plant-visiting bats in tropical or subtropical habitats. These studies have taken me to tropical forests in Panama, Costa Rica, and Australia, and, more recently, to the lush Sonoran Desert of northwestern Mexico and Arizona. During the course of this work, my family, colleagues, and I have had many adventures. These adventures have arisen because science, particularly field biology, is a very human endeavor. As scientists, our immediate goal may be to answer a particular question or test a specific hypothesis, but the pathway toward that goal is seldom straight. Nature and humankind always provide distractions along the way. Some of these distractions are delightful; others are frustrating. Some distractions are totally unexpected and lead to new scientific discoveries. Unexpected discoveries can be euphoric. They provide the only reason most of us need to continue the study of nature and its secrets. As we will see, some discoveries also open up new fields of investigation, leading us into unexpected research directions (the "El Duende" effect). But the euphoria of discovery rarely appears in a scientific paper reporting the results of our studies, not even when the molecular structure of DNA was being described for the first time.

The aim of this book, then, is to provide a personal account of over thirty years of biological fieldwork, primarily with tropical bats. I will describe some of the background behind my studies—Why did I conduct this or that particular study? What was our state of knowledge when I began a scientific study?—and some of the notable results. At the same time, I wish to convey the sights, smells, and sounds of everyday life in the field. What were my joys and concerns? And how did my family and I deal with my life as a field biologist? Finally, this book is about the conservation of bats. What do we need to do to make sure these captivating animals are around for future generations of curious naturalists to study?

Acknowledgments

I owe a large debt of gratitude to many people who directly or indirectly helped me write this book. They include my mentors Clara Dixon, Emmett Hooper, and Charles Handley; my collaborators Don Wilson, Ray Heithaus, Don Thomas, Frank Bonaccorso, Merlin Tuttle, Hugh Spencer, Leo Sternberg, Jim Hamrick, John Nason, Sandrine Maurice, and Jerry Wilkinson; my graduate students Rick Williams, Larry Herbst, Renee Borges, Randy Breitwisch, Peg Horner (officially a student at Texas A&M), Vinnie Sosa, Cathy Sahley, Andy Mack, Deb Wright-Mack, Gerardo Herrera, Sophie Petit, Jafet Nassar, Nat Holland, and Lyndsay Newton; and my field assistants par excellence Juni Barrett, Orlando Barbosa, and Liz and Rich Chipman.

Many institutions provided essential support for my research. These include the Smithsonian Institution, the Smithsonian Tropical Research Institute, the Universities of Michigan, Missouri (St. Louis), and Miami, the Organization for Tropical Studies, the Costa Rican National Park Service, and Bat Conservation International. Financial support for my studies has come from the U.S. National Science Foundation, Earthwatch, Fulbright Foundation, National Geographic Society, National Fish and Wildlife Foundation, Arizona Game and Fish Department, and the Ted Turner Endangered Species Fund.

Making my family and me feel "at home" while we were away from home on research leaves or sabbaticals were the following people and institutions: Karl Johnson and the Middle America Research Unit (Panama); Ruth and Steve Stevens and Zenel and Carmen Coto (Costa Rica); Chris Perrins, Nick Davies, John Krebs, and the Edward Grey Institute (Oxford); Dick White and the Department of Botany (Duke University); Helene Marsh and the Department of Zoology (James Cook University, Australia); Judie Bronstein, Goggy Davidowitz, and the Department of Ecology and Evolutionary Biology (University of Arizona); and Gary Nabhan and the Arizona–Sonora Desert Museum.

I also wish to acknowledge the warm friendship and camaraderie of my Mexican colleagues Hector Arita, Gerardo Ceballos, Rodrigo Medellin, and Francisco Molina and their students, whose enthusiastic interest, help, and joie de vivre have made my work in Mexico truly enjoyable.

Bill Deiss kindly allowed me to work with the extant field notebooks of Edward Goldman and the correspondence of Alexander Wetmore in the Archives of the Smithsonian Institution. I thank Keir Sterling and Oliver ("Paynie") Pearson for their insights into the history of explorations by late nineteenth- and early twentieth-century mammalogists. I wish to thank Luther and Betty Goldman specially for sharing with me stories about their lives and that of Luther's father, Edward Goldman.

I thank Jim Findley, Nat Holland, Marleta Nemire-Pepe, Peter Scott, and Don Wilson for reading and commenting on the entire manuscript. Special thanks go to Christine Henry, the biological sciences editor at the University of Chicago Press, for her early interest in and encouragement of this project, and to Doris Kretschmer for shepherding my manuscript into publication.

My greatest thanks go to my wife, Marcia, who has provided unfailing support at all times, good and bad. I could not have written this book without her love, encouragement, and help, which included correcting my memory lapses and editing early drafts of all chapters. We both shared a vision of travel adventures early in our marriage but never imagined how extensive those adventures would be. She has cheerfully accompanied me from the top of Panama to the tip of Baja California; from Monteverde, Costa Rica, to El Verde, Puerto Rico; from the Fox Glacier in southern New Zealand to Lake Patzcuaro in the highlands of Mexico. She does draw the line at bat caves, however. Being claustrophobic, she cannot stand to be in dark, narrow spaces. When I slither into bat caves, I'm on my own!

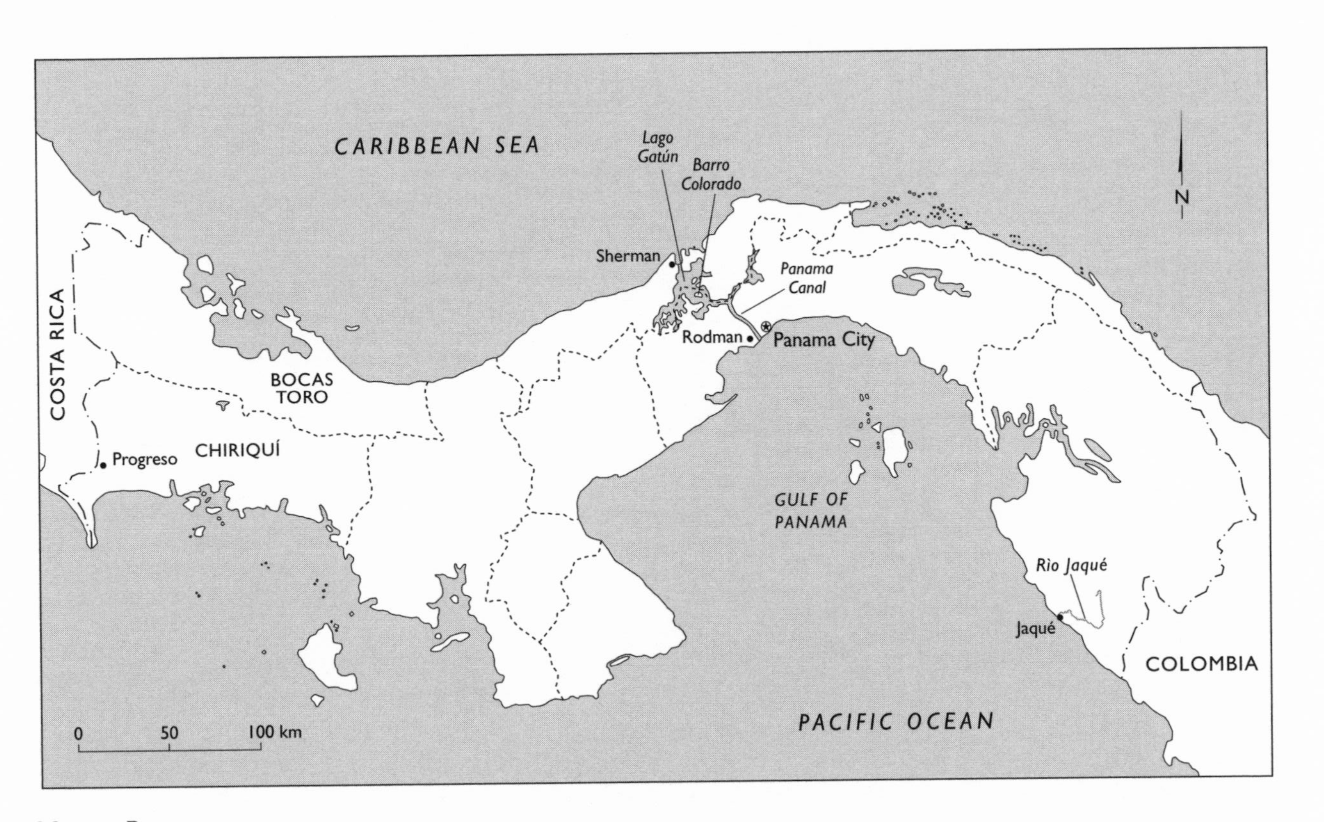
CARIBBEAN SEA
Lago Gatún
Barro Colorado
N
Sherman
Panama Canal
COSTA RICA
Rodman
Panama City
BOCAS TORO
CHIRIQUÍ
Progreso
GULF OF PANAMA
Rio Jaqué
Jaqué
COLOMBIA
PACIFIC OCEAN
0
50
100 km

Map 1. Panama

Map 2. Costa Rica

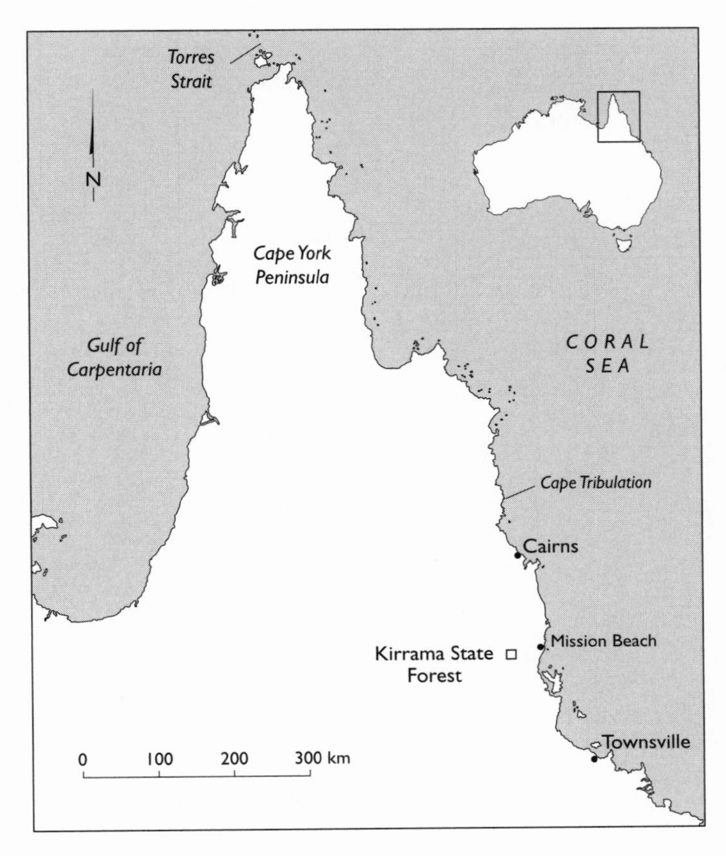

Map 3. Northeastern Australia

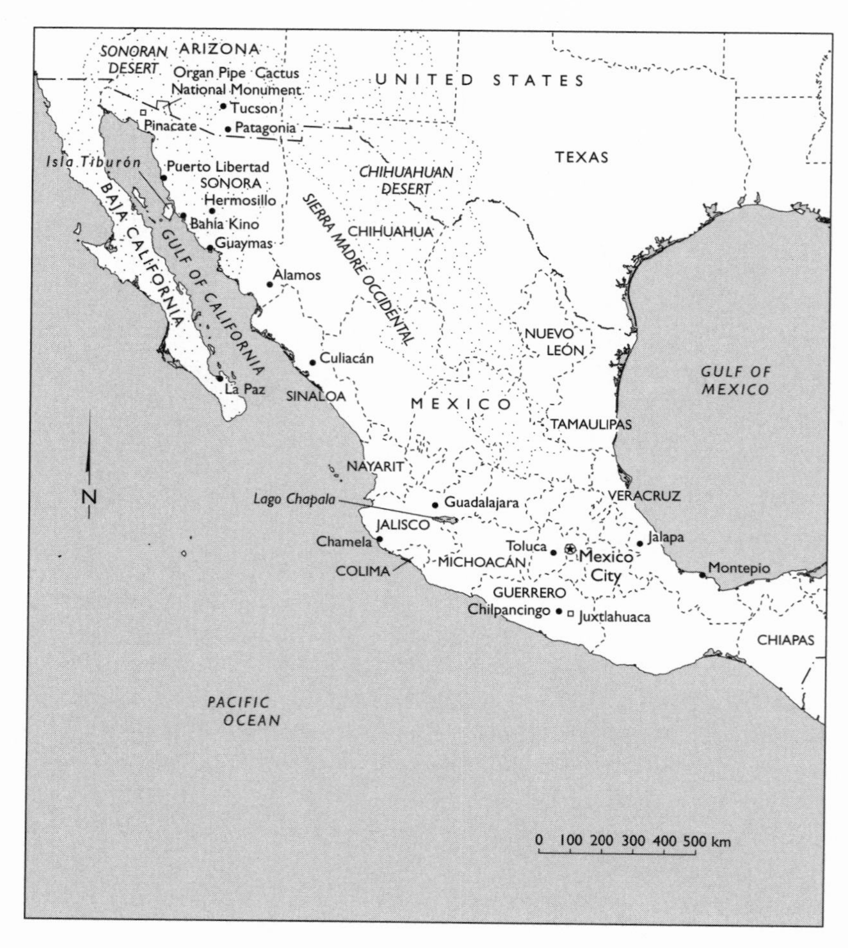

Map 4. Mexico and the Southwestern United States

1 Up a *Quebrada* without a Paddle

On a beautifully clear morning in late January 1966 I found myself seated behind the pilot in a small twin-engine plane that had just taken off from a small airfield outside Panama City, Panama. In the plane with me was Frank Greenwell, who worked for the Smithsonian Institution's Division of Mammals, and eight hundred pounds of field gear and a mountain of food. Our flight would take us an hour south along Panama's Pacific coast to the village of Jaqué, where we planned to hire Choco Indians to transport us in a boat up the Río Jaqué (see map 1). Once upriver, Frank and I would make collections of as many different kinds of mammals as possible as part of the Smithsonian's Mammals of Panama project. Directed by Charles Handley, the project aimed to produce a detailed account of the taxonomy and distribution of Panamanian mammals. Handley's volume would be an update of an account of Panamanian mammals published in 1920 by another Smithsonian mammalogist, Edward A. Goldman.

Both Goldman and Handley are scientific descendants of the irrepressible C. Hart Merriam, the founder of modern faunal studies and the head of the U.S. Biological Survey at its inception in the 1890s. Merriam and his

Baby howler monkey. Drawing by Ted Fleming.

colleagues were the first naturalist-scientists to systematically document the biological richness of much of North America, including Mexico. They set the stage for modern studies of biological diversity by describing hundreds of new species of plants and animals and placing them within a new biogeographic framework. Over the course of my research career, my path would cross with that of Edward Goldman on a number of occasions. Our paths were about to cross for the first time in Panama, a land Goldman described in his Panamanian monograph as

> [a] region of surpassing biological interest, owing to peculiar configuration, varied topography, and geographic position, forming as it does a slender artery blending the complex elements or converging life currents of two continents, and through which countless migrations of non-volant terrestrial animals probably passed during the Tertiary or early Quaternary ages. But of recent migratory movements in the region we have no evidence, and how effective a barrier the completed Panama Canal may prove to be in limiting the distribution of species remains to be determined. The country, said to have been named "Panama" from an Indian word meaning rich in fish, might with equal propriety have received an appellation meaning rich in mammals, or birds. (Goldman 1920, 4)

The plane purred along several hundred meters above the Panamanian coast. I peered down and watched the passing landscape—a mostly unbroken expanse of forest bordered by a thin strand of white beach and the azure Pacific. This was the moment that I had been waiting for most of my life. As a child growing up in suburban Detroit in the 1940s and 1950s, I often dreamed of going on expeditions to capture animals in foreign lands. Those dreams were inspired, no doubt, by grainy black-and-white television images of Frank ("Bring 'em back alive") Buck and Osa and Martin Johnson in Africa. The wildlife adventures of Marlin Perkins on *Zoo Parade* ("Quick, Jim, get this damn anaconda off my neck!") also thrilled me. In my spare time I read all the Raymond Ditmars books about tropical snakes that I could find in our local library. The closest I ever came to realizing my dreams, however, was when I chased after snakes, frogs, and turtles in the shrinking wildlands of Michigan's lower peninsula. Herpetology, not mammalogy, had been my early passion. Now I was about to begin my own adventure of exploration and discovery—an adventure that would continue for more than thirty years. I could hardly wait!

How did I end up realizing my life's dream in 1966? A large part of the answer turns out to be serendipity. I was simply in the right place at the right time. The place was graduate school in the Department of Zoology at the University of Michigan. I was in my third semester of graduate school when my academic adviser, Emmett Hooper, told me that Charles Handley, one of his former students, was looking for a student to study the natural history of Panamanian mammals for a year and a half. Charles had written to Hooper: "The fact that so little is known about seasonal variation in the biology of tropical mammals makes this especially attractive as a thesis problem." Hooper asked me, "Are you interested in applying for Handley's position?" I replied, "You bet!" Three months later I was on my way to Panama.

Soon, Jaqué's old World War II landing strip was in sight. The plane circled low over the village, a maneuver that sent my stomach to churning. Despite the tropical heat and humidity, as the plane bumped down the crumbling asphalt runway I broke out in a cold sweat. This was a pretty shaky way to begin my first tropical adventure, I thought.

Once on the ground, we were greeted by a ragtag crowd of young boys and men accompanied by an old, beat-up Japanese car and a noisy Honda motorcycle. Everyone was in a festive mood while we unloaded the plane and hauled our gear along dusty footpaths into town. Little boys gave Frank and me juicy oranges to keep us hydrated as we worked in the hot sun. To my city boy's eyes, Jaqué, a town of about a thousand people, was pretty uncivilized. Its houses came in two basic models: tin-roofed, shabbily painted wooden shacks, or thatch-roofed, palm-walled huts. Garbage was everywhere, and outhouses were few and far between. The town did have electricity, provided by a generator that ran for a few hours after sunset, but other amenities of urban life seemingly were absent.

Shortly after arriving, Frank and I met Francisco Pruente, a Protestant minister who had been teaching the local Choco Indians to read, write, and speak Spanish. Pruente arranged for us to stay for two nights in the town's "hospital," an unfurnished cement building run by a gruff-looking but warm-hearted old nurse from Colombia, whom the townspeople called La Señorita. Pruente also contacted some of the local Indians and arranged for Hombría, chief of the village of Mamé, several hours up the Río Jaqué, to transport us to our campsite. He even managed to find us a cook: a pudgy, cheerful twenty-three-year-old named Germán Rodriguez. Germán had learned his trade and a bit of English at Piñas Bay, a nearby gringo fishing camp.

Germán's father, also a cook, had worked for Alexander "Doc" Wetmore, former secretary of the Smithsonian Institution and, in 1966, dean of American ornithologists, when Wetmore collected birds in the Jaqué region in 1946–47. Frank and I had socialized with Doc Wetmore and his wife, Bea, in the Panama Canal Zone before departing for Jaqué. Eighty years old but still spry, Doc was about to embark on his last collecting trip to gather additional material for his multivolume treatise on the birds of Panama.

Jaqué was much different when Wetmore had worked there. For one thing, the decaying airstrip where we had just landed was the site of a World War II auxiliary airfield of the Army Air Forces, complete with screened wooden barracks, cement sidewalks, and a post exchange. None of these structures was present in 1966. In 1946, Wetmore spent a month collecting 170 species of birds within ten kilometers of the airfield. The following year he returned to Jaqué and, with permission from the Choco Indians, traveled nearly a hundred kilometers up the Río Jaqué with four assistants, including Germán's father, in two wooden piraguas.

In a letter dated 12 June 1947, Alexander Wetmore described his experiences up the Río Jaqué:

> [T]he whole region is still in primitive forest. The trails are poorly marked tracks such as these forest Indians use everywhere. There is little food in this country, except for a fair amount of game. We killed the big pacas, which the natives call Conejo [rabbit], the kinkajou among the animals, the curassow and the tinamou among birds for food. There were a good many peccaries but these are seldom good eating as the natives never remove the musk gland before cleaning. When we ran out of other meat we shot a few of the abundant big-billed toucans which are a little tough, but are not bad eating at that. From the Indians it is possible to buy only green bananas (Plantain) or once in a while yuca [*sic*].
>
> The Choco Indians are friendly but do not welcome strangers. We did not have the slightest difficulty with them. In fact, I am now "compadre" with all the Indians on this river! They told me that they were getting poisoned arrows ready to use on some witch doctors over on the headwaters on the Sambu to the north. The story was that these brujos had gone over into Colombia, had learned some new witchcraft and that people were dying because of it so they were going to get rid of the witch doctor as the easiest way to settle the difficulty.
>
> There is much malaria in this country and a great deal of other sickness. By proper care and guarding against mosquitoes, Mr. Perrygo [his longtime assistant] and I both came out in good health. I used quite a bit of medicine among the Indians. (Wetmore Papers)

After lunch on our first day in Jaqué, Frank and I walked a short distance out of town along the broad Río Jaqué. With our machetes we cut down a series of saplings for mist-net poles. In preparation for a night of bat netting, we set up four Japanese mist nets across a shallow stream that emptied into the Río Jaqué. This was only the second time I had ever worked with mist nets, the first time being in a bird-banding exercise in a vertebrate natural history course as an undergraduate at Albion College. Since then, I have set up thousands of mist nets but have never tired of the routine. To me, a well-set mist net, its gossamer nylon netting forming four shallow bags for entangling bats and birds, is a thing of beauty. Folded up in their plastic bags, the black nets look like human scalps. Once set, they form a delicate, nearly invisible wall up to twelve meters long and two meters high.

Since their introduction to the scientific community in the 1930s, these elegant capture devices, used for centuries in Japan for snaring birds for the cook pot, have revolutionized the study of bats and birds. Charles Handley was one of the first researchers to make serious use of mist nets to capture bats in his tropical faunal and systematic studies. Edward Goldman, of course, did not have mist nets when he collected mammals in central and eastern Panama in 1911–12. Instead, he and his assistants employed only shotguns and hand nets to collect bats. As a result, in his 1920 monograph Goldman covered only thirty-three of the eighty-nine species of bats that were included in Handley's 1966 checklist of Panamanian mammals.

Just before sunset we returned to open our nets only to discover that we had a slight problem. We had failed to note that the water level in the Río Jaqué and its small tributary fluctuated with the Pacific tide. We had set our nets at low tide, when the water was only ankle deep, but now the tide was in and the stream was chest deep on me and eyeball deep on Frank. Instead of abandoning the operation, we decided to open the nets for an hour or so to see what we could catch. Being over six feet tall, I was elected to wade out into the stream to remove the bats, my belt and six-volt headlamp battery hanging around my neck. The water was pleasantly warm, so my job was enjoyable, once I adjusted to the fact that the stream doubled as the local toilet.

In our hour of netting we managed to capture eleven bats of four species. The most notable was the large, foul-smelling greater fishing bat. Covered by a coat of short orange fur, this bat has a broad naked muzzle, long pointed ears, and large hind feet tipped with sharp claws, with which it gaffs small fish at the water's surface. Less bizarre were three members of the New World leaf-nosed bat family, Phyllostomidae, the bats that I have studied intensely during most of my research career. Most members of this family have a flexible, triangular flap of skin called a nose leaf surrounding and projecting above their nostrils (figure 1). The nose leaf is thought to serve as a parabolic reflector for increasing the directionality of the ultrasonic sounds these bats emit through their noses. (See appendix 1 for an overview of the classification and diversity of bats, appendix 2 for a list of the scientific names of species of amphibians, reptiles, mammals, and plants whose common

Figure 1. A phyllostomine, *Phyllostomus hastatus* (greater spear-nosed bat). Photo © Merlin D. Tuttle, Bat Conservation International.

names are mentioned in the text of this book. Only the common names of birds are used in this book.)

Containing about 150 species, the Phyllostomidae exhibits perhaps the greatest variety of food habits of any mammalian family. It is certainly the most ecologically diverse of all eighteen bat families. Among its species are blood feeders (the vampire bats), insect eaters (the ancestral feeding habit), and large carnivores that eat frogs, lizards, snakes, birds, other bats, and rodents. Most phyllostomids, however, are plant visitors that eat nectar, pollen, fruit, and occasionally leaves. The three species of phyllostomids that we caught included two ubiquitous species—the short-tailed fruit bat and the common long-tongued bat—and the less common Heller's broad-nosed bat. We kept our bats alive in cloth bags overnight and killed and prepared them as museum study skins the next morning. Catching, killing, and preserving bats would be an important part of my job during my stay in Panama. Although these collections would yield valuable information about tropical bats and their reproductive cycles, I wouldn't really study bats as *living* creatures until I began working in Costa Rica in 1970.

We departed for our field camp located about twenty-five kilometers up the Río Jaqué two days after arriving in Jaqué. Hombría had agreed to take us upriver in his piragua, a seven-meter boat carved from a single tree trunk and powered by a small Johnson outboard motor. Hombría knew a few English phrases but spoke mostly Spanish to non-Indians. Dressed in Western clothes—shorts and a tee shirt—when he was away from his village, he loved to kid around and had his Indian helpers in a good mood as we loaded our pile of gear into the narrow boat. We shoved off at 8 A.M. and had a smooth ride up the broad river under a bright sun for four hours. Then it clouded over, and we traveled in a steady downpour for another four hours. By midafternoon we had reached Hombría's village of Mamé, where we sought shelter from the rain. The Chocos were hospitable, feeding us roasted corn and grilled meat and providing us with a place to unroll our sleeping bags for the night in one of their houses. Away from Jaqué the Indians dressed in their traditional clothes—long skirts and no tops for women and loincloths for men. They made their living as subsistence farmers, hunters, and fishermen. Their houses were simple open-sided platforms raised about a meter and a half off the ground and covered with a palm-thatch roof. Before sunset Frank and I set out a few wire-mesh traps baited with fruit in a nearby banana plantation. As I set my traps in the gathering darkness, I was amazed to see small bats flitting effortlessly around and through the thick vegetation. If you didn't know they can echolocate, it would be easy to view these creatures as highly mysterious, I thought.

The next morning dawned clear and cool. Our rain-soaked traps contained a common opossum, three mouse opossums, and a spiny rat. The Indians contributed a forest rabbit and a night monkey to our collection. After packing up, we continued upriver for another couple of hours to the junction of the Ríos Jaqué and Chicao, which we had selected as our campsite upon the recommendation of the Indians. An abandoned Indian "stilt" house located on a small bluff above the Chicao was available for our use. A family of Chocos with three small children lived a bit farther up the Jaqué.

Our campsite was located in a broad valley surrounded on three sides by forested hills. The vegetation immediately surrounding camp was an abandoned banana plantation. Thick with shrubs, vines, and fallen banana trunks and emanating an earthy and slightly fermented smell, this habitat easily fit my preconception of a jungle. The real jungle—intact evergreen forest—was within easy walking distance of camp across the two rivers. This forest contained emergent trees three meters or more in diameter, more than thirty meters tall, and festooned with vines and epiphytes. The understory was moderately dense and filled with low palms and shrubs. High densities of philodendrons and monsteras on and near the ground and a profusion of moss, bromeliads, and orchids above ground indicated that this region receives substantial rainfall, though we were now well into the dry season.

Once we unpacked, life settled into a comfortable routine, much as it had, I imagine, for Edward Goldman, Doc Wetmore, and Charles Handley at their Panamanian campsites. With the arrival of George Barrett, Jr. ("Juni"), who would be my field assistant for my entire time in Panama, and Emiliano Segura, a Choco Indian whom we hired to collect larger mammals, our field party contained five people, all housed under the thatched roof of our five-square-meter palm trunk platform. Germán did the cooking and laundry; Juni, who was fluent in (Jamaican) English and Spanish, helped Frank and me preserve specimens and set out traps and mist nets; and Emiliano, who spoke only Spanish, hunted for food and specimens every day. Barefoot and wearing only cutoff shorts and a battered Panama hat, Emiliano would go off into the forest each morning armed with an old shotgun; he seldom returned empty-handed. He kept us continuously supplied with tinamous, chicken-sized relatives of ostriches and rheas, as well as an occasional crested curassow, turkey-sized terrestrial birds, for food. He also collected collared peccaries, four species of opossums, tamanduas (a kind of anteater), tayras (large semiarboreal weasels), a river otter, agoutis, pacas, a kinkajou, an olingo (a relative of the kinkajou), and four species of monkeys for us.

My daily routine consisted of checking and rebaiting our eighty to a hundred live traps shortly after sunrise. Our capture rate usually was low, and I would bring the few opossums and rodents back to camp for killing and processing. After breakfast (dry cereal with powdered milk, a fried egg, some fried tinamou, or, on special occasions, fried freshwater shrimp), Frank and I measured, skinned, and made museum specimens of opossums, rodents, and bats. We pinned the cotton-filled skins onto cardboard trays for drying and hung the labeled skulls out in the sun to dry. After lunch (rice, beans, fried plantains, and tinamou), we would take a break for writing notes or exploring before setting out traps and mist nets in different locations near camp in the late afternoon. After dinner (usually a repeat of lunch but often including *carampalas,* a sweet fried bread), we would open the mist nets for several hours. Then we would wash up in the stream and sleep under our mosquito nets for a few hours before beginning the routine again.

The weather at this time of the year was lovely. It was hot in the sun, but midafternoon temperatures in the shade were around eighty degrees Fahrenheit. Early mornings were clear and cool, the only rain occurred as short afternoon showers, and night skies were clear and star-filled. Civilization (and my wife, Marcia, who was still in grad school in Ann Arbor) seemed thousands of miles away as I concentrated on enjoying the life of a field biologist. I constantly stopped to watch birds during my morning and afternoon work. Common birds around camp included two species of tinamous, yellow-rumped tanagers, chestnut-headed oropendolas, several species of hummingbirds, toucans of two sizes, three species of kingfishers, golden-collared manakins, black-chested jays, northern orioles, and a large, red-headed woodpecker. Early each morning a squawking flock of yellow-headed parrots would fly over camp. Some birds, such as the oropendolas, called *linguas* (tongues) in Panama, made sounds I had never heard before. A colony of these large orioles was nesting in an orange-flowered tree across the Río Jaqué, and small groups would fly over camp with loud wingbeats, making strange clucking and gurgling sounds with their throats. The mournful whistles of tinamous were common in the late afternoon. The strangest sounds, however, came from manakins, small fruit-eating birds that feed in the forest understory. The first time I heard the loud *cr-r-r-r-ack* followed by intermittent short *cracks* produced by the wings of males, I could not guess what was making that sound. I finally discovered that male manakins produce these sounds before flying rapidly just above the ground between small trees that served as their display perches. I later learned that pairs of males make these noises to attract females to traditional mating sites, called leks.

Although it was dry season, frog calls punctuated the night air. Several individuals of a small tree frog sang in a loose chorus from rocks along the edges of rivers. Individual calls were *cricks,* and frogs would be silent for long periods, only to begin cricking in a chain reaction when one frog gave the "go" *crick.* These rapid-fire bursts of male exuberance probably serve as an antipredator adaptation. The predator in this case undoubtedly is the fringe-lipped bat, a species that specializes in capturing singing male frogs. Under threat of death from the sky, a male's best mating strategy probably is singing its heart out in a chorus rather than as a soloist. Issuing a loud *br-r-r-r-rup* from shrubs along the water's edge was a large, solitary tree frog. A third tree frog produced a machine-gun-like burst of sound at irregular intervals. In addition to these noisemakers, we often encountered large marine toads around camp and a large, copper-colored *Leptodactylus* frog that Juni called spring chicken in small *quebradas* (streams) in the forest at night.

We saw relatively few reptiles at this site. Arboreal *Anolis* lizards were common in the banana grove, and terrestrial *Ameiva* lizards were constantly poking through the leaf litter in search of food in the plantation and forests. My favorite lizard quickly became the basilisk, or *jesu cristo* lizard. I would commonly see the striped females along stream edges; the larger, frilled adult males were usually higher up in trees. When frightened, basilisks would sometimes dash across water in upright, bipedal fashion without sinking. I once saw a male jump off a log onto a swiftly flowing stream, which it traversed without sinking or being swept away. Long, fringelike scales on its hind toes allow this lizard literally to walk on water.

We encountered few snakes and, contrary to my expectations, no venomous species. I missed seeing (nonpoisonous) snakes, because as a youngster I had loved to catch snakes. In the 1940s and 1950s, city lots in Detroit still contained garter snakes and northern brown snakes. Lakes and streams had banded water snakes. Fortunately for me, poisonous snakes were uncommon in southern Michigan. Only the massasauga rattlesnake, an inhabitant of swamps and bogs, occurs there. I say "fortunately" because I am extremely nearsighted; my uncorrected vision was once measured as 20/800 in a government physical examination. I began catching snakes long before I wore glasses, and in my frenzy to grab them, I was often bitten. Poor myopic Teddy couldn't tell the head end from the tail end! Except for the ones that bit me, I sometimes couldn't even distinguish snakes from sticks. Despite a high likelihood of being bitten, I tried to catch every snake I encountered.

Our main task along the Río Jaqué, of course, was capturing and preserving as many mammals as possible. In the five weeks we were there, we preserved over seven hundred specimens of sixty-two species, including thirty-three species of bats. This diversity of bats, equivalent to three-quarters of the total species living in the United States and Canada, is only moderately high by neotropical standards, where sites containing from fifty to seventy species are known. But to capture thirty-three species in our mist nets in about one month suggests that this region of Panama is very rich in bats. A longer, more intensive effort, including searching for roosts and using additional capture or detection methods, certainly would have increased the length of our species list. Using a variety of methods, mammalogists from the American Museum of Natural History, for example, have detected seventy-eight species of bats in a couple of square kilometers of similar forest in French Guiana. Over 50 percent of the mammalian diversity in most neotropical habitats is made up of bats.

In contrast with the high bat diversity, the number of rodent species we captured was disappointingly low. My traps yielded only a handful of rodents and marsupials each day, and many traps went days without catching anything. Tome's spiny rat, called *macangué* in Spanish, was by far the most common rodent in our traps. Its scientific name is *Proechimys semispinosus,* and it is related to agoutis, guinea pigs, and porcupines of South American ancestry. Gentle in disposition, this handsome rodent, whose adults weigh two hundred to three hundred grams, has large eyes and a rich, rust-colored pelage containing many stiffened, spinelike hairs. It normally has a long tail, but many individuals lose their tail, probably as a result of close encounters with predators. Like many lizards, one of its tail vertebrae has a weak centrum, which breaks easily when the tail is grabbed. As much as I like spiny rats as living creatures, I disliked dealing with them as museum specimens, because their skin is soft and tears easily during skinning, a common feature of many spiny-skinned mammals. It became a real challenge to produce intact study skins of this species.

The only other common rodent in our traps was the Talamancan rice rat (known at the time as *Oryzomys capito*). Resembling woodland species of deermice *(Peromyscus)* back in Michigan, this slightly larger rich-brown, white-bellied mouse seemed to be relatively common in both disturbed habitats and intact forest in this region. Along with the spiny rat, it turned out to be one of the three species of rodents whose populations I studied in my doctoral research.

More interesting to me than the rodents were the opossums. We caught or shot six species, including the beautiful black and gray water opossum with its webbed hind feet and, in both sexes, a fur-lined pouch held tightly shut with special muscles. The *perro de agua* was common in larger streams, where it fed heavily on the freshwater shrimp we so enjoyed for breakfast. The terrestrial opossums included a mouse opossum, three medium- to large-sized semiarboreal kinds (the common or *Didelphis* opossum, the gray four-eyed opossum, and the brown four-eyed opossum), and the strictly arboreal woolly opossum. *Didelphis* and gray four- eyed opossums turned out to be real nuisances around mist nets at night. Unless we checked our nets frequently, we were likely to lose low-caught bats to one of these critters. The only animal I killed in anger in Panama was a large *Didelphis* that insisted on raiding our nets despite our efforts to shoo it away. In exasperation, I finally backed up a dozen paces and blasted the poor *zorro* with my shotgun. Because its behavior irritated me, this animal now permanently resides in the mammal collection in the U.S. National Museum.

Like Edward Goldman and Charles Handley, we relied on hunting to secure the larger species of mammals. Emiliano and his friends who would occasionally drop by camp were responsible for obtaining most of these specimens, which provided us with skins, skulls, and meat for the table. Emiliano had a rather distinctive way of deploying his shotgun, as I learned when I went hunting with him one night. Instead of shouldering his 12-gauge, sighting, and firing, he held his gun out in front of him in one hand and fired it like a pistol. I presume he used a more conventional style during the day, when he didn't have to aim in the narrow beam of a headlamp.

Depending on your point of view toward killing monkeys, Emiliano's greatest triumph (or disaster) came in late February when he and a buddy shot twelve of them—five mantled howlers, four white-faced capuchins, and three Central American spider monkeys. They must have traveled quite a ways from camp—in the hills to the east, they said—to obtain these animals, because we never heard howlers roaring in the forest around camp. They returned to camp in late afternoon carrying their load of monkeys. Two of the howlers turned out to be nursing females, and their babies were still alive. We quickly decided that we would adopt the babies and try to keep them alive, but our more immediate concern was preserving the dead adults as fast as possible. We worked late into the night skinning the monkeys under the light of our lanterns. It took me two hours to skin one animal. The long fingers and toes with their soft, ridged skin were particularly difficult to skin. We spent most of the next day cleaning and salting the skins and defleshing the skulls.

Lacking a baby bottle, we fed the two baby howlers powdered milk from a rubber glove. To our pleasant surprise, they flourished under our care. To keep them from completely disrupting our camp, we fashioned a collar from a handkerchief for each of them and tied them to one of the house's support poles with string leashes. When we weren't busy, we often untied and played with them. They loved to cling to us like the affectionate babies they were and acted like little "Curious Georges" when let loose to explore (figure 2). They were champion climbers and often used their prehensile tails to hang from various parts of our house during their explorations. At night they slept with their arms and long tails wrapped around each other in a cardboard box lined with one of our towels.

The two baby howlers were our second and third camp pets. Our first pet was a baby tamandua whose mother had been killed by our hunters. "Tammy," as we called her, was also affectionate and loved to be held and scratched. She would nuzzle us with her long snout. Her basic color was tan, and she had a handsome "jacket" of coarse, dark brown fur covering her back, shoulders, and sides. We initially fed her powdered milk but quickly discovered that she preferred termites, which she lapped up with her long pink tongue. To feed her, all we had to do was cut off part of a termite nest from a nearby tree. She would immediately rip into the carton nest with her large front claws and begin slurping termites out of their tunnels.

Mist-netting bats quickly became my favorite activity of the field trip. We set our nets in a variety of different locations—across streams, in the banana plantation, in intact forest—and were always anxious to see how many species we could capture. While we checked our nets at night, we often saw other mammals—opossums and rodents on the ground, an occasional

Figure 2. Ted Fleming with a baby howler monkey. Photo by F. Greenwell.

kinkajou or olingo high up in trees. These last two, arboreal relatives of raccoons, both gave the same call—*wick-up, wick-up*—and often fed in fruiting trees in mixed-species groups. I wondered what ecological and behavioral differences separated these two species.

Our collection of thirty-three species of bats included representatives of five families, (sac-winged bats, fishing bats, leaf-nosed bats, free-tailed bats, and plain-nosed or vesper bats), but the majority of our captures were leaf-nosed bats, as is the usual case in the lowland New World tropics. One night we set our nets over and around a small *quebrada* in tall forest and caught 160 bats of twenty species. But nightly totals of eight to ten species and thirty to forty individuals were much more common.

I quickly learned to recognize members of four subfamilies of leaf-nosed bats: the Phyllostominae, Glossophaginae, Carolliinae, and Stenodermatinae. Considered to be the most primitive members of the family because of their insectivorous or carnivorous food habits, phyllostomines were rela-

Figure 3. A fringe-lipped bat *(Trachops cirrhosus)* attacking a singing male frog. Photo © Merlin D. Tuttle, Bat Conservation International.

tively uncommon among our captures. Our phyllostomines covered a broad size range, from the nine-gram tiny big-eared bat to the ninety-gram greater spear-nosed bat, the second largest bat in the New World. Two interesting phyllostomines included the fringe-lipped bat and the woolly false vampire bat. The fringe-lipped bat gets its name from a cluster of fleshy projections on its chin pad and along its lips. This bat is now famous as the "frog-eating bat" as a result of behavioral studies conducted by Merlin Tuttle and Michael Ryan on Barro Colorado Island, in central Panama's Lago Gatún, in the late 1970s and early 1980s (figure 3). Almost as large as the greater spear-nosed bat, the woolly false vampire bat gets its common name from its long, fluffy, light-colored fur. This handsome carnivore eats lizards, mice, and other bats.

Much more common than the insectivorous or carnivorous phyllostomines in our captures were plant-visiting members of subfamilies Glossophaginae, Carolliinae, and Stenodermatinae. The ten-gram brown long-tongued bat, with a small nose leaf and a moderately elongated rostrum for poking into flowers, was one of our most common captures (figure 4). The most common bat among our captures was the short-tailed fruit bat, an eighteen-gram, gray-brown frugivore specializing on the spikelike fruits of

Figure 4. A glossophagine, *Glossophaga soricina* (common long-tongued bat). Photo © Merlin D. Tuttle, Bat Conservation International.

Piper shrubs. Generally considered to be a "trash" bat by most mammalogists because of its abundance, the short-tailed fruit bat would eventually become the subject of an intensive study by my research group in Costa Rica. Members of the Stenodermatinae were also common, especially the geographically widespread Jamaican fruit-eating bat, the subject of a detailed ecological study by Charles Handley and associates on Barro Colorado Island beginning in 1975 (figure 5).

Stenodermatine bats tend to be more colorful than other phyllostomids. Most species have white facial stripes and white- or yellow-edged ears; some also have a white middorsal stripe. The stripes help camouflage these bats as they roost by day in the foliage of trees rather than in caves or hollow trees, the usual roosts of other leaf-nosed bats. The common tent-making bat and a few other small stenodermatines live in small groups under the folded blades of palm leaves and other broad-leafed plants. One or more members of the group chew the leaves to form the "tent" that protects them from rain and predators. A common nonstriped stenodermatine was the little yellow-shouldered bat. Males of this golden brown bat, which is about the same size as *Carollia,* have dark brown tufts of fur on their shoulders. Glands in these shoulder tufts emit a strong, pleasant odor that reminded

Figure 5. A stenodermatine, *Artibeus jamaicensis* (Jamaican fruit-eating bat). Photo © Merlin D. Tuttle, Bat Conservation International.

me of citronella. Years later I would smell this same odor coming from males of the Queensland tube-nosed fruit bat, a member of the flying fox or pteropodid family, in tropical Australia.

The common vampire bat, the most specialized of all phyllostomids, was also fairly common in this area (figure 6). This was one species whose feeding behavior I could actually watch. The Choco family near us had two pigs that slept under our house every night. Shining my headlamp between cracks in the floor, I observed several vampires settling down to feed on the pigs one night. The bats were wary and would skitter like large spiders into the shadows of a pig when I shone my light on them. They would hop to the ground whenever the pigs stirred. The next morning, one of the pigs had a bloody ear indicating where the bats had finally fed. Knowing that vampires were feasting away about a meter below me made me glad that I was sleeping under a mosquito net. Mosquitoes were not a problem at this time of the year, but mammalian "mosquitoes" certainly might have fed on us if given the chance.

We also shared our house with another species of bat. A group of black myotis, the common vesper bat of neotropical lowlands, lived in our thatched roof. This turned out to be a maternity colony, and females were each nursing a single young when we captured a few in early March. We could hear

Figure 6. A desmodontine, *Desmodus rotundus* (common vampire bat). Photo © Merlin D. Tuttle, Bat Conservation International.

the bats squeaking and moving around overhead every day as we sat at our skinning table preparing study skins. The bats would dart out of the roof in small groups at sunset. We never captured them in our mist nets. In 1968 a young biologist named Don Wilson, from the University of New Mexico, later to become director of biodiversity studies at the Smithsonian Institution, studied this species on Barro Colorado Island for his doctoral research.

Although at times it felt as though we were a long way from civilization, we of course were not. Our camp was located near a trail that went to Colombia, only about a six-hour walk away. Scruffy-looking men dressed in ragged shorts and shirts, carrying a rifle and invariably accompanied by a skinny dog, passed by our camp once or twice a week. They would usually stop to see what we were doing and then continue. Juni told us these men often were fugitives from Colombian law, which made us a bit nervous. One day two such men decided to camp a short way up the Chicao from us. They were well-armed, and none of us liked their looks. We prominently displayed our firearms around camp and kept our shotguns near us as we slept that night. Nothing remotely threatening ever ensued, however, and the men moved off toward Colombia in a couple of days.

Much more enjoyable were the visits of the Choco Indians. In addition to our neighbors, who visited us every day and with whom we shared meat supplied by our hunters, we were regularly visited by Pechandé, the chief of a village located about half an hour down the Río Jaqué and a good friend of Hombría. Taller than most of the Chocos, Pechandé was about forty years old. He had a calm, intelligent demeanor and was an excellent hunter. In addition to occasionally bringing us animals he had shot, he also brought us our mail and supplies from Jaqué.

Our most dramatic interaction with the Chocos occurred one afternoon when the two older children of our neighbors, a boy and girl about six or seven years old, came to watch us work. In visiting us, they left their year-old baby brother alone at home. Somehow, the baby fell into the river and was washed downstream. When the parents discovered their baby was missing, they raced downstream in a piragua and found him on a gravel bar. Frank and I were unsuccessful in our attempts to revive the child by artificial respiration. For two days the family mourned the loss of their baby by singing dirgelike songs in Choco day and night. The mournful singing created an eerie atmosphere around our camp after dark. After two days, the family took their baby down to Jaqué for a Christian burial. Sadly, the brother and sister were never allowed to visit us alone again.

By the end of February our supplies were running low, and we had collected reasonable series of most of the common mammals in the area. Our dinners had degenerated to a couple of slices of Spam, crackers spread with little dabs of peanut butter, and a cup of hot chocolate. We had saved a small, rum-soaked fruitcake that Bea Wetmore had given to us for these lean times, so meals weren't a total bust. But it was obvious that it was about time to return to civilization. We sent word downriver that we needed transportation back to Jaqué, and on 2 March two boatloads of Chocos arrived to transport us the next morning. They brought some fresh paca meat—a real delicacy because of its soft texture and sweet flavor—and we had a minifeast that night. It rained steadily for several hours during the night, and all the Indians crowded into our house to sleep.

Frank and I spent ten days in the Canal Zone before heading west to Chiriquí Province on our next collecting trip. During that time Charles Handley spent a few days with us on his way back to Washington, D.C., from visiting his field crews in Venezuela. After completing most of his intended fieldwork in Panama, Handley initiated a much more ambitious survey of the mammals and their ectoparasites of Venezuela, a much larger and more diverse country than Panama. During the period I was in Panama, he had two field crews collecting mammals there. One crew was headed by two brothers, Arden and Merlin Tuttle, whose collecting abilities were legendary. Whereas Handley and his workers collected about ten thousand mammals in field trips spanning seven years in Panama, in three years the Venezuelan collectors preserved nearly forty thousand individuals of 270 species of mammals, including 137 species of bats. Merlin acquired what for most people would be a lifetime's experience with bats during his Venezuelan work, but that was to be only a small part of his total bat experience. He was just warming up, back in the mid-1960s.

Our next collecting trip was at the opposite end of Panama near the town of Progreso, about three kilometers from the Costa Rican border. Within Panama in the 1960s, Progreso and the lowlands of Chiriquí Province were about as far as you could possibly get geographically, culturally, and ecologically from Jaqué and the heavily forested Darién lowlands. The first difference was accessibility. Whereas the only way to get to Jaqué was by boat or plane (which is still true), we could easily drive to Progreso, about 550 kilometers from Panama City, along the Pan American Highway. True, part of that highway was unpaved and dusty in 1966, as I discovered while riding in the back of our truck for several hours. But we were able to drive, rather than walk or boat, to most of our collecting sites around Progreso and Puerto Armuelles, an hour's drive south of our field quarters.

This accessibility, of course, was the result of the encroachment of civilization into much of western Panama. Whereas in Goldman's time most of western Panama had been forested, by the mid-1960s much of the lowlands of Chiriquí Province and the rest of southwestern Panama had been converted to agricultural and grazing lands. Our field base near Progreso, for example, was located on an abandoned banana plantation owned by the Chiriquí Land Company. A subsidiary of "El Pulpo" (the octopus), the derogatory name applied by Latin Americans to the United Fruit Company, the Chiriquí Land Company had been established in 1927 after Panama disease (a fungus) devastated thousands of hectares of banana groves along Panama's Atlantic coast. In the mid-1960s, its banana fields between Progreso and Puerto Armuelles encompassed about fifteen hundred hectares. The local manager of the plantation kindly let us stay at Finca Quira in an unoccupied two-story wooden house, complete with electricity, a refrigerator, a kerosene stove, and indoor plumbing. Compared with our accommodations on the Río Chicao, our new quarters were almost luxurious.

Another major difference between the two field trips was the absence of Indians on this one. The Chocos had had a marked presence on the previous trip, but we saw no Indians on this trip, not even the Guaymi, who lived in the mountains of Chiriquí. Instead, our neighbors and contacts in Progreso and Puerto Armuelles were Hispanics. We found the Panamanians in Chiriquí to be friendly and interested in our work. They would wave to us as we drove around the countryside looking for places to set our traps and mist nets. I had let my beard grow on the Jaqué trip, and now whenever Chiriquí cowboys saw my black hair and beard and my thick, black-rimmed glasses, they would yell out an enthusiastic "Fidel!" Actually, they thought I was an American army officer, and Juni Barrett led them on by saying that, yes, I was an 'alf-capitan in the U.S. Army.

On numerous occasions people let this strange crew of biologists set up mist nets around fruiting plants in their backyards to catch bats at night. Small crowds of people would gather around us to see what we had caught. They were well-acquainted with bats; after all, most thatch-roofed houses had bats (and rats) living in the roof, and bats sometimes entered houses to capture insects or eat fruit. But people generally feared bats and thought that all were *"vampiros,"* a common belief throughout most of Latin America. We tried to explain to people that nearly all the bats flying around their yards were beneficial because they ate insects, pollinated flowers, and dispersed the seeds of fruiting plants. But the canine teeth on some of our bats, especially those of the Jamaican fruit-eating bat and the greater spear-nosed bat, were so large (for impaling hard-skinned fruit) that it was hard to convince people that these bats weren't dangerous.

The final obvious difference between Jaqué and Progreso was the depauperate bird and mammal fauna resulting from the absence of intact forest. The habitat around our house was either fallow banana fields or newly plowed cornfields, not pristine forest. The closest trees were growing along an irrigation canal a bit east and south of our house. Most of those trees were early successional species such as *Cecropia* or tall thickets of bamboo. All along the road from Progreso south to Puerto Armuelles, forest was being cleared and burned for farming or ranching. Owing to its high annual rainfall, the land must have been covered with tall evergreen forest prior to the late 1920s, but now forest patches were few and far between.

Despite the grossly human-modified landscape, certain kinds of birds and mammals were still abundant around Progreso. Our common avian neighbors included the beautiful scarlet-rumped tanager, groove-billed anis, common ground doves, several species of small, seed-eating finches, fork-tailed flycatchers, roadside hawks, black and turkey vultures, white-throated crakes (rails), and, feeding on the roads at night, the pauraque (a nighthawk). Conspicuously missing were the mournful whistles of tinamous and the gurgling calls of *linguas.* Blue-headed motmots sat quietly on branches along the irrigation canal, swinging their raquet-shaped tails in pendulum-like fashion, and I watched a couple of large brown squirrel cuckoos hopping along branches searching for lizards and large sphingid moth caterpillars.

As a result of forest destruction, our mammal trapping and netting was much less productive than it had been near Jaqué—a vivid introduction for me to the loss of biological diversity caused by deforestation. Our collection here included only four species of opossums, of which the common and gray four-eyed opossums were by far the most common kinds, four species of rodents, and twenty-four species of bats. The rodents included spiny rats and the volelike dusky rice rat, which lived in banana groves and along irrigation canals, and two grassland species—the hispid cotton rat and common cane rat. In all, we preserved only thirty-nine species of mammals, a little over half the number of species that we had collected in the Darién.

Although we had preserved the skins and skulls of five species of primates at Jaqué, I had not yet seen live monkeys (except for our baby howlers) in the wild. The three larger species (capuchin, howler, and spider) originally lived in the forests around Progreso, but we saw none of them in the woodlots that we visited. However, I did see red-backed squirrel monkeys along the irrigation canal near our house. My first of several encounters with a troop of about seven adults occurred while I was bird-watching one afternoon. A harsh bark coming from a tall tree caught my attention while I was observing a pair of tanagers. When I turned around to locate the sound, I discovered that a group of monkeys, scattered over several trees, was barking at me. I began watching the monkeys as they moved around the trees, stopping to peer down at me and bark. In addition to the barks, they also made birdlike twittering sounds. They leaped gracefully and deliberately from one tree to another with their long, black-tipped tails trailing behind. Their bright rusty backs made them conspicuous in the green foliage, and I noted that one of them had a youngster clinging tightly to her back.

After that, we often encountered squirrel monkeys along the irrigation canal and in nearby banana groves. One time we watched a group cross the canal by leaping from a tall *Cecropia* tree into a shorter one on the other side. The monkeys launched themselves into space and quickly scrambled when they hit their target. They also leaped from bamboo trees into rather flimsy banana plants without losing their balance on the swaying leaves. Although they seemed afraid of us, their curiosity usually compelled them to stop for a look, accompanied by a series of harsh barks, before moving away. We learned from Juni that these monkeys usually slept in the bamboo groves at night.

As at Jaqué, most of our preserved specimens were bats. Three plant-visiting phyllostomids—the common tent-making bat, the common long-tongued bat, and the ubiquitous short-tailed fruit bat—were our most common captures. Especially common in banana plantations were pygmy fruit-eating bats, Heller's broad-nosed bats, orange nectar bats, and pale spear-nosed bats. The dusty covering of pollen on the fur of the latter three species indicated they were frequent visitors to banana flowers. We did not capture Jamaican fruit-eating bats in the banana fields but frequently netted them in people's yards around fruiting plants.

We caught few free-tailed bats in our nets but discovered that four species were living under the corrugated tin roofs of the buildings around us. Tin or tile roofs are typical roosting places for these "house" bats throughout Latin America. Black mastiff and little mastiff bats lived in our attic. The Sinaloan mastiff bat was roosting in the attic of a nearby schoolhouse, and the large, blunt-eared Wagner's bonneted bat, whose northern distributional limit is in Coral Gables, Florida, a few kilometers from my house, lived in the attic of the *finca*'s dispensary. The presence of bats in our attic was the probable cause of my getting a mild case of histoplasmosis, which gave me a low-grade fever and weakened me for a couple of days. I didn't discover that I had been exposed to "histo," a fungus whose spores cause serious lung damage in some people, until I later tested positive for it in the Canal Zone.

Along with rabies, histoplasmosis, which dwells in guano-filled roosts, is one of the occupational hazards of bat researchers. There are essentially two ways of avoiding contracting histo: avoid histo-plagued caves altogether (local people usually know which caves these are) or wear a respirator when entering a contaminated roost. Rabies, of course, is a fatal disease and is much more serious than histo. Most (but not all) bat workers take a series of rabies immunization shots before handling bats. Additional booster shots are needed should you be unlucky enough to be bitten by a rabid bat (which has a very low probability).

Although our ecological and domestic setting in Chiriquí was very different from that in the Darién, our daily routine was pretty much the same: check the trap lines before breakfast, prepare our specimens after breakfast, set out new trap lines and mist nets in the late afternoon, and net bats for several hours after sunset. Again, Germán Rodriguez was our cook, and Juni was our all-around field assistant. Juni also became our hunter in Emiliano's absence, but he shot few large mammals because there were few to be hunted.

Specimen preparation time was a bit different from our situation in Jaqué. There we had worked to the sounds of a flowing river, whining cicadas, and birdsongs along the Río Chicao. Now we skinned our mammals to the sounds of a radio playing popular Latin American music. I really enjoyed listening to Latin rhythms—mambos, salsas, meringues, boleros, and rumbas—performed in bright, infectiously bouncy or floridly romantic fashion, depending on the style. The brassy, upbeat pieces with their emphasis on trumpets, trombones, and percussion reminded me of my high school and college years when I played trumpet in dance and jazz bands. The music made me want to play along. Instead, however, I had to settle for bouncing around on my chair in time with the music. We also heard music that was more typical of the rural Panamanian countryside. This music usually featured an acoustic guitar, violin, accordion, at least two percussionists, and a female vocalist singing a highly repetitious, wordless melody in unison with the violin and accordion. We watched one of these local *conjuntas* perform one night at a fiesta in Progreso between rounds of checking our bat nets. Beer was flowing freely, and it was easy to get caught up in the festivities here, but our obligation to the nets forced Frank and me to remain sober. Germán, our cook, was under no such obligation. After several beers he got into a shouting match with the local police and spent the night in jail. It cost us five dollars to bail him out the next morning.

Our diet was also different in Progreso. Instead of a steady supply of tinamou for breakfast, lunch, and dinner, we now had to settle for common iguana as a major source of meat. These large (three feet or more in total length) arboreal herbivores were still quite common in the area, especially in *Cecropia* trees. Although they tried to hide on the back side of tree trunks or branches as soon as they spotted us, iguanas were easy to shoot. Fried, their meat tasted "just like chicken," of course. Caimans were also fairly common in the irrigation canals and permanent streams, and we shot a couple for dinner. Their flaky white tail meat did *not* taste like chicken but had a fishy flavor. We even resorted to eating a few spiny rats after we had skinned them. Their meat was soft and tasty, just like the meat of their cousins, agoutis and pacas, which we had eaten in Jaqué. I've never eaten guinea pig but imagine that its meat is similar to that of its tropical relatives. I suspect that the ancestors of the Incas did not domesticate guinea pigs just to keep them as friendly but stupid house pets.

The four weeks that we had planned to spend in Chiriquí passed quickly. As we moved into April, afternoon thundershowers became more frequent and harder, indicating that the dry season, the best time to preserve study skins of birds and mammals, was over. We packed up all of our gear on 10 April and drove thirteen hours back to the Canal Zone the next day, Easter Sunday. After repacking all of our specimens for air shipment to Washington, Frank and I said a friendly good-bye and went our separate ways. We had enjoyed each other's company but were philosophically divided over the Vietnam War, which was rapidly escalating in 1966. An army veteran, Frank was an ardent hawk, whereas I was a dove. We often discussed our differing views on that faraway conflict while we prepared museum specimens, but we were never able to come to any agreement. Our low-keyed discussions contrasted sharply with the violent clashes between hawks and doves that rocked large parts of U.S. society, especially on university campuses, during the late 1960s and early 1970s.

Now that my collecting duties were over, I had to make serious plans for my year-long study of seasonality in populations of tropical mammals. I needed to choose my study sites and find permanent living quarters for Marcia and me. Marcia's second semester of graduate courses in mathematics was nearly over, and she was scheduled to arrive in Panama at the end of April. The two collecting trips had been a great introduction to tropical diversity and the cultures of Panama, and I had finally realized a long-held dream of capturing animals in an exotic land. But I knew my real job was just beginning.

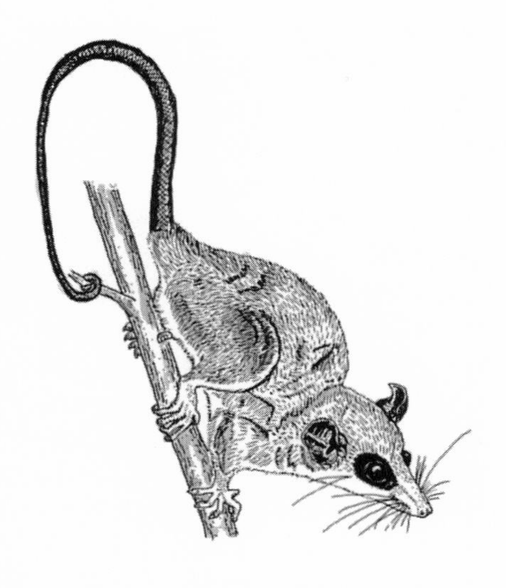

2 Year of the *Marmosa*

In his first letter to me, Charles Handley had indicated that little was known about "seasonal variation in the biology of [neo]tropical mammals." In a way, this was paradoxical. Here was the richest mammal fauna in the world containing a myriad of different lifestyles, but it had barely been studied by ecologists and natural historians. Whereas the natural history and rudiments of the population ecology of many species of temperate small mammals, including a number of bats, were quite well-known by the mid-1960s, little was known about the lives of their tropical counterparts.

The main reason for this, of course, is "location, location, location." Modern animal ecology developed in temperate lands, at places such as Oxford University in England and the Universities of Chicago and Illinois in the American Midwest, beginning in the early part of the twentieth century. Consequently, most of its early studies were based on temperate or arctic species. Even today, despite their greater accessibility and increased attention from scientists, the tropics still take a backseat to temperate regions as

Robinson's mouse opossum. Redrawn by Ted Fleming, with permission, from an illustration by F. A. Reid, *A field guide to the mammals of Central America and southeast Mexico,* Oxford University Press.

a source of material for most ecology textbooks. Until recently, textbooks in tropical ecology did not exist, and most ecology students in Latin American universities still read translations of temperate zone ecology texts. Energy flow in Cedar Bog Lake, Minnesota, niche partitioning in boreal forest warblers, and population cycles of Canadian lynx and snowshoe hare are far removed from the interests of tropical students, but this is what they read about in their ecology texts.

Despite the absence of detailed ecological knowledge, many ecologists and evolutionary biologists in the 1960s believed that "things were different in the tropics." The writings of nineteenth-century tropical explorers such as Alfred Russel Wallace, Henry Bates, and Thomas Belt clearly supported this viewpoint. The influential evolutionary biologist Theodosius Dobzhansky summarized many of these beliefs in a famous essay entitled "Evolution in the Tropics," published in 1950. With the following hyperbole, Dobzhansky set the stage for his thesis that, indeed, things *were* different in the tropics:

> Becoming acquainted with tropical nature is, before all else, a great esthetic experience. Plants and animals of temperate lands seem to us somehow easy to live with, and this is not only because many of them are long familiar. Their style is for the most part subdued, delicate, often almost inhibited. Many of them are subtly beautiful; others are plain; few are flamboyant. In contrast, tropical life seems to have flung all restraints to the winds. It is exuberant, luxurious, flashy, often even gaudy, full of daring and abandon, but first and foremost enormously tense and powerful. (Dobzhansky 1950, 209)

Although Dobzhansky did not single out any particular group of plants and animals in his essay, he could easily have used bats to support this generalization. With a few notable exceptions, such as the beautiful black and white spotted bat of western North America and the frosted-orange eastern red bat, most temperate zone bats are rather nondescript. Usually weighing less than an ounce (twenty-eight grams), most are plain brown or gray in color. Except for members of the Old World leaf-nosed bat families, which are tropical immigrants in Eurasia, they lack fancy facial appendages such as nose leaves. In strong contrast, tropical bats come in a wide variety of sizes, shapes, and colors. There are white bats (the northern ghost bat and Honduran white bat), orange bats (certain species of funnel-eared, thumbless, and mustached bats), and bats with white facial stripes (many stenodermatines) or white shoulder patches (African epauletted fruit bats). The faces of four tropical families—hipposiderids, rhinolophids, megadermatids, and phyllostomids—are adorned with nose leaves. And members of two families—disk-winged and sucker-footed bats—have independently acquired suction cups on their wrists and ankles for clinging to the inner surfaces of rolled-up banana-like leaves that serve as their day roosts. Compared with temperate bats, many tropical bats are flamboyant in appearance, and many are large (flying foxes and hammer-headed bats) and powerful (the Australian ghost bat and the New World false vampire bat).

In his essay on the tropics, Dobzhansky pointed out that a high diversity of species—the hallmark of the tropics for most groups of plants and animals—creates many adaptive opportunities as well as strong interspecific competition. Furthermore, he noted, tropical species' demographic adaptations, favored by the absence of strong physical challenges from a benign climate, should be very different from adaptations in temperate species. He wrote,

> The process of adaptation for life in temperate and especially in cold zones consists . . . primarily in coping with the physical environment and in securing food. Not so in the tropics. Here little protection against winter cold and inclement weather is needed. In the rainforests, . . . relatively little effort is necessary . . . to secure food, and it seems that the amount of food is less often a limiting factor for the growth of populations of tropical animals than it is in extratropical zones. (Dobzhansky 1950, 220)

Finally, he postulated,

> Physical factors, such as excessive cold or drought, often destroy great masses of living beings, the destruction being largely fortuitous with respect to the individual traits of the victims and the survivors. . . . [This] indiscriminate destruction is countered chiefly by development of increased fertility and acceleration of development and reproduction. . . . Where physical conditions are easy, interrelationships between competing and symbiotic species become the paramount adaptive problem. The fact that physically mild environments are as a rule inhabited by many species makes these interrelationships very complex. (220)

Ideas about life-history differences between temperate and tropical animals were formalized in 1967 by Robert MacArthur and Edward O. Wilson in their famous book *The Theory of Island Biogeography,* one of the landmark publications in modern ecology. In addition to developing a general model to predict the number of species that co-occur on islands—a model that would have an enormous impact on conservation biology in the 1970s and beyond—these influential American ecologists developed the concepts of *r* and *K* selection in their book. The symbols *r* and *K* refer to a population's growth rate and its maximum size or environmental carrying capacity, respectively. Expected differences between temperate and tropical mammals were seemingly (and naively) clear-cut. Under the influence of *r* selection, which favors maximum population growth rate in a seasonally uncrowded environment (or on a newly colonized island), temperate and arctic mammals would reach sexual maturity rapidly and would produce large litters during a short lifetime. Tropical mammals, in contrast, would be slower to mature and would produce smaller litters over a longer lifetime as a result of *K* selection, which favors maximum adult survival in a crowded environment (or island). Despite Dobzhansky's glowing but incorrect picture of bountiful food in tropical environs, interspecific competition for food would be intense among mammals in species-rich tropical communities; it would be much less keen in species-poor communities at middle and high latitudes.

MacArthur and Wilson's theory of *r* and *K* selection had not yet been published when I began my thesis research on rodent population ecology in 1966, so I didn't go about my Panamanian fieldwork chanting the mantra "*K* selection, *K* selection, . . ." Instead, I was more concerned with fulfilling my original objective of documenting seasonal variation in the biology of tropical mammals. As Handley had indicated, information about the natural history of Panamanian mammals was scarce. One source of this kind of information was the species accounts in Edward Goldman's 1920 monograph. Not surprisingly, most of those accounts were anecdotal. In describing the behavior of Panamanian monkeys, for example, Goldman quoted observations made by Lionel Wafer way back in 1681:

> There are great Droves of Monkeys, . . . most of them black; some have Beards, others are beardless. They are of middle size, yet extraordinary fat at the dry Season, when the Fruits are ripe. . . . In the rainy Season they have Worms in their Bowels. I have taken a handful of them out of one Monkey we cut open; and some of them 7 or 8 Foot long. They are a very wagish Kind of Monkey, and played a thousand antick Tricks as we marched at any Time through the Woods, skipping from Bough to Bough, with the young one's hanging at the old one's Back, making Faces at us, [and] chattering. (Wafer 1729, 330)

In 1966, the best source of information about seasonal variation in the lives of Panamanian mammals came from the studies of Robert K. Enders, a long-time professor of biology at Swarthmore College. Early in his career Enders had studied the reproductive biology and embryology of Robinson's mouse opossum. To gather specimens and field data for his research, he worked in Panama, mostly on Barro Colorado Island, between 1929 and 1932. During the course of this work, he conducted general trapping surveys of small terrestrial mammals and kept many species in captivity for behavioral observations. One of Enders's most important general observations was how quickly many mammals responded to changes in the locations of food. Thus he noted: "[O]ne may be led to believe that a certain species is all too abundant at one time and then to wonder why all have disappeared when a change in food supply or season or both brings about a shift away from what the observer has come to consider their usual haunts" (Enders 1935, 390). When the almendro *(Dipteryx panamensis)* crop failed in March 1932, for example, white-lipped and collared peccaries, white-nosed coatis, common opossums, and other species moved from the forest into the laboratory clearing on Barro Colorado Island and ate anything that was available. "Then conditions changed almost overnight. No small mammals entered the traps, neither did large mammals disturb them. Coati, peccary, opossums, and fruit-eating bats were nowhere to be seen. Only small rodents were left about the clearing; everything else, apparently, having moved. Trap lines along the streams yielded nothing but Spiny Rats, and few of them. Imperfect bunches of bananas discarded in the clearing were not consumed. The finca was deserted" (391).

Except for Enders's work, no detailed studies of the population ecology of Panamanian rodents, marsupials, and bats existed in 1966. My research plan was therefore to systematically study small terrestrial mammals and bats at two locations in the Canal Zone for one year. My study sites included tropical dry forest at the Rodman Naval Ammunition Depot on the Pacific side of the Zone and tropical moist forest at Fort Sherman on the Atlantic side of the Zone (see map 1). The Rodman site receives about 1,750 millimeters of rain annually, but its forest canopy is mostly bare during the dry season, which runs from late December to late April. The Fort Sherman site receives about 3,250 millimeters of rain annually, and its forest canopy is intact during the dry season.

In May, Juni Barrett and I set up a five-hectare grid of one hundred trap stations in each forest. This job was slow going at Rodman, where thick vegetation limited visibility to a distance of about twenty-five meters, the distance between our grid stakes. With machetes we chopped narrow trails through a wall of greenery for our grid lines. Vines were everywhere, and I quickly discovered that some of them were armed with stout spines capable of penetrating the soles of my rubber boots as well as my scalp. I still bear the tip of one of those spines in my skull. It took us nearly a week to complete the Rodman grid. Then we repeated the process at Fort Sherman. Fortunately, the understory there was much more open, and the grid went in quickly.

In addition to putting in the trapping grids, Juni and I had to find permanent living accommodations. I helped Juni and his girlfriend, Carmen, who were from the town of Almirante in Bocas del Toro Province in northwestern Panama, move into a small apartment in Río Abajo, on the far eastern side of Panama City. Because he was black, Juni was not free to settle just anywhere in Panama City; he had to live in a black housing area. Most blacks moving to Panama City from Bocas settled in Río Abajo to be with friends and relatives, and Juni did the same. Living in this barrio meant that Juni had a long bus ride to and from the Canal Zone each day when we worked on the Pacific side. I would pick him up in front of the Hotel Tivoli, the classic Canal Zone guest house, at 7 A.M. each morning, and we then drove across the Thatcher Ferry Bridge, completed in 1955 as the "bridge between the Americas," to the Rodman site on the western side of the canal.

Marcia arrived in Panama from Michigan at the end of April. After spending two weeks at the termite-ridden Tivoli, we rented "vacation quarters" for two months in Balboa—half of a fully furnished two-story duplex owned by a Canal Zone employee—before moving into a small efficiency apartment in a housing development near Albrook Air Force Base. While I scurried about getting my research under way, Marcia did her own job hunting and ended up as an employee of the Panama Canal Company. She was hired to teach eighth and ninth grade math at Curundu Junior High, a beautiful new school near Fort Clayton.

Life in and around Balboa was quite idyllic for us, mainly because we lived close to many of the amenities available to Pan Canal Company employees. These included the "clubhouse," which contained a theater showing first-run movies, commissaries where we could buy U.S. groceries and hardware and department store items, an Olympic-size swimming pool, lighted tennis courts, and a gym for basketball, volleyball, and square dancing. Pan Canal employees had all of this (and more) plus paid furloughs back to the States every couple of years. In addition, they received a "hardship" allowance on top of their regular salary for having to endure life in a disease-ridden tropical country. It was easy to understand why Zonians were worried about losing their comfortable lifestyle as a result of the new Canal Zone treaties being negotiated by the United States and Panama in 1966 and 1967. But those treaties were never ratified. It wasn't until 1977—the year a new Canal Zone treaty was successfully negotiated—that the cushy lifestyle of the Zonians began to fade into history, a process that was completed in 1999.

As a field biologist who routinely came home from work in muddy, sweat-stained clothes, I hardly resembled the accountants and engineers who lived in our sedate, carefully manicured Canal Zone neighborhoods. But what really set me apart was our motor vehicle. Whereas most Zonians drove late-model U.S. or foreign cars, we drove a World War II–vintage Dodge army-weapons carrier. Nicknamed "the Blue Bomb," this vehicle was on loan from the Canal Zone Biological Area, the forerunner of the Smithsonian Tropical Research Institute. Distinctive features of the Bomb included large tires that cost $195 apiece, a twenty-four-volt electrical system (miserable to repair), and canvas coverings over the cab and truck bed. The bed itself was covered by a large sheet of plywood and quickly became a breeding ground for a tremendous population of cockroaches. In short order, the Blue Bomb became a mobile roach motel. In addition, after our fieldwork began in earnest, our net poles usually bore clusters of partly cleaned mammal skulls, which dangled out the back of the truck. These skulls must have caught the attention of bag boys when we went grocery shopping at the Balboa commissary. For some reason, we never had to tell them which vehicle was ours. They automatically made a beeline to the Bomb.

Before he left Panama in March, Charles Handley arranged for me to have office space in the Middle America Research Unit (MARU), a unit of the Walter Reed Army Hospital on Ancon Hill across from the sprawling Gorgas Memorial Hospital. The director of MARU was Karl Johnson, an expert on tropical diseases such as Bolivian hemorrhagic fever and Venezuelan equine encephalitis. In 1976 Karl would become famous for naming the Ebola virus while working for the Center for Disease Control in Atlanta. About thirty-five years old and sporting a ready smile behind his scraggly brown goatee, Karl was a relaxed and friendly scientist and administrator. From the start, Karl treated me as a colleague, not as a neophyte field biologist. He and Merle Kuns, another senior scientist at MARU who had trapped rodents all over South America (and even had a genus of mice, *Kunsia,* named after him), were supportive both scientifically and socially during our year in Panama. In "exchange" for use of an office, I provided MARU scientists with blood and kidneys for a survey of leptospirosis in Panamanian rodents and opossums (figure 7).

After all the preliminaries had been completed, fieldwork began in early June. The basic design of my study was very simple. Juni and I worked at each field site for ten days every lunar month. Expecting low daily capture rates, I operated my trapping grid for nine nights. The grid contained a hundred live traps set on the ground as well as a series of traps tied to trees and vines a meter or so above the ground. The traps were baited with slices of bananas and sunflower seeds. They were sensitive enough to capture juve-

Figure 7. Juni Barrett and Dr. James Gale (left), of the Middle America Research Unit, drawing blood from a spiny rat, Rodman Naval Ammunition Depot. Photo by T. Fleming.

nile rodents and bats weighing less than twenty grams but were large enough to capture adult *Didelphis* opossums weighing up to about a kilogram. All the animals I captured were given a numerical code on the basis of toe clipping and then examined and released. This was the tried-and-true method of studying small mammal populations developed by a generation of ecologists in north-temperate habitats, beginning in the 1930s. It had seldom been employed, however, in the tropics.

While I ran the grids, Juni operated transects of a hundred or more live traps at other locations in our study areas. All of the animals that Juni caught were killed and preserved. The skins were prepared "flat," that is, pinned out to dry on cardboard trays for the study of molt seasonality. Skulls were partly cleaned and dried for the study of tooth wear, which I used to determine survivorship and the age structure of my populations. The occasional death of animals of known age allowed me to "calibrate" each species' tooth wear curve. Reproductive tracts were preserved in formalin for later histological examination; stomachs were similarly preserved. We also operated several mist nets at each site for five nights a month. The nets were open for three or four hours beginning at sunset. All of the bats we captured were quickly killed and preserved in formalin for detailed analyses of reproduction and diet. To characterize the vegetation on my two grids, I also collected and preserved flowering and fruiting specimens and made crude estimates of the amounts and kinds of seeds and fruits that were available for consumption by mammals each month.

This, then, was the basic daily routine that we followed each month for one year: check the traps, preserve the animals caught in Juni's "supplementary" lines, set up mist nets, write up field notes, open the nets for several hours, preserve the bats, and fall into bed. When we worked at Rodman, we lived at home, but at Fort Sherman we lived on the army base a few kilometers north of the live-trap grid. On days that we weren't going to net bats, we sometimes took the Panama Canal train back across the isthmus to spend a night at home. The eighty-kilometer train route skirted the east side of the canal for much of its length, and the ride was always relaxing. I enjoyed seeing the ships heading north or south in Lago Gatún and Barro Colorado Island in the middle of the lake. The next morning we took an early train back to the Atlantic side to check our traps.

We always rode in second-class cars, equipped with wooden benches, rather than in more comfortable first-class cars to avoid any incidents in response to Juni's race. The two-class railroad system struck me as an anachronism harking back to the days when the Panama Canal Company adhered to the "gold" and "silver" form of segregation. During the early years of the Canal Zone, North American workers were paid in gold, and all other employees were paid in silver. Until the mid-1940s, Panamanian employees of the company even had to live in separate towns from North American employees, and public restrooms and drinking fountains were also segregated by nationality. While overt signs of racial and ethnic discrimination were gone by the mid-1960s, Marcia and I found that the attitudes of many North American Canal Zone employees toward non-Anglos were still quite bigoted.

Considering the effort that it took to reach the train station by bus from Río Abajo, it amazes me that Juni was on the train, hanging out a window with a big smile and waving to me, each morning when we returned to Fort Sherman. But Juni was a loyal and conscientious assistant. The only times he did not show up for work were real emergencies: once when he went to the hospital with Carmen when she had a medical problem and twice when he was briefly put in jail in Panama City for an altercation with "some Panamanian guy" (i.e., a Hispanic). On the latter two occasions I had to run around in Panama City, first to find out where Juni was being held and then to arrange for his release. He was always apologetic in these instances.

Juni and I were about the same age, and we got along well. Somewhat shorter than I am, he was lean, muscular, and athletic. His hair was close cropped, and he wore a thin goatee. His favorite sport was baseball, not soccer, which fit his rangy build. He probably had a grade-school education but had spent much of his youth in Bocas playing outdoors and hunting and fishing. As a result, he was a keen observer and always spotted more animals in the forest than I did. We spent many hours talking about differences between life in Bocas and the United States. I never tired of listening to his Jamaican patois and his colorful way of describing things. I've forgotten much of our conversation, but I do remember that he would always describe a snake sticking out its tongue by saying, "Dat snake, 'e was longin' out 'is tongue."

Until I got used to his dialect, I sometimes had trouble understanding what Juni was saying. For example, one day he told me he had just seen an 'awk chasing a squirrel. I first pictured an extinct arctic seabird flopping around the forest after a squirrel before I finally caught on to what he was saying. On another occasion he described the first time he met Carmen: "Den one day I met dis girl from I-T." I asked, "Where's I-T?" to which he replied, "Oh, it be one of dose islands [Haiti] near Jamaica."

Juni and I worked together day and night collecting data and specimens. In one year we captured twenty-one species of rodents, five species of opossums, and thirty-five species of bats. My two grids yielded nearly thirty-four hundred captures and recaptures of rodents and opossums, and Juni's trap lines produced about eleven hundred specimens of rodents and opossums. We also preserved twenty-three hundred bats. Just in walking my grids ten times each trapping session for a year, I covered at least six hundred kilometers, stopping every twenty-five meters to check and bait a trap. We accumulated about eighteen hundred net-hours (each mist net open for one hour) in our bat work.

These are the raw numbers that I eventually used to write a series of papers on the biology of Panamanian mammals, but what was it actually like gathering those data? What did I experience and learn on a daily basis, and did this work ever become tediously routine? In one sense, the work *was* tedious, because it involved a very systematic set of actions that were nearly invariant from day to day. In such a situation, it was easy to think ahead and imagine what it would feel like when the fieldwork was done, to be impatient to complete the immediate task at hand (gathering the data) and get on with the fun part (the analysis of data and making sense of the results). In another sense, there was always some uncertainty involved in the work. Would capture rates fall off, so that I would end up with too few data to conclude anything meaningful about the lives of Panamanian mammals?

Where should Juni put his trap lines, and where should we put our bat nets to get good capture rates each month? How long would it be before the Blue Bomb had another breakdown? And most immediately, when was it going to rain? How wet would I get checking the grid that day, and would we be able to net bats that night?

We began the study in the wet season, which in 1966 was extremely wet on the Pacific side of the Canal Zone. Nearly 750 millimeters of rain fell in October alone, making that month the wettest in Balboa's history. Almost every day at Rodman we would get soaked with rain while checking our traps. Because of the warm air temperatures, I didn't feel cold in the rain. But my glasses were constantly drenched, and it was hard to keep them dry enough to see clearly. I vowed that if I ever returned to the tropics to work, I would get contact lenses. I carried a golf umbrella with me and huddled under it each time I had to examine a captured animal and record data. To keep my captives dry, I covered each of my traps with a tarpaper "sleeve." Heavy rains and wind in October knocked down branches and trees on the Rodman grid. Deep pools of water accumulated in low-lying areas, and water streamed over my boot tops as it ran off the grid during and after rains. The one small stream on the grid was full of roiling water at the height of the rainy season. Surprisingly, although the Atlantic side of the Zone receives nearly twice as much rain as the Pacific side, I managed to stay drier (in a relative sense; I was always thoroughly soaked with sweat even when it wasn't raining) at Sherman, because the rain fell mainly at night or just before dawn rather than well after sunrise.

The heavy Pacific coast rains in the latter part of 1966 literally and figuratively dampened my enthusiasm for tropical fieldwork. I couldn't wait for the arrival of the dry season in late December. At the depths of my rain-caused depression, I read George Schaller's *The Year of the Gorilla,* which recounted in popular form his experiences in central Africa during his doctoral research on the mountain gorilla. Reading about the wet and cold conditions that George endured made me feel a lot better about my own situation. At least I had a warm shower to look forward to each day. Schaller's book gave me renewed enthusiasm for my work.

Although it seemed as though it took forever, the dry season finally did arrive pretty much on schedule, just after Christmas. The ensuing four months of little or no rain at both field sites resulted in dramatic changes in the vegetation. Pools of water and streams dried up. By March the ground was hard and cracked and covered with a thick layer of crisp, dry leaves. No longer could I (or other terrestrial mammals) walk silently through the forest. Many of the canopy trees lost their leaves at Rodman, but only a few did so at Sherman. Dry season was prime time for the flowering of canopy trees. Brilliant yellow flowers of poro-poro and bright red flowers of palo santo were everywhere at Rodman. The more subtle yellows and whites of *Cordia* and *Luehea* flowers also graced the canopy. The hills at Rodman and elsewhere in the Zone were splashed with purple *Jacaranda* and orange *Erythrina* flowers. Brilliant blue skies and strong northeast trade winds replaced the heavy clouds and muggy air of the wet season. Fieldwork again became enjoyable.

The rainy season returned with a series of hard afternoon showers at the end of April. Almost overnight vegetation in the Rodman forest turned a deeper, richer shade of green. The layer of dead leaves on the forest floor became water-logged and matted. Small pools of water formed in low-lying areas, and Túngara frogs, the main food of the fringe-lipped bat, began breeding in them almost immediately. Leaf cutter ants, which had switched to nocturnal foraging during the dry season, again began to carry crescent-shaped slices of leaves along their well-worn paths during the day. It was time to get out the golf umbrella again.

Despite the physical discomfort of the wet season and the routine nature of our data collection, each day was just different enough to make fieldwork enjoyable. It was always interesting to see how many and which animals were going to show up in the traps each day. The bulk of the captures at Rodman came from three species of rodents: the Panamanian spiny pocket mouse, *Liomys adspersus* (of the kangaroo rat family Heteromyidae), and two species that I knew well from the two collecting trips, the Talamancan rice rat and the spiny rat. The next three most common species at Rodman were Robinson's mouse opossum, the common opossum, and the gray four-eyed opossum. My most common captures at Fort Sherman were rice and spiny rats; common opossums and the gray four-eyed opossum were also relatively common there.

My capture rates started out slowly on both grids, but by the latter part of the wet season, in October and November, I was capturing about thirty-two and twenty-five animals per night at Rodman and Sherman, respectively. My all-time record was fifty-three captures (in 120 traps), including twenty-four spiny pocket mice, twelve spiny rats, fifteen rice rats, and six mouse opossums in October at Rodman. Charles Handley was pleased to learn about my "phenomenal" trap success, because the prevailing wisdom at the time was that many species occurred in the tropics, but they had low population densities. My grids contained respectable numbers of rodent species (sixteen at Rodman, thirteen at Sherman), and at least three species could be considered "common," even by temperate zone standards.

Other than data on reproduction, survivorship, and home range sizes, the

Figure 8. Three Panamanian rodents: Tomes spiny rat *(Proechimys semispinosus)*, Talamancan rice rat *(Oryzomys talamancae)*, and spiny pocket mouse (*Liomys adspersus;* from left). Last two species redrawn, with permission, from illustrations by F. A. Reid, *A field guide to the mammals of Central America and southeast Mexico,* Oxford University Press.

grid captures provided me with relatively little information about the behavior of the different species of rodents and marsupials. Once I released them, most individuals immediately dashed out of sight. This certainly was true of individuals of rice and spiny rats (figure 8). Rice rats were especially anxious to get away from me. They often dashed off quickly, wildly crashing into vegetation in their desire to escape. In contrast, spiny pocket mice usually were calm upon release. These gray and white animals would slowly walk or hop off a ways until they reached an obviously familiar spot on the ground, probably part of their invisible (to me) trail system, and then headed off with speed and confidence. Unlike their desert-dwelling relatives—pocket mice and kangaroo rats—male spiny pocket mice (and their wet-forest relatives, forest spiny pocket mice) have very large testes and epididymides (masses of sperm storage tubules that project into the scrotum) during the breeding season. The occurrence of grossly enlarged testes suggests that the mating systems of tropical heteromyids involve considerable sperm competition among males as a result of several individuals mating with the same sexually receptive female. This hypothesis, however, has not yet been tested. When moving faster than a walk, male *Liomys* held their rear ends high to raise their bulging scrotum and testes off the ground. Both males and females of this species often ran directly to inconspicuous burrows and disappeared. With time, I was able to locate many burrows used by frequently captured individuals. Though heteromyid rodents of the arid U.S. Southwest are asocial and solitary animals, my spiny pocket mice seemed to share burrow systems. Different individuals sometimes entered the same burrows on successive days, and their home ranges overlapped extensively. Intraspecific tolerance, rather than aggression, seemed to rule their lives.

Of all the animals that I encountered in my traps, my favorite was Robinson's mouse opossum, *Marmosa robinsoni,* the subject of Robert Enders's reproductive research. Tan in color, this little marsupial has large, flexible "Mickey Mouse" ears, large black eyes surrounded by a narrow black mask, a pointed snout, and a long naked tail. As Enders had observed, *Marmosa* was much more common in the thick, viney forest at Rodman than in the taller, more open forest at Sherman. I marked and released sixty-five individuals at Rodman compared with only nine at Sherman.

Although adults usually weighed less than a hundred grams, *Marmosa* acted as if it were a large, ferocious beast. Whenever I approached a trapped individual, it invariably reared back on its hind legs, opened its mouth wide to display its numerous sharp teeth, and hissed, just like its much larger relatives. Much of this fierce display was bluff, even in the large opossums, and individuals rarely attempted to bite when I removed them from a trap with a bare hand. When released, most individuals tried to climb the nearest object, even if it was only a foot tall. One male dashed up a short, thin palm stump, continued "climbing" in thin air when he overshot the top, and then crashed back to earth at my feet. Shaking itself off, the creature ran to a sapling and successfully ascended into the subcanopy. Mouse opossums climbed the spiny trunks of *Bactris* palms with impunity as well as most other vertical surfaces. After climbing to a height of about two meters, individuals would stop for several minutes, seemingly to catch their breath, before proceeding.

We watched *Marmosa* climb along vines on several occasions at night. Individuals were always solitary, but Juni captured an adult male and female together in one trap in November. We never found any *Marmosa* leaf nests in the forest, but in July through September an adult female nested in the plywood shack where we kept bananas and skinned our supplementary animals near the Sherman grid. Her nest in July was about thirty centimeters in diameter and consisted of an outer layer of fresh green leaves and an inner layer of dead leaves. When I disturbed the nest, the small opossum stuck her masked face out of the entrance before quickly ducking back inside. In August she had moved her nest to a darker corner of the shack and was nursing four well-furred babies. Since females give birth to eleven to thirteen embryonic young, this female had lost a considerable portion of her litter by August. In September, her litter had decreased to three nearly independent young.

The fourth day of December was a notable date on the Rodman grid, because it marked my first capture of a supposedly "rare" rodent. On that day I caught not one but three males of the vesper rat in tree traps located near each other. Gentle in disposition, this rodent superficially resembles Robinson's mouse opossum: it is about the same size, its dorsal pelage is a rich tawny brown, and its large black eyes are surrounded by a black mask. Unlike *Marmosa,* however, its long tail is covered with a brush of short dark fur. Charles Handley had collected only a handful of vesper rats in Panama and was astounded to learn that I apparently had a colony of these arboreal rodents living in the middle of my grid. He wanted me to collect every individual that I caught and make the first decent taxonomic "series" from Panama. Being an ecologist rather than a taxonomist, I didn't do this and instead marked and released thirty-four individuals, capturing and recapturing them nearly eighty times over the next seven months.

The vesper rat was truly a denizen of the forest canopy. The few individuals that I caught on the ground hopped slowly to the nearest tree and quickly ascended into the canopy. Individuals ran much more swiftly along vines and branches than they did on the ground. One female ascended a hog plum tree toward a leafy nest, which I presumed was her home. Another climbed into a small fruiting tree bearing white berries. It deftly moved along thin branches, nosing different berries before taking one in its mouth. Then it climbed up a vine, where it ate the berry and its single large seed. From the locations of captures and recaptures, I learned that the home ranges of vesper rats were large—up to a hundred meters in diameter—and overlapped extensively.

Why did this apparently common rodent take nearly six months to begin entering my traps? It is highly unlikely that vesper rats suddenly "invaded" the grid in December. They were undoubtedly living on the grid when we began trapping in June. Their sudden interest in traps most likely was caused by a general decrease in the availability of fruit at the end of the wet season, as witnessed by Robert Enders on Barro Colorado Island thirty years earlier. Not only vesper rats but also a variety of other mammals, including bats, began showing up in the traps in numbers in the late wet season and early dry season, a time when my monthly surveys indicated fruit was particularly scarce.

This fruit scarcity made *Didelphis* opossums a downright nuisance on both grids. Starting in October, individuals began to raid my traps and steal the banana bait without getting caught. On some days, nearly every trap that didn't catch something was closed and baitless by morning. One evening, Juni and I staked out a series of traps at Sherman and watched half-grown opossums crawl partway into traps, eat the banana bait, and then back out to visit other traps. Nearly half of the traps in two grid lines were stripped of their bait and closed by 7:30 P.M. From then on we alternately put pieces of carrot and banana slices in traps to reduce the bait raiding. Opossums bothered only the banana-baited traps, and their raids decreased considerably in April, when fruit levels were on the upswing.

In addition to the rodents and marsupials that routinely showed up in our traps, we encountered many other animals on the grids. The mammals I saw most frequently during the morning trap checks were agoutis, red-tailed squirrels, and coatis. At least eight agoutis lived on the Sherman grid. Usually solitary, these hare-sized, stub-tailed, rusty red rodents would sometimes bound away from me with a surprised "bark." After traveling a short distance, they would freeze until I was out of sight. More often, however, they would quietly slip away as inconspicuously as possible when they saw me. I often spotted them only when sunlight passed through their thin, naked ears, turning them bright pink. On other occasions, their loud chewing of hard seeds, which they held in their front paws while sitting on their haunches, would catch my attention.

Red-tailed squirrels and female coatis were often conspicuous as a result of their group activities. The red-tailed squirrels sometimes formed noisy groups of six to eight individuals that chased each other on the ground and in low vegetation, all the while grunting and squealing. These groups probably contained several males that were intent on mating with a sexually receptive female. However, I usually encountered solitary male coatis on the Rodman grid. These animals were a nuisance because they regularly killed animals in my traps. They attacked rodents and mouse opossums as well as larger species such as *Didelphis* and four-eyed opossums.

Unlike the solitary males, female coatis are gregarious and live in groups containing several females and their offspring year-round. In August, Juni and I encountered a group of at least forty female coatis and their young feeding in two fruiting trees on the Sherman grid. We were drawn to the trees by loud squeaking and rustling coming from the canopy. As we approached, the coatis began to "panic" and quickly descended the tree trunks headfirst. When they got within five meters of the ground, adults and young jumped the rest of the way, hitting the ground with a *thump!* and then scattering in all directions. At one point at least twenty coatis seemed to pour out of one of the trees. The undergrowth was alive with scurrying, grunting *pizotes,* dashing away with their long dark tails held vertically like periscopes. Youngsters tried valiantly to keep up with their faster moms but temporarily became separated from them. As quickly as it had begun, the coati melee disappeared, and the forest became quiet once again.

Monkeys and tamanduas were other mammals that I occasionally encountered on the grids. The thick second-growth forest at Rodman was perfect habitat for Geoffroy's tamarin, the smallest of Panama's primates. We sometimes spotted small groups crossing over the Rodman roads along overhanging palm leaves. We more frequently encountered western night monkeys while netting bats. Fruiting *Cecropia* trees were good places to find night monkeys as well as woolly opossums and kinkajous. One night we were watching a woolly opossum feeding in a *Cecropia* when I spotted two night monkeys, twittering like birds, in an adjacent *Cecropia*. They were quickly joined by a third individual of this monogamous, family-dwelling species. The first two monkeys appeared to be mating and then separated and moved out of the tree. The third monkey remained behind, silently leaping from branch to branch and eating fruit.

The monkey most frequently found on the Sherman grid was the white-faced capuchin, the "organ grinder monkey" of my childhood days in Detroit. A small troop of this medium-sized, black and white monkey moved onto the grid to feed in mango trees during their fruiting season. Well-scattered in the canopy, these monkeys kept in contact with each other through doglike barks. Whenever they spotted me, they maneuvered about in a tree to get a better look and sometimes broke off branches and dropped them to the ground. I never had the impression that they were hurling the branches at me, however.

We encountered tamanduas by day and night in both forests. One morning when I heard rustling noises in tall grass near a trap, I stood still, expecting to see a *Didelphis* opossum or an armadillo ambling along. Instead, a female tamandua with a baby clinging to her back emerged from the grass. Head down, she walked right up to me, raised up slightly when I softly spoke to her, grunted once, and quickly retreated into the grass. In January along the road at Rodman, Juni and I spotted an adult female, her nipples still enlarged from recent lactation. Foolishly thinking that we were cowboys, we tried to lasso her with a short bit of rope. Sitting back on her haunches, she lashed out at us with forepaws armed with long, sharp claws. She managed to nick both of us before we gave up the struggle and let her retreat in peace. I later read in Phillip Bole's account of the birds and mammals of Panama's Azuero Peninsula how a wounded adult tamandua put its claws through his leather shoe, narrowly missing his toes (Aldrich and Bole 1937). Tamanduas are not animals to be messed with by mere mortals!

On another occasion, I again seriously underestimated the power of another toothless edentate—the silky anteater. A Panamanian had brought an individual to MARU, and people wanted to know if I would care for it. A little larger than a mouse opossum, this golden-furred, woolly anteater lives in the forest canopy, where it is seldom seen. Knowing the creature wouldn't bite me, I casually reached into the box to pick it up. Before I realized what was happening, the anteater sank one of its razor-sharp foreclaws into the base of my left thumbnail and nearly ripped it off. After this rude awakening, I treated the animal with much more respect while keeping it in a mesh cage at home for a couple of days. Its doll-like appearance and slow-motion climbing behavior made it seem like a wind-up toy. I tried to feed it a mixture of hamburger, egg, and crushed ants. When it refused to eat, I released it in the Rodman forest.

I went to Panama expecting to see lots of snakes in the jungle, but this was not what I experienced. As was the case during the two collecting trips, I actually encountered rather few snakes in the two forests of the Canal Zone during my field study. Thick-bodied common boas and the long, black and cream rat snake *Spilotes pullatus* were the species I saw most commonly. Giving in to my boyhood interests, I usually caught these snakes and held them briefly before releasing them. Actually, the boas, which sometimes reached a length of three meters, were as likely to have a firm a grip on me with a coil or two as I was on them ("Quick, Juni, get this damn boa off my neck!"). Their muscular power was impressive. *Spilotes* was a less powerful but still impressive snake. My usual way of catching them was to step on their tail to detain them until I could grab them by the neck. Upon being pinned by the tail, one three-meter individual reared up in cobra fashion and struck me on the chest before I could grab it. Both of these species undoubtedly were important predators of my rodents and opossums (probably only the boa preyed on the latter), but I never gained any direct evidence of this.

I encountered only one pit viper during my eighteen months in Panama. This snake, a two foot-long individual, was coiled on top of a trap containing a rice rat. I spotted the rodent first and didn't notice the snake until I had picked up the trap to look at the mouse. Muttering an automatic "whoops," I gently set the trap back down without agitating the snake. Then, thinking that it was merely a fer-de-lance, I pinned its head with my umbrella and picked it up. When I showed the snake to Juni, he chastised me for handling a poisonous species so casually. I preserved it, and it was added to the U.S. National Museum's reptile collection.

Two decades later, the Cornell University herpetologist Harry Greene, who was studying the diets of carnivorous snakes on the basis of museum collections, asked me how I happened to capture a bushmaster in Panama. "That little snake was a bushmaster?" I asked incredulously. Having since learned how dangerous bushmaster bites are from working in bushmaster country in Costa Rica (they are usually fatal), I briefly shuddered at the thought of "what if . . ." had happened way back in Panama. But then I relaxed. That snake, like the much bigger ones I have encountered in Costa Rica, was sluggish and unaggressive. It was easier to capture than most of the harmless garter snakes of my childhood.

During my year of fieldwork in the Canal Zone bats were very much a side issue with me because of the time we devoted to livetrapping and preserving rodents and marsupials. Whenever possible, however, we netted bats at both sites and caught quite a few species and individuals with this effort. But we saw relatively few bats away from our nets, and the bats we captured were quickly killed and preserved. Although I didn't gain a particularly deep understanding of their behavior from this work, I did learn a great deal about their general diversity, reproduction, and diets.

One of the enjoyable aspects of working at night as well as during the day was seeing life in a tropical forest from an entirely different perspective. The transition from daylight to night, when diurnal species went to their sleeping quarters and the night shift became active, was a particularly pleasant time to be in the field. My field notes recorded at least two of these transitions:

> 8 January 1967 at Rodman: The sun went down in a beautiful blaze of bronze. I watched bats flying in the clearing above the Emergency Air Raid Shelter [near our grid]. A medium-sized bat (probably a mustached bat) flew around issuing regular, sharp clicks. As it chased an insect, the clicks speeded up. Small molossids flew around in rapid, steady flight. A *Myotis* dodged and darted around in its feeding territory. The bats moved out of the clearing to feed elsewhere as it got darker.
>
> 12 April 1967 at Sherman: A sac-winged bat *Saccopteryx bilineata* leaves its roost [the buttress of a large tree] a little before 1800 and flies silently around the forest understory hawking insects. It's interesting to sit in the forest in the dark and hear things moving around in the underbrush and the trees. The sound of falling fruit is common, especially espave. We saw few of the animals we heard. One *Proechimys* ducked under a log near a pool of water. An olingo . . . jumped gracefully from one tall cativo *[Prioria copeifera]* to another and continued along its arboreal highway. Along a limestone ridge we saw several scorpions, which would duck into cracks in the stone when our lights hit them. A small wren is nesting in a small, shallow tunnel in the limestone.

Being out at night attending our bat nets gave us fleeting glimpses of many other kinds of nocturnal mammals. In trees and vines we spotted four kinds of marsupials as well as kinkajous, olingos, night monkeys, and two species of arboreal rodents. On the ground we saw our three common rodent species *(Liomys, Proechimys,* and *Oryzomys capito)* as well as pacas, three species of marsupials, nine-banded armadillos, tamanduas, raccoons, and white-tailed deer. Though we never saw any cats, we sometimes heard their screams in the distance and found ocelot and puma tracks in the mud.

> 27 April at Rodman: [This] was a good night for spotting animals. On our first walk up the road [to check our nets], we saw a large spotted owl perched on a tree limb near the road. We checked the entrance to an ammunition bunker and saw a *Carollia perspicillata* eating an espave fruit, which it dropped when we disturbed it. There were many such fruits on the ground under this night roost. I saw a couple of northern pygmy rice rats [small, grassland-inhabiting relatives of *Oryzomys capito*] around the truck. They kept popping into and out of the "bush." We spotted a kinkajou high in an *Anacardium* tree. An adult male *Didelphis* was caught raiding one net. It had killed several bats in the bottom shelf and was devouring an *Artibeus lituratus* [a big bat] when we chased it off. Then we saw a *Philander* opossum hurry up a steep bank and disappear. Frogs of at least three species were chorusing in the pools of water that had formed in today's rain [at the beginning of the wet season]. *Engystomops [Physolaema]* and *Leptodactylus* frogs were quite vocal along with a tiny, light-colored "cricket" frog. We watched a female *Didelphis* with a great bulging pouch [of babies] run off near one of the water holes.

As in the case of our livetrapping, our success in catching bats varied seasonally, probably in response to the same environmental factor—changes in the availability of food. Whereas our trap captures *increased* late in the wet season as food became scarce, our bat captures *decreased* then. Rodents and marsupials entered our traps (or raided them) as natural food became scarce, and bats probably moved to different areas in search of new sources of fruit and flowers. In support of the food shortage hypothesis, I even captured bats in my banana-baited live traps in October, November, and December. At Sherman in November, for example, I caught seven short-tailed fruit bats in ground traps on one night, four the next night, and two on the third night. In October and November, bats entered our skinning shack at Sherman to eat the bananas we stored there and also entered the back of the Blue Bomb whenever it contained bananas.

The response of certain kinds of bats to local concentrations of food—usually fruiting fig trees—was truly impressive, as my notes from Sherman on 27 September attest:

> What a fantastic night for *Uroderma bilobatum!* All but a handful of the [fifty-six] bats were taken in the net set under a short fruiting fig

> tree. When we arrived at the net, shortly before 1830, many bats were flying into and out of the fig tree, and figs were falling to the ground; this was before it was very dark. With the aid of my headlamp, I watched *Uroderma* feed: the bats would fly up to a cluster of figs, which were closely attached to a branch, land for an instant on the cluster, and then bite off one fruit and fly away. The bats were sensitive to the light and often fled from the fruit as soon as the light hit them. It seemed like scores of bats were flying around the tree. When they grabbed one fruit, the bats often knocked others to the ground. The bats were silent as they fed, and I saw no conflicts between feeders. Each bat was in the tree for only a few seconds, and then it was flying again.
>
> At first the net failed to catch anything, but within 10–15 minutes after it was opened, it began to catch *Uroderma*. As I began to empty the net more bats flocked around, particularly when a captured bat began to squeal. When *Uroderma* squealed, many bats arrived at the net, many avoided the net, some bounced off of it, and a few got caught. Bats didn't seem to converge directly on the "squealer" but just all around the area in general. They never flew into me, but I was really surrounded by bats a couple of times. So many bats were in the net that I could only clear about one-third of it before it was full again. The net was only open for half an hour, and we caught more than 65 bats of three species. When we left the area, at around 1930, the bats were still actively feeding in seemingly undiminished numbers.

Over the years, I have run into other "swarms" of stenodermatine bats visiting fig trees. In his work with the Jamaican fruit-eating bat and other species on Barro Colorado Island, Charles Handley has reported that bats arrive and leave fruiting fig trees in "pulses." I have often wondered how so many bats, roosting in small groups throughout the forest, locate these scattered fruit sources. Do bats living in different roosts converge independently at these trees? What sensory cues do they use to locate fig trees; do fruiting fig trees create a substantial odor plume? From how large an area do these trees attract bats, and how do the bats coordinate their pulse-like arrivals and departures? Working out the details of how fig-eating bats track their ever-changing food supply would be a fascinating research project.

All too soon my year of trapping and netting was over. At one point, Charles Handley had indicated that he might send a replacement to continue the trapping for another year. But in April 1967 he indicated that his grant funds simply were stretched too thin to pay for another year of data from Panama, as valuable as those data would be. Therefore in June I reluctantly shut down my grids by sacrificing all of my marked animals, many of whose lives I had followed for months, for tooth wear and reproductive data. The project had had its low points (principally, difficult weather, annoying trap disturbances, and a finicky field vehicle), but I was enthusiastic about how things had turned out. The Rodman and Sherman field sites had been good choices for conducting the first intensive population studies of small mammals in Panama. I had more than enough data for a thesis and was anxious to begin the next phase of the project: analysis and preparation of the data for publication. This next phase ended up taking far longer than the actual fieldwork.

Part of the success of the project undoubtedly resulted from working on military property that was off bounds to civilian disturbance and hunting; Rodman was entirely so, and Sherman was partly so. But proximity to the military had its trying moments. In July 1966, for example, I arrived at my Sherman trapping site to find a squad of fifty army recruits—soon to be on their way to Vietnam—setting up a jungle camp, complete with thatch-roofed "hootches," along the eastern edge of the grid. They had cut down a fair number of small trees and had removed the large leaves of many *Corozo* palms on the grid. Thatch collecting continued near my grid in October, and when I asked one of the soldiers how much longer they would be cutting palm leaves, he replied, "Until we destroy the whole f—— jungle!" His attitude toward tropical diversity obviously didn't exactly coincide with mine. The final environmental insult came in May 1967 at Sherman when two men from the U.S. Army's Tropic Test Center told me they were about to test a new (secret) weapon on my grid. I persuaded them to conduct the test, which was supposed to destroy an area with a radius of thirty meters, south of my grid. This was, I suppose, a small victory for me but a substantial loss for the forest.

Not all my encounters with the military were quite so negative. A place called Devil's Beach was one of our most productive netting sites at Fort Sherman. We had just opened our nets there in November when a squad of soldiers arrived and politely asked us to abandon the area for an hour. They were setting up a mock ambush along the beach road. At 7:30, all hell broke loose in the dark. Automatic-weapon fire and mass confusion rocked the area. Acrid smoke filled the air. Suddenly, all was quiet again, the army quickly departed, and we resumed checking our nets. As you can imagine, with all that noise and disturbance, however, we caught few bats. I'd be willing to bet that the firefight sent lots of bats winging their way to quieter parts of the Canal Zone that night. Because of the intensifying Vietnam War, these kinds of training exercises were common throughout this part of the Canal Zone, and, as a soldier once told me, firefights sometimes ended up with unplanned participants. Panamanian hunters sometimes got caught in the middle of mock battles. The trouble was that, whereas the soldiers were firing blanks at each other, the Panamanians, who were undoubtedly scared witless, were firing live ammunition.

Fieldwork was much calmer at Rodman. The only military personnel that we encountered there were Air Force military police patrolling the ammunition depot day and night in trucks. I emerged from the jungle one night after checking my bat nets and ran into one such patrol. The two servicemen, both African Americans, were amazed to see me. One exclaimed, "You mean to tell me you go into that jungle at night without a gun or anything? Man, you couldn't pay me enough to do something like that! And you do it just to catch bats? Ma-a-a-n, you must be crazy!" Then he proceeded to tell me that he was much more scared driving around this place than he had been at any time during his two-year hitch in Vietnam. I thought to myself, "Ma-a-a-n, *you* must be crazy!"

After packing up my data and specimens and our belongings and saying good-bye to all of our friends, including Juni, Marcia and I left the Canal Zone aboard the S.S. *Cristobal*, a Panama Canal steamer, at the end of June. Because she had been an employee of the Pan Canal Company, Marcia qualified for boat passage from Cristobal on the Atlantic end of the Canal Zone to New Orleans. During the four-day trip, we reminisced about all that we had seen and done in the last fourteen months. Although our work had dominated our lives, we also managed to make a lot of friends from the Balboa Union Church, the armed services, and the scientific community and had a fun social life. We also enjoyed being tourists in Panama by exploring the city, including the ruins of Panama Viejo, where Edward Goldman had collected *Carollia* and *Glossophaga* bats in a vaulted cellar in 1912, attending the February carnival festivities in downtown Panama City, and taking weekend trips away from the Zone.

Once we were back in the States, we visited both of our families and gave the first of numerous slide shows of our adventures in Panama. Then we resettled in Ann Arbor to resume our lives, I as a graduate student and Marcia as a junior high school math teacher. In addition to course work, I began acquiring two basic tools that I needed for my research: elementary computer programming and histology. In the late 1960s, academic computing was done on IBM mainframe machines, and data were read into the computer from punch cards. There were no remote terminals at Michigan yet, and decks of cards had to be taken to the computer center to be run. The turnaround time on data analyses usually was several hours, and it often took several trips to the computer center before all of the typing errors and other "bugs" had been eliminated from the command cards and data set so that the analysis would run. Compared with today's desktop computing, the mainframe system of the 1960s was slow and cumbersome.

I needed to learn histological techniques to study details of the reproductive biology of Panamanian mammals. I planned to make slides of the ovaries of females to document cycles of egg production and ovulation and slides of the testes and epididymides of males to document cycles of sperm production. Lois Lowenthal, the zoology department's histologist, enthusiastically adopted me and taught me basic histological procedures in her lab. When I wasn't attending classes, I spent hours dehydrating my preserved rodent and marsupial tissues, embedding them in wax, slicing the organs at a thickness of ten microns, mounting my thin slices on glass slides, and staining the tissues purple and pink with hematoxylin and eosin. Then I spent hours viewing my slides and recording data from them. After we moved to St. Louis in 1969, I taught these techniques to Marcia and an undergraduate student, and they prepared all of the slides of bat tissues for me.

So what did my year of Panamanian fieldwork reveal about seasonal variation in the lives of tropical mammals? It is important to remember that, despite its proximity to the equator, Panama still has a very seasonal environment. Daily and monthly fluctuations in air temperature are minimal, but rainfall in central Panama is strongly seasonal. The alternation between a four-month dry season and an eight-month wet season has a strong effect on the lives of Panamanian plants and animals. For example, my botanical observations plus those of other researchers indicated that the dry season was a major flowering time at all levels of the forest, from grasses and shrubs at ground level up to the top of the canopy. Closely following the flowering peak was a fruiting peak. Fruit availability was greatest from March through September and was scarce at the end of the wet season and early in the dry season. Given these strong seasonal shifts in food availability, it was not surprising to find that many aspects of the lives of Panamanian mammals, including reproduction, population dynamics, molt, and behavior, all showed significant seasonal variation.

One of the most obvious seasonal responses was in reproduction. A review of the reproductive cycles of Panamanian mammals based on my data and published observations revealed that twenty-three of forty-five species, including all marsupials, many bats, a few rodents, most primates, and the coati, were strictly seasonal breeders. The spiny pocket mouse is a good example of a strongly seasonal breeder. I captured pregnant females during

December through April, and most adults produced only one litter a year. Peak testis size and sperm production in males occurred during October through April. In contrast, my other two common rodents, the rice rat and spiny rat, reproduced year-round. Females in these species produced up to five litters a year, and adult males had enlarged testes and produced sperm year-round.

My monthly collections of bats also provided clear evidence of reproductive seasonality in these animals. Frugivorous species such as the Jamaican fruit-eating bat, the common tent-making bat, and the short-tailed fruit bat showed two annual pregnancy and lactation periods resulting in two births per female per year. One birth occurred in March or April toward the end of the dry season, and the other occurred in July or August in the middle of the wet season—times when fruit levels were high. This reproductive pattern, technically known as bimodal polyestry, turns out to be common in many species of fruit-eating bats, in both Old and New World tropics. It contrasts sharply with the pattern of seasonal monestry (one pregnancy per year) found in virtually all species of temperate insectivorous bats as well as in many species of tropical insectivorous and carnivorous bats. An exception to seasonal monestry occurs in the black myotis, studied by Don Wilson on Barro Colorado Island. Like many plant-visiting phyllostomids, this small insectivore is polyestrous, and females undergo two or three pregnancies each year. The general message that emerges from these reproductive studies is that, in contrast with the glowing picture painted by Theodosius Dobzhansky, life is far from benign and food-rich in the tropics for most species of mammals. Supplies of insects, fruit, flowers, and seeds undergo substantial seasonal fluctuations in the tropics. This, in turn, has favored the evolution of seasonal reproductive cycles in many tropical mammals.

Careful scrutiny of my histological slides revealed a real surprise—my first novel scientific discovery—regarding the reproductive cycle of the Jamaican fruit-eating bat. Whereas females of other fruit bats showed no evidence of ovulation after their July or August births, my slides revealed that many females of *Artibeus jamaicensis* conceived another baby at this time. Microscopic examination of the reproductive tracts of females captured in September through November indicated that most were carrying a single tiny embryo firmly implanted in their uterus. However, instead of growing at their normal rates, which would have resulted in births occurring in October or November, when food levels were low, these embryos remained tiny and did not begin growing until late November. By December they were macroscopically visible. These babies were born in March or April after a gestation period of about seven months compared with a "normal" gestation period of about four months.

Long after my discovery of delayed embryonic development in the Jamaican fruit-eating bat, Paul Heideman found another example of delayed development in a frugivorous bat—this time in a small pteropodid bat, *Haplonycteris fischeri,* that lives in rain forests in the Philippines. As in the Jamaican fruit-eating bat, delayed development allows late pregnancy, lactation, and weaning in *Haplonycteris* to occur at the end of the dry season and early in the wet season, when fruit and flower levels are at their annual peak. As we will see in later chapters, flowering and fruiting rhythms of tropical (and subtropical) trees and shrubs have a profound effect on the lives of plant-visiting bats. In order to understand the ecology and behavior of these bats, one has to pay close attention to the reproductive rhythms of their plant communities.

Finally, I examined Panamanian rodents to see if they were more *K*-selected than their temperate zone relatives or ecological counterparts, as predicted by the MacArthur-Wilson *r* and *K* selection theory. To answer this question, I analyzed all of the available data on the demographic characteristics of temperate and tropical rodents. Results of these analyses were equivocal. Although females of temperate murid rodents (i.e., typical forest and field mice) had higher monthly reproductive rates than tropical murids, these two groups did not differ significantly in the size of their litters, number of litters produced per female per year, or annual survival rates. Therefore, contrary to prevailing theory, tropical murids did not appear to be more strongly *K*-selected than their temperate relatives. My analyses, however, did reveal significant differences in demographic characteristics among rodent families. For example, murid rodents, regardless of latitude, have higher reproductive rates and lower annual survival rates than heteromyid rodents. I tentatively concluded from this that phylogenetic history probably has a stronger influence on the life history traits of rodents than latitude does, a conclusion that has been supported by many subsequent studies.

Unlike tropical rodents, bats as a group turn out to be quintessential *K*-selected animals. Although they are about the same size as many rodents, bats are much slower to reach sexual maturity (often in one or two years), have much smaller litters (usually one or rarely two young per birth; a few species of tree-roosting temperate bats produce up to five pups at a time), and have much longer life spans (up to thirty years in hibernating little brown bats but as long as seven years in the tropical black myotis and eighteen years in the common vampire). Be they temperate or tropical, bats have life histories whose pace is much slower than that of similar-sized terrestrial mammals. As we will see, two consequences of this life history are stable population sizes and complex mating systems.

Just as Charles Handley had predicted when we first met, our Panamanian association turned out to be mutually profitable. I learned a tremendous amount about tropical mammals from Charles and from the opportunity he had given me to work in Panama. In return, I began to produce a series of papers on the ecology and biology of tropical mammals and their responses to tropical seasonality. Unfortunately, because of his involvement in other projects, Charles never got around to writing an updated version of Goldman's "Mammals of Panama."

In May 1969 I received my Ph.D. degree from the University of Michigan and accepted a job as assistant professor of biology at the urban branch of the University of Missouri in St. Louis. Before I left Ann Arbor, Emmett Hooper and I began to plan my next trip to the tropics. Costa Rica would be the site of our new project. Except for brief trips to visit Karl Johnson at MARU in the fall of 1970 and to a scientific meeting in 2002, I have never returned to Panama. I sometimes wonder what Rodman and Sherman and their mammal populations are like today, more than thirty years after I studied them. And what ever became of my friend and assistant Juni Barrett?

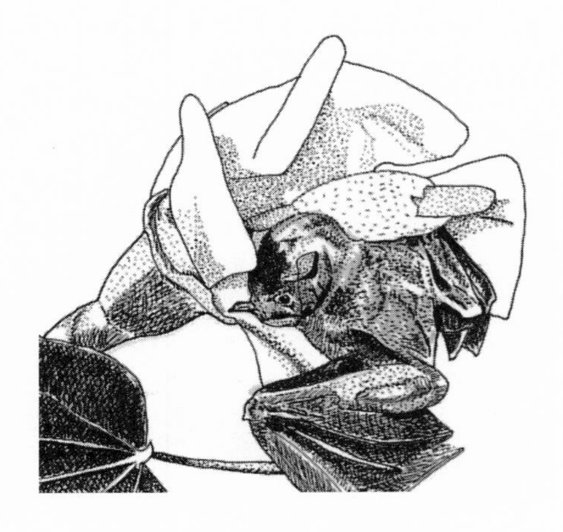

3 Along the Río Corobici

I'm sure that my wife Marcia's women friends consider her to be a saint. Who else but a saint would put up with a field biologist who annually disappears for months at a time and who from time to time asks his entire family to pull up stakes and move to a new location for a year? To be fair, Marcia didn't necessarily know that this was going to be our lifestyle when we were married in the summer of 1965. But she received a strong preview of our coming life when I flew off to Panama without her four months later. By now, after over thirty years of my itinerant lifestyle, she is resigned to my absences. But she is still is not thrilled with them and all the domestic responsibilities they place on her. Who can blame her? I wish that, like me, she had been bitten by the "El Duende" bug early in life. Then perhaps she could understand why the lure of foreign creatures and places is so strong for me.

We were barely settled in St. Louis when we again packed up all of our belongings and stored them with family and friends, sold our car, took a leave of absence from the University of Missouri, and moved to Costa Rica for

---

Jamaican fruit-eating bat in a balsa flower. Redrawn by Ted Fleming, with permission, from a photo by Merlin D. Tuttle, Bat Conservation International.

fourteen months. In the spring of 1970 Emmett Hooper and I received a grant from the National Science Foundation to continue my study of the population ecology of tropical rodents. This time my objective was to explicitly test predictions of the MacArthur-Wilson *r* and *K* selection theory, which, as we've seen, was the reigning theory to explain the evolution of life history strategies in plants and animals. This theory predicted that species living in strongly seasonal environments have higher reproductive rates and lower survival rates than those living in weakly seasonal environments.

Using the same field methods as in Panama, Hooper and I planned to compare the demography of closely related tropical rodents living in habitats that differed strongly in climatic seasonality. These habitats included tropical dry forest, located in the northwestern corner of Costa Rica, and tropical wet forest, located in the Atlantic coastal lowlands. Our proposed site in Guanacaste Province received about one and one-half meters of rain annually but had an intense six-month dry season (see map 2). The site in Heredia Province, in contrast, received about four meters of rain annually and had a modest dry season lasting no more than two months. Given these strong climatic differences, we were confident that our sites would harbor species whose demographies differed to a much greater extent than the sites I had studied in Panama.

I went to Costa Rica expecting to find evidence for *r* selection in Guanacaste and *K* selection in Heredia. This is what I eventually found, but along the way I also discovered that I enjoyed working with bats, especially plant-visiting species, much more than rodents. My ecological studies in Costa Rica thus marked the beginning of my becoming a bat man and not a rat man. And in Costa Rica, Marcia and I found a foreign setting in which we felt perfectly comfortable. So comfortable that I would spend the next sixteen years conducting research almost exclusively in that country.

If knowledge about the natural history and ecology of Panamanian mammals had been scanty when I began working there in 1966, the situation was even worse in Costa Rica. To be sure, European naturalists had been living and working in Costa Rica for about a century, but most of them were "Victorian collectors," in the words of Luis Diego Gomez and Jay Savage (1983). Extensive knowledge of the kinds of plants and animals occurring in Costa Rica was now at hand, but little was known about their natural history and ecology. The first modern account of the mammals known or suspected to occur in Costa Rica was written by George G. Goodwin, of the American Museum of Natural History, and appeared in 1946. Goodwin's study was based on the examination of dry and dusty museum specimens and contained little of the first-hand "feel" for the country and its fauna that permeated Edward Goldman's "Mammals of Panama," published a generation earlier. In his report, Goodwin treated 189 species of land mammals, including 84 species of bats and 50 species of rodents. These numbers compare favorably with the current checklist of Costa Rican land mammals, which contains 200 species, including 103 species of bats and 45 species of rodents. Goodwin's guesses placing three species of bats in Costa Rica—the ghost-faced bat, the lesser long-nosed bat, and the Mexican long-tongued bat—were wildly off. These three bats are inhabitants of much more arid lands to the north and south of the "rich coast."

In 1970, Costa Rica was the site of a growing research effort by U.S. scientists interested in tropical ecology. American involvement in tropical research had a much longer history in Panama, mainly because of the well-established facilities run by the Smithsonian Institution on Barro Colorado Island. Now a similarly strong effort was beginning in Costa Rica under the auspices of the Organization for Tropical Studies (OTS). OTS had been incorporated in 1963 as a result of meetings and discussions among several U.S. universities and the University of Costa Rica. The U.S. universities wanted to develop field stations where their faculty and graduate students could study tropical biology, and Costa Rica wanted a stronger liaison with American scientists. Initial goals of the fledgling organization were to teach U.S. and Latin American graduate students the fundamentals of tropical ecology, to help foreign scientists conducting research in Costa Rica, and to promote the sustainable use of tropical resources. These goals still form the core of a much-expanded OTS mission today.

OTS ended up in Costa Rica because of that country's strong commitment to democracy, education, and social justice. By 1970, Costa Rica could boast the most European population, the highest literacy rate, the largest per capita gross national product, and the longest period of democracy in Central America. Firmly established in a stable and progressive political setting, OTS began its first attempt at "big science" in 1968. The late 1960s and early 1970s were halcyon years for ecological research in the United States. Supported by generous grants from the National Science Foundation and the Atomic Energy Commission and as part of the International Biological Program (IBP), teams of ecologists began to study ecosystem processes in tundra, temperate grassland, and eastern deciduous forest biomes. Other ecologists pursued comparative studies in climatically similar habitats in different parts of the hemisphere, including chaparral in California and matorral in Chile and the Sonoran Desert in Arizona and the Monte in Argentina.

In 1968, infected by the IBP spirit, a loosely organized coalition of U.S.

researchers initiated comparative studies of biological processes at two OTS sites that differed strongly in climatic seasonality—Palo Verde in Guanacaste Province and La Selva in Heredia Province. Scientists from the University of Washington focused on various aspects of forest ecology, soil nutrients, and micrometeorology. Herbert Baker and his postdoctoral student Gordon Frankie at the University of California–Berkeley began their classic work on the reproductive biology of forest trees. And Dan Janzen of the University of Chicago was simultaneously studying all manner of insect-plant interactions, including predation by bruchid beetles on seeds of legume trees, the dominant tree family in most neotropical forests.

Emmett Hooper and I joined this coalition in 1970 to study rodent ecology. Dan Janzen urged us to become involved in the two forest comparisons to document the impact of rodents as predators on tropical forest seeds. I personally was much more interested in testing demographic theory but would make a modest effort to satisfy Dan by feeding different kinds of seeds to captive rodents and noting their gross physiological response—did they live or die? I made a much stronger link with Dan's new postdoc, Don Wilson, who had recently completed his doctoral dissertation on the ecology of the black myotis bat on Barro Colorado Island. Don and I planned to study the ecology of Costa Rican bats in our "spare" field time. As it turned out, Don had little spare time, because Dan Janzen always demanded that his employees work "half time," that is, twelve hours a day, for him. Don and I did collaborate but were seldom in the field together. I eventually formed a much stronger collaboration with two other bat enthusiasts.

Our arrival in Costa Rica—accompanied by nearly four hundred pounds of luggage—in early July was a bit different from my arrival in Panama four and a half years earlier. In contrast with Panama City, which is located at sea level and is perpetually hot and humid, San José is located at an elevation of about 1,150 meters in a broad, volcano-ringed valley called the Meseta Central. Its climate is often described as perpetually springlike. We stepped off the LACSA airplane into typical rainy-season weather—heavy afternoon showers. We hired a large taxi to take us to the Holland House on the east side of San José near the University of Costa Rica. The Holland House was a sprawling, one-story pension that was a favorite hotel of visiting biologists. Although not as classy as the Tivoli in the Canal Zone, the Holland House had the true ambience of Costa Rica—comfortable, a bit spartan, and unpretentious. Except for the honeymoon suite, all of its rooms had single beds. Its showers often issued cold, rather than hot, water.

After dinner on our first night in Costa Rica, Marcia and I walked around the neighborhood near the Holland House, just as Frank Greenwell and I had done in the Canal Zone. Instead of the warm, humid air of Panama, the San José air was cool and misty, and the only animal sounds I heard were crickets and the bell-like calls of a tree frog. The streets were full of Toyota Landcruisers and Land Rovers, the two vehicles that dominated the rough roads in rural parts of Costa Rica. Compared with our Panama experiences, it hardly felt like we were in the tropics.

The next day we walked to the OTS office, which was located at the University of Costa Rica. With an enrollment of about eleven thousand students, the UCR campus was attractively landscaped; its modern two-story buildings gave it a prosperous appearance. At the OTS office we met Jorge Campabadal, the resident director of OTS operations. Dark-haired, of medium height, and always dressed in a short-sleeved shirt and tie in the office, Jorge was an affable, immediately likable guy. A former travel agent, he had many connections in the maze of Costa Rican bureaucracy and was invaluable in helping gringo scientists get their equipment into the country. Jorge helped me set up a research account with OTS, put out the word that I wanted to hire a field assistant for a year, and gave us some advice about finding an apartment at a reasonable price and in a reasonable part of town.

Our domestic plan was to live in San José, like the rest of the long-term U.S. scientists. I would rent an OTS jeep to travel to my field sites each month. Otherwise we planned to travel by foot, bus, and taxi around San José. Marcia's plans weren't completely settled. She wanted to study Spanish and possibly teach math in an English-language school in San José.

In mid-July, I accompanied Gary Hartshorn, a doctoral student from the University of Washington studying the growth and life history of tropical trees, on a trip to Finca La Selva, the OTS wet-forest site. It took us about four hours to drive the hundred kilometers from San José to Puerto Viejo, where we parked our jeep and were taken four kilometers up the Río Puerto Viejo in a piragua. This drive, which I would make once a month for the next year, took us from the Meseta Central northward along a narrow paved road through dairy country. From Alajuela, twenty kilometers west of San José, the road climbed steadily in elevation along a saddle between two volcanoes, Poás and Barva. We passed through sparsely populated temperate zone habitat containing cypress and alder trees. Wooden houses with tin roofs were clustered into hamlets along the road. Their inherent plainness was brightened by pots and coffee cans full of colorful flowers, hanging around doorways and fastened under windows. The road crested at Varablanca, a cold and misty village where we stopped for a mug of hot chocolate, a routine I would follow each time I drove to La Selva.

Beyond Varablanca the road was unpaved and began its winding descent into the Atlantic lowlands. We quickly left the cool temperate zone behind and passed a lovely waterfall plunging under a narrow wooden bridge en route to the warm tropics. The descent followed the heavily wooded valley of the Río Sarapiquí and reached the extensively deforested lowlands at the village of La Virgen. From there we sped along a paved road past oil palm plantations and cattle pastures toward Puerto Viejo, so named because of its historic role as the terminus of river traffic coming inland from the Caribbean Sea via the broad Río San Juan and its smaller tributary, the Sarapiquí.

At Puerto Viejo we parked the jeep where the road crossed the Río Puerto Viejo and waited for Rafael Chavarria, the manager of Finca La Selva, to pick us up in his piragua, powered by an outboard motor. A knowledgeable naturalist, Rafael had been the manager of La Selva when it was owned by Leslie Holdridge, a tropical forester and creator of the modern life zone system used to classify vegetation formations throughout the world.

Holdridge had purchased about six hundred hectares of primary rain forest in 1953, when this part of Costa Rica was sparsely settled and most of its forest was intact. He sold his property to OTS in 1968. During the fifteen years he owned La Selva, he left most of the virgin forest alone but planted experimental stands of economically important trees in disturbed areas along the Río Puerto Viejo. These plants included laurél, a source of high-quality hardwood, the pejibaye palm, a source of heart of palm and edible fruits, as well as cacao and robusta coffee. Holdridge built a rustic two-story farmhouse on a bluff overlooking the Río Puerto Viejo and used the property as a weekend retreat. He encouraged other scientists to conduct research on his property. One of these people was a young ornithologist named Paul Slud, who conducted his doctoral research on the bird communities of Finca La Selva in the late 1950s. While in the Canal Zone I had read Slud's monograph and studied intently his black-and-white photographs of undisturbed tropical rain forest. La Selva seemed like a much wilder, more remote place than any of the forests I had seen in the Canal Zone, including those on Barro Colorado Island. Little did I know then that I would be working at La Selva three years later.

We loaded our gear into Rafael's piragua and enjoyed the trip up the broad, tree-lined Río Puerto Viejo. The river's water was clear and relatively shallow. We passed turtles basking in the sun on logs and watched the silhouettes of fish in the water. Occasionally those silhouettes turned out to be large tarpons. At other times, as I would learn, the Puerto Viejo would rise nearly ten meters during periods of torrential rains and become a chocolate-brown expressway full of uprooted trees acting like enormous, out-of-control eighteen-wheeler freight trucks. The La Selva property began at the junction of the Ríos Sarapiquí and Puerto Viejo, so about half of the boat ride was spent traveling along the forested boundary of the OTS property. The other (east) side of the river was less forested, and cattle pastures crept up to the river's edge opposite La Selva. In time, these pastures would threaten to make La Selva a forest island in a sea of cattle-infested grasslands.

We arrived at the field station in about twenty minutes and lugged our gear up a series of cement steps to the still-rustic wooden field station. Its first floor consisted of four rooms: a dining room/workroom, a kitchen with a propane stove, the cook's quarters, and a storage room that housed a chest freezer. Outside stairs led to the second floor, which contained both a large dormitory with bunk beds of stretched-canvas and three small bedrooms. Screened windows encircled the upper and lower stories. An outbuilding with indoor plumbing and cool (cold) showers was a few paces behind the station, and a small generator was housed behind the outbuilding. The noisy generator ran a few hours each day to keep the freezer cold and to provide electricity for lights for a couple of hours after sundown. Except for the modest houses of a few workers and Rafael and his family a bit farther upstream, the property was "vacant," occupied only by plants and animals.

Over the next couple of days Gary introduced me to La Selva's forest and helped me choose a site for my rodent grid. I was surprised to learn that La Selva was not an impenetrable jungle—the impression I had gained from Slud's photos—but was relatively easy to traverse along a system of well-marked, muddy trails radiating from the field station. One could walk along the Río Puerto Viejo through plantations of cacao and laurél and inland through primary forest over hilly and gradually ascending terrain. Trails crossed over several *quebradas* by way of rough wooden bridges or large logs.

My lasting first impression of this habitat was of a collage of green in many shades, shapes, and sizes. Everywhere you looked—up, down, sideways—there were plants. Huge canopy emergents were few and far between in this forest. Instead, a matrix of relatively thin gray boles formed the skeleton that supported the greenery. Attached to this matrix was a myriad of vines, epiphytes, lichens, ferns, mosses, and leaves. Standing trees were gardens of plants. Fallen trees were gardens of plants. And individual leaves were gardens of plants. Often these gardens could be viewed only through the graceful slotted leaves of palms. *Geonoma* and *Asterogyne* palms dominated the understory, and *Iriartea, Socrotea,* and *Welfia* palms were common in the subcanopy. Although the La Selva forest contained over three hundred species of trees, the feathery pinnate leaves of one species, the legume *Pen-*

taclethra macroloba, dominated the canopy. Baby *Pentaclethra*s formed thick patches in the understory. Gary Hartshorn was studying the demography of this plant for his dissertation research.

The exuberance of the La Selva plant life was hardly matched by a similar exuberance of animal life, at least at first glance. Neither birds nor mammals were conspicuous during my initial daytime wanderings. This was in strong contrast to my first hike through tropical moist forest on Barro Colorado Island in 1966. Mammals were everywhere on the island, where it was easy to spot agoutis, coatis, peccaries, and howler monkeys. It would take many hours of trail walking at La Selva to equal the mammal sightings one could accumulate in a couple of hours on the island. This impression of much lower mammal densities in Costa Rican forests than in forests in the Panama Canal Zone was to persist throughout my work in Costa Rica.

After orienting me to the property and some of its conspicuous plants and animals, Gary took me to one of the three intensive study areas where researchers from the University of Washington were inventorying tree species diversity, which he thought might be a good site for a trapping grid. Intensive Area I was located about fifteen minutes from the field station. Relatively flat in topography, the four and one-half hectare site was already gridded with steel reinforcing rods spaced at twenty-meter intervals. As I walked through it, the plot felt very "ratty" and struck me as a good place to catch rodents.

After a couple of days back in San José, Gary and I next drove to Guanacaste Province, in western Costa Rica, to visit the OTS dry-forest site at Palo Verde. The two-hundred-kilometer trip took nearly five hours, and we traveled along the two-lane Inter-American Highway through far different country from what we had traveled en route to Puerto Viejo. The first part of the road wound through the Meseta Central past large coffee plantations and sugarcane fields and through a series of small towns—Alajuela, Sarchi, Naranjo, and San Ramón. At San Ramón the road descended from the plateau along thirty kilometers of curves with congested bus and truck traffic until it reached Esparta, near sea level. At Esparta, it was imperative to take a *refresco* break—a cool drink rather than hot chocolate—to recover from the tedium of the mountain road.

From Esparta we drove another hundred kilometers to the town of Cañas through relatively flat land that originally had been covered with tropical dry forest. Houses that we passed along the highway—mostly brightly painted stucco structures with red barrel tile roofs—were much more prosperous looking than those along the road to Puerto Viejo. To the north rose the Cordillera de Guanacaste, part of the volcanic backbone that bisects Costa Rica and separates the Pacific lowlands from the Atlantic lowlands. By 1970 much of the forested lowlands of the Pacific coast provinces of Puntarenas and Guanacaste had been cleared for cattle pastures, but stretches of the Inter-American Highway were still shaded by large ceiba (or kapok) and guanacaste trees. The former species, often a canopy emergent in neotropical forests, including La Selva, flowers biennially in the dry season and is pollinated by bats. The latter species, the national tree of Costa Rica, is a large legume with a spreading, umbrella-shaped canopy. It graces many pastures as a shade tree in dry parts of Costa Rica and elsewhere in Central America.

Though it had been paved since 1955, the Inter-American Highway in western Costa Rica in 1970 was full of deep potholes that could break an axle if a vehicle traveled too fast. The tropical forest ecologist Steve Hubbell, who worked in Guanacaste Province in the mid-1970s before moving to Barro Colorado Island to continue his studies of forest structure and dynamics, once described to me his impression of this highway before it was paved and the land adjacent to it cleared. He had accompanied his dad, an eminent entomologist who was the director of the Museum of Zoology at the University of Michigan during my early graduate student years, on a trip to Costa Rica in the mid-1950s. To young Steve, the road through Guanacaste had seemed an unending tunnel formed by large trees. This certainly wasn't the situation in 1970, after extensive deforestation.

Once we reached Cañas, we continued west for a short distance to the Río Corobici, a river that flows year-round from the Cordillera de Guanacaste south to the Gulf of Nicoya, and our final destination for the day, Finca La Pacifica. La Pacifica was an eight-hundred-hectare farm and ranch managed by Werner and Lily Hagnauer. The Hagnauers and their four children had emigrated to Costa Rica from Switzerland in the late 1950s to manage La Pacifica for the pharmaceutical giant Ciba-Geigy. Ciba had planned to grow tropical medicinal plants on the land but failed to realize that this part of Costa Rica suffered from a severe six-month dry season. Werner Hagnauer, a well-trained agricultural engineer, told Ciba that the plants they wanted to grow at La Pacifica could not tolerate such a prolonged drought. He suggested other uses for the land to Ciba and proceeded to implement his plans by growing cotton, sorghum, and cattle. By 1970 cotton growing was no longer economically or ecologically feasible because of its susceptibility to insect attack. But Hagnauer was a determined and resourceful man, and he managed to find other ways to make a living (and a profit for Ciba) from this land.

Although a very businesslike applied scientist, Werner Hagnauer was also interested in the natural history of La Pacifica and, beginning in the mid-1960s, allowed OTS ecology courses to live and work on his land whenever they were in Guanacaste. Lily Hagnauer loved wildlife and began to build a collection of orphaned Costa Rican mammals and birds into a private zoo, the best in Costa Rica in 1970. Her speciality was nursing injured ocelots and margay cats back to health and breeding them. Hard-working, warm-hearted, and very hospitable, the Hagnauers became the hosts for many foreign scientists over the years.

The next day we hired a small plane in Cañas to fly us into Palo Verde, about thirty-five kilometers to the northwest. It was possible to drive to Palo Verde over dirt roads from the Inter-American Highway during the dry season, but during the wet season the OTS field station was accessible only by plane. We landed on a grass strip between the single-story cement block field station and a large freshwater swamp formed by the Río Tempisque. Full of fulvous tree ducks and wading birds such as roseate spoonbills and jabiru and wood storks at certain times of the year, the swamp bred hordes of mosquitoes that made sleeping at the field station difficult during the wet season, according to Gary. Most of the land around the field station had been cleared during previous ranching activities, and intact forest was over a kilometer's walk or horseback ride away. Seeing the isolated nature of this field site and anticipating logistical problems in running supplementary trap lines and mist nets at a variety of sites in the absence of a field vehicle, I began to have serious reservations about using Palo Verde as my dry-forest field site.

Back at La Pacifica, I talked with don Werner, as he was respectfully called by his Costa Rican employees, about the possibility of conducting my research on his land. Especially attractive to me was the presence of the Río Corobici, which looked like a good place to net bats. With its canopy of tall trees, many of which were hollow and housed colonies of bats, the Corobici was a shady oasis in the midst of mostly cleared land. I spent a day or so exploring La Pacifica, looking for potential trapping sites, before making a decision about my study site. After weighing the pros and cons of both sites, I decided in favor of La Pacifica. In the end, the convenience of living and working at La Pacifica outweighed any advantages that the remote but still highly disturbed Palo Verde seemed to offer. Werner and Lily Hagnauer said that my field assistant and I could live in one of their spare houses. They even offered to feed us if we didn't want to cook for ourselves. With that kind of hospitality, it was hard to consider working anywhere else.

In late July and early August my new assistant and I set up trapping grids at La Selva and La Pacifica. OTS had hired a young gringo as my assistant. The work at La Selva went quickly, because all we had to do was extend the grid on Area I by a couple of acres. At La Pacifica, we had to start from scratch after I selected a relatively large block of forest a bit over half a kilometer's walk from the main housing area. Over the years this patch of forest became known as Fleming's Woods, and it housed a fair amount of wildlife, including rodent predators such as snakes, owls, and hawks as well as opossums, agoutis, raccoons, coatis, skunks, peccaries, white-tailed deer, and a troop of howler monkeys.

As I was settling down to my work, Marcia was getting her new life in order. For fun, she began taking guitar lessons from a classical musician who came to our apartment each week. She and Kate Wilson, Don's wife, signed up for two hours of Spanish lessons a day at a nearby language institute. And she was hired by the Country Day School, the largest English-language school in San José, to teach fourth-grade reading, spelling, and grammar for two hours each day beginning in mid-August. She would have plenty to pursue when I was in the field.

All of these plans came under serious reconsideration, however, when we learned on 14 August, our fifth wedding anniversary, that Marcia was pregnant. Our baby was due the first week of March 1971. We celebrated our good news and anniversary by having a special dinner at the Hotel Royal Dutch, one of San José's few elegant downtown hotels. The little band played "Happy Birthday" for us. But our joy was tempered by the trauma associated with Marcia's first pregnancy less than a year ago. Without any warning, Marcia had gone into premature labor and gave birth to a baby boy on 26 December 1969. The five-month-old fetus died shortly after birth, and I was given only a brief glimpse of him, dead in a wax bucket, before he was gone forever.

Could this tragedy happen to us again? we wondered. Should Marcia reduce her activities in anticipation of a difficult pregnancy? Marcia's new doctor, Arturo Esquivel, was fully aware of her medical history and assured us that he would closely monitor her condition. In his usual calm and confident fashion, this U.S.-trained doctor told us not to worry, that everything would turn out all right this time. Ultimately he was right, but preventing our second son from being born fatally early turned out to be a difficult task.

My monthly trapping sessions began in late August 1970 and continued through mid-August 1971. When the trapping began, I had a new assistant named Orlando Barboza, a soft-spoken *tico,* or Costa Rican. Relatively tall and stockily built, Orlando proved to be an intelligent and able field assistant. He didn't speak much English, so I was forced to use Spanish regularly for the first time in my life. Orlando was a patient teacher, and most of my rather meager Spanish skills, as well as my Costa Rican accent, can be attributed to him. We spent eight days per lunar month at each site, and traps were open for seven nights. Each morning I checked from 140 to 150 grid traps while Orlando checked another 100 supplementary traps. As in Panama, my grid animals were marked by toe-clipping and released; the supplementary animals were sacrificed and preserved. We baited the traps only with corn kernels to avoid attracting opossums. This bait strongly restricted the number of species we captured at both sites. Although we captured or observed ten species of rodents at La Pacifica, the bulk of our captures came from Salvin's spiny pocket mouse *(Liomys salvini)*, a close relative of the spiny pocket mouse I had studied in Panama *(L. adspersus)*. Similarly, although we captured or observed fourteen species of rodents at La Selva, most of our captures came from the forest spiny pocket mouse *(Heteromys desmarestianus)*, which I had captured in low numbers at Fort Sherman in Panama. Absent from our Costa Rican captures were the large numbers of rice rats and spiny rats that had filled our traps in Panama. I also missed capturing marsupials, especially mouse opossums, in Costa Rica.

However, a trapping study based on only two species of heteromyid rodents was perfect for my purposes. Since I was interested in testing demographic theory, I needed to be studying two closely related species, one that was adapted to a highly seasonal environment and one that was adapted to a relatively aseasonal environment. My heteromyid rodents were ideally suited for this comparison because of their habitat specializations. When I had studied *Liomys, Oryzomys,* and *Proechimys* in Panama, I was comparing mangoes with oranges and avocados. Any demographic differences among them were more likely to have resulted from their vast phylogenetic differences (they were members of different rodent suborders) than from different adaptations to tropical seasonality. In the parlance of modern comparative biology, my Costa Rican heteromyid comparison controlled for phylogenetic effects and directly examined demographic and other responses to different environments.

Besides the scientific merit of such a comparison, I was aesthetically pleased to be working with tropical heteromyid rodents. I had enjoyed working with *Liomys adspersus* in Panama because it was a pretty animal. With its sleek, gray and white spiny pelage and relatively large, bright eyes, Salvin's spiny pocket mouse was just as handsome as its southern relative. About one-third larger than *Liomys,* the forest spiny pocket mouse was black dorsally and white ventrally. However, its smaller eyes reduced its aesthetic appeal, and its bumbling behavior when I released it from traps suggested that it wasn't as "bright" as the more agile *Liomys.*

Results of my demographic study would indicate that *Liomys* was a much more strongly seasonal breeder and that it had lower annual survival rates, produced larger litters, and reached sexual maturity at an earlier age than *Heteromys.* As in Panama, most *Liomys* babies were born during the long dry season, the time when most of the edible seeds fell from trees, shrubs, and grasses into the deep leaf litter on the forest floor. In contrast, *Heteromys* reacted to its short and often unpredictable dry season at La Selva by reducing its reproductive activity. All of my demographic data supported the hypothesis that *L. salvini* was more *r*-selected than *H. desmarestianus.* Additional behavioral studies in the field and back in the lab in St. Louis further indicated that *Liomys* was socially more aggressive and a more intense burrower and seed hoarder than *Heteromys.* These results were consistent with the idea that *Liomys* lived in a physically harsher environment than *Heteromys.*

As in Panama, my daily fieldwork in 1970 and 1971 settled into a sometimes tedious routine. But life at each of the two field sites was far from boring. There were always new things to see, new organisms to learn, and new physical conditions to contend with. Because of its tremendous biological diversity, La Selva was an endlessly fascinating place, if you liked working in the rain. It didn't rain all the time, of course, but more often than not I got soaked checking my traps. I hated to get up at 6 A.M. each morning to the sound of rain pounding on the station's tin roof. Almost as bad was having to listen to Perry Como singing Schubert's "Ave Maria" on the cook's radio at breakfast and lunch every day. After a breakfast of Kellogg's Corn Flakes ("¿Porque corn flakes para desayuno?" I asked Francisco, our cook. Perhaps because he had seen too many old U.S. programs and commercials on Costa Rican TV, Francisco answered, "¡Porque todos los gringos comen corn flockies!"), Orlando and I would slog through the mud to our traps. In Panama I had vowed to purchase contact lenses before my next trip to the tropics, and I did this before we moved to Costa Rica. My contact lenses were great for working in the rain, but on at least three occasions a twig knocked a lens out of my eye while I was checking traps. I always carried my glasses along for such emergencies.

November and December were extremely wet months at La Selva. The Río Puerto Viejo rose steadily in the incessant rain and threatened to isolate the field station from the rest of the property. In November we had to cut our stay one day short, because all the trails and part of my grid became covered with water. Rafael took Orlando and me to check our traps by boat on our last day there. December was even worse. It rained constantly each day I was at the station. The Puerto Viejo rose about ten meters, and the field station became an island surrounded by muddy brown water. Again Rafael had to take us to our traps in his boat. On our fifth day at La Selva, he evacuated us to the town of Puerto Viejo, which was slowly becoming inundated. Nearly one and one-half meters of rain, by far the most rain ever recorded in one month at La Selva, fell in December 1970. Water levels had receded when we returned at the end of December. Some of my traps that had been under water earlier in the month had rodent skeletons in them.

The heavy rainfall at La Selva was a boon to at least one group of organisms—frogs. The number of species of frogs known from the station is currently forty-one. This list includes fourteen species of *Eleutherodactylus* and six species of *Hyla* tree frogs. Everywhere I went in the forest, frogs, usually "eleuths," jumped out of my way. *Eleutherodactylus fitzingeri* was a streamlined, rapid jumper. *E. bransfordii,* a highly polymorphic, toadlike species and by far the most common member of its genus, merely hopped away. The visually and vocally most conspicuous frogs were strawberry frogs. Members of the "poison dart" family, this species is bright red with dark blue legs. Unlike their more palatable cousins, male strawberry frogs eschewed courting under the cover of darkness. Instead, they brazenly sang their short, insectlike buzzes from low perches everywhere in the forest during the day. Harder to spot but nearly as colorful was the red-eyed tree frog. A skinny frog with large red eyes, orange toes, and purple thighs, it has become an icon of neotropical rain forests.

La Selva also turned out to be snake city. It has a high density of very dangerous snakes, including the fer-de-lance, or *terciopelo* (velvet), and the bushmaster, or *mata buey* (ox killer). Without going out of my way to find snakes at La Selva, I managed to see about one fer-de-lance a week. I undoubtedly walked by many more of these beautifully cryptic serpents than I saw. For example, during a two week period in July 1986, a particularly sharp-eyed Earthwatch volunteer walking behind me along La Selva's trails spotted one fer-de-lance *a day!* I rarely saw the snakes until she pointed them out to me. In 1970–71 only one individual ever parked itself by a trap containing a rodent. As was the custom at La Selva at the time, I dispatched this snake with a couple of machete blows to the neck.

Far more deadly but much less common than fer-de-lances were bushmasters. Another of my 1986 Earthwatch volunteers was Dr. Dave Hardy of Tucson, an anesthesiologist by vocation and an expert on tropical poisonous snakes by avocation. Dave grew up chasing snakes in the Panama Canal Zone, where his father was chief of surgery at Gorgas Memorial Hospital, during the 1940s. He told me that, given injections of antivenin, the mortality rate from fer-de-lance bites was only about 2 percent, whereas the mortality rate from bushmaster bites was over 80 percent. In 1970–71, I encountered only one individual bushmaster, stretched out to its full length of two meters across a grid trail. Unlike the fer-de-lance, which is a relatively slender snake with a large, lance-shaped head, the bushmaster is thick-bodied and small-headed. Its basic body color is tan, and it has a series of dark-brown inverted triangles edged with light spots along its back. I saw the snake in plenty of time to avoid it, but I made the regrettable decision to kill it. The snake was totally unaggressive, and I had to chase it to break its neck with a large branch. I severed its head, equipped with fangs two and one-half centimeters long, and put it in a trap for transport back to the field station. I carried the heavy, headless body draped over my shoulder the rest of the way around the grid. It took me fifteen minutes to stop shaking after this encounter.

Why did I kill this beautiful but deadly snake? I guess my main reason was that if it had lived, I would have worried about its whereabouts every time I walked the grid. As it was, even with the snake dead, I had a case of the runs each morning for the next two months before checking my traps. I was amazed to see what a strong psychological effect encountering that snake had on me. Shortly after I killed my bushmaster, another one of similar size was found on the same part of the grid. The *tico* workers killed that one, too. Years later Harry Greene and Dave Hardy radio-tracked several bushmasters at La Selva and reported that they spend most of their time waiting along trails to capture spiny pocket mice. I wonder how strongly the "kill fer-de-lance and bushmaster" policy in the early days at La Selva (it was discarded years ago) affected the survival rates of the rodents I was studying.

Just for the record, in 1986 my Earthwatch volunteers and I conducted an unplanned snake census at La Selva as we systematically censused *Piper* plants in large belt transects. We encountered two fer-de-lances in a total of forty-nine carefully searched transects; these snakes were within a hundred meters of each other but were in different transects. This represents a rough density, undoubtedly an underestimate, of about one fer-de-lance per two and one-half hectares (or at least 240 snakes in La Selva's original six hundred hectares of primary forest). We also carefully searched thirty-one treefall gaps for *Piper* plants and found one fer-de-lance. According to these data, the probability of encountering one of these snakes in a gap is about .03. I'm sure this value is far lower than the encounter rates experienced by plant ecologists regularly working in treefall gaps at La Selva.

We found no bushmasters in our belt transects or gaps but did run across one (literally) in a line transect near a gap. That morning I had a brand-new crew of nervous Earthwatch volunteers out in the field to collect data for the first time. I was busy censusing a gap with some of them, and my assistant had others laying out a line transect nearby. Suddenly one woman came running up to me saying, "Come quickly. Our line just passed over a big snake!" Suspecting what they had encountered, I told everyone to step back from the snake. Sure enough, their line had gone right over a two-meter bushmaster that was beginning to form a defensive coil when I arrived. We all watched the bushmaster from a safe distance, and it soon slowly crawled up against a fallen log and proceeded to form a sleeping coil. It was still asleep when I brought some friends to see it that afternoon.

My biological surroundings at La Pacifica were totally different from those at La Selva. I had no pristine forest to work in, far less biodiversity to learn, and fewer dangerous snakes to worry about. The tropical dry forest is home to the very venomous South American rattlesnake, or *cascabel*, but we never saw one at La Pacifica. Barbed-wire fences, the scourge of Guanacaste in my opinion, and stinging plants were about the only noxious things we encountered on a regular basis. Despite its extensive conversion to agricultural fields and pastures, Guanacaste still had its charm, and I quickly came to love the wide open, savanna-like spaces and views of mountains that were lacking at La Selva. I suppose this supports the evolutionary biologist E. O. Wilson's contention that humans have a deep-seated preference for habitats that resemble their ancestral East African homeland. Whatever the reason, I always felt much more comfortable working in tropical dry forest, initially at La Pacific and later at Santa Rosa National Park than in the claustrophobic tropical wet forest.

One of the pleasures of working at La Pacifica was the contact I had with other young gringo biologists. It wasn't that I was working alone at La Selva. Larry Wolf and Gary Stiles were regularly working on hummingbirds and their food plants, and Gary Hartshorn was studying tree demography there. Larry and Gary S. broke the ice when we first met by jokingly telling me that the water cistern we all drank from was contaminated with dead bats. Could I please devise a way to keep bats from drowning in our drinking water? they pleaded. Gary Hartshorn was always friendly and taught me a lot about the natural history of La Selva. Back in San José, Gary's wife,

Lynne, who was taking care of their three young children, was very helpful to Marcia as she settled into her new life.

But, somehow, it was more fun to be working at La Pacifica. Larry Rockwood, one of Dan Janzen's doctoral students at Chicago, was studying the effects of herbivory on flower and fruit production in a series of dry-forest trees. It was fun to climb into bushy calabash trees and help Larry and his girlfriend, Jane, play herbivore by stripping off all of the leaves. Ken Glander, also a doctoral student at Chicago, was just beginning his long-term studies of the social organization and behavior of howler monkeys in riparian (riverine) forest along the Río Corobici. Ken is the only biologist I've known to routinely go into the field wearing a pith helmet and carrying a comfortable lawn chair.

Years later I learned first hand why Ken wore that helmet. It was midday, and I was sitting on a rock closely examining a bat that I had just hand-netted in a cave along the nature trail at Santa Rosa National Park, farther west in Guanacaste. Suddenly I felt something warm and wet, smelling slightly sweet, falling onto my bare head. I looked up to see a large male howler sitting on a low branch directly above me. Had he silently sneaked into that tree for no other reason than to pee on me? I had to laugh. I probably would have done the same thing if I had been a howler with a full bladder and such an inviting target. I'm just thankful he wasn't diarrhetic.

The people I interacted with most extensively at La Pacifica turned out to be Paul Opler and Ray Heithaus. Paul and his wife, Sandy, and their two young sons arrived in Costa Rica shortly after we did. Paul had just received his Ph.D. in entomology from the University of California–Berkeley and was now Herbert Baker's postdoc, replacing Gordon Frankie, in their study of the reproductive biology of tropical trees. After he learned an extensive new flora, Paul continued monitoring the flowering and fruiting biology of dry- and wet-forest trees and began a new study of the reproductive biology of tropical understory plants. Paul was a keen entomologist who had been catching and studying insects all his life. He and another young insect enthusiast, none other than Dan Janzen, had corresponded about butterflies as teenagers, but they didn't meet until Paul came to Costa Rica.

Ray Heithaus showed up at La Pacifica in early 1971. A doctoral student at Stanford under the supervision of Peter Raven, Ray planned to study the community ecology of pollinating insects and their food plants in different habitats in Guanacaste. He wanted to learn how flowering plants and insects divide up their biotic resources. Questions motivating his research included: How much ecological similarity or "niche overlap" in pollinators do different plant species experience and vice versa for their pollinators? How much competition for pollinators do plants experience, and how much competition for nectar and pollen do insects experience?

Looking for evidence for competition as a major factor structuring communities was all the rage in the 1970s. As in the case of *r* and *K* selection, this kind of research was inspired by the Princeton ecologist Robert MacArthur, whose provocative theories dominated much of American ecology in the 1960s and early 1970s. Unfortunately, MacArthur died of cancer at the young age of forty-two in 1972. By the end of the 1970s, his influence was waning as a new generation of American animal ecologists began to question the importance of competition as a factor behind community structure. Although competition has never completely disappeared from discussions in community ecology—after all, both plant and animal ecologists have amassed considerable experimental and observational evidence for its existence—its role as a dominant force in determining the coexistence of species has been downgraded since the heady days of "MacArthurian ecology." Ecologists currently advocate a much more pluralistic approach to the study of community ecology. Predation, mutualism, physical disturbance, and the influence of history are now viewed as major contributors to community structure to a much greater extent than was the case in the early 1970s.

By the time Paul and Ray arrived at La Pacifica, I was regularly netting, banding, and releasing bats there in collaboration with Don Wilson. We had established a series of netting sites along a two-and-one-half-kilometer stretch of the Río Corobici and at one site along the Río Tenorito, about four kilometers west of the Corobici, at which we opened five to ten nets for several hours each night. Most of these sites were on dry land, but two of them included nets stretched over water. I always loved wading in the cool water to clear nets of their bats. Before being released, each bat that we captured was marked with a numbered aluminum forearm band, a technique that bat biologists in North America and Europe had been using to study longevity and seasonal movements since the 1930s. Our study was the first large-scale banding project in the New World tropics. In addition to obtaining data on population structure, survivorship, and reproductive cycles, we hoped to learn something about the local movements and home-range sizes of tropical bats.

Out of curiosity, Paul began to join me on many of our netting efforts. He had never worked with bats before and was keen to learn more about them. I taught him how to remove bats alive from mist nets. Don Wilson had recently taught me how to do this, because in Panama we had always killed our captures by cervical dislocation before removing them from the nets. It took me a little time to learn how to handle a squirming, snapping bat with one gloved and one bare hand. I learned that you hold and manipulate the bat with your gloved hand and remove the net from the bat with your bare hand. The trick to quickly removing a bat turned out to be working from the hind legs forward. Once the legs were free, it was often a simple matter to roll the net over the wings and head to release the bat. Once I got the knack of the manipulations, I could remove most bats in a matter of seconds, relying more heavily on a sense of touch than on sight.

Shortly after Paul joined us, Ray joined our bat team. He had enjoyed netting bats during an OTS tropical ecology course and told me that he really wanted to study bat pollination for his dissertation research but was discouraged from doing so by his adviser. Bright and easy- going, Ray and I quickly became good friends and really enjoyed working together. We spent many happy hours gossiping about tropical biologists and talking about science while we captured bats over and beside the Río Corobici. Dan Janzen's unconventional, workaholic behavior and his stimulating ideas (for example, why are mountain passes higher in the tropics?), which were now appearing regularly in prominent journals, often were the subjects of our discussions.

Paul and Ray added an important botanical dimension to our bat study. Don Wilson and I were both classically trained zoologists and viewed bats primarily from a zoological perspective. But Paul and Ray were heavily involved in botanical and plant-animal studies, so it was natural that they would want to know more about the botanical side of the lives of plant-visiting bats that we were capturing. Our bat-plant focus had become much sharper early in Guanacaste's six-month dry season, which lasts from late November to mid-May, when several species of bat-pollinated plants began flowering. In mid-December 1970, for example, we set a net under a pair of flowering balsa trees next to the Hagnauer's house. In six hours we caught fifty-six bats of five species. Most of our captures were pale spear-nosed bats, which were coming in large numbers to drink nectar from balsa's robust, creamy white flowers. Many of our captured individuals were heavily covered with white balsa pollen, as were a few individuals of the greater spear-nosed bat, the common long-tongued bat, and the great fruit-eating bat. Weighing ninety grams and ranking as the second largest bat in the New World, the greater spear-nosed bat did not appear to be visiting balsa trees solely to drink nectar. Instead, individuals flew around the tree giving out loud screeches and chasing other bats, including their smaller relative. Were they trying to capture smaller bats for dinner? I wondered. At any rate, the air was marvelously alive with bats, and it was exciting to watch bats and plants "interact" at close range. Slowly but surely, I was becoming hooked on members of the leaf-nosed bat family as fascinating study animals.

During the dry season we commonly captured phyllostomid bats covered with pollen, even far away from flowering trees. Particularly dramatic were individuals of Jamaican fruit-eating bats that were covered from head to toe with the yellow kapok pollen. I had never seen this species covered with pollen in Panama. At Ray's suggestion, we began to routinely "dust" each bat that we captured with a small cube of gelatin impregnated with purple fuchsin dye. Ray was using this technique in his insect-plant studies. The gelatin picked up pollen grains and stained them. We placed each cube in a glassine envelope until the next morning, when the cube was melted on a microscope slide and covered with a cover slip for permanent storage. Paul began to assemble a reference collection of pollen from trees and shrubs so that we could identify the pollen we obtained from bats.

We added a final dimension to our bat-plant study in April 1971, when bats began switching from visiting flowers to eating fruit. I had frequently noted that phyllostomid bats defecated seeds when they were handled but had not bothered to save them. Now it was a simple matter to put the seeds in labeled envelopes and later identify them with the help of another reference collection assembled by Ray and Paul. We were finally studying phyllostomid bats and their diets in a nondestructive fashion.

We conducted our monthly netting sessions at La Pacifica for two years. Ray and Paul continued to collect data after I left Costa Rica in late August 1971. I then picked up where they left off when we returned to Costa Rica in the summer of 1972. I also banded and released bats at La Selva in 1970–71 but not with the same enthusiasm or intensity as at La Pacifica, for several reasons. First, it rained a lot at La Selva, and mist nets are ineffective capture devices when they are dripping with water. Second, on the nights that Orlando and I did net bats, our capture rates were relatively low, especially in primary forest. Either the bats were flying too high to encounter nets set at ground level or bat densities were chronically low in primary forest. In an intensive netting effort at La Selva in 1986, I was to discover that capture rates at ground level were about two and a half times higher in secondary forest than in primary forest, reflecting the greater resource density for bats in disturbed forest. Finally, without Ray and Paul, netting was not particularly fun at La Selva. Sitting in the dark waiting for bats to fly into mist nets can be pretty boring, especially when capture rates are low. I have found that it really helps to have some company during the slow times to make the time go faster. Orlando was a fine field companion, but we really couldn't communicate deeply, so we often sat in silence, fighting sleep, between rounds of checking the nets at La Selva.

What did our pollen-covered and seed-defecating bats tell us about their lives in the tropical dry forest at La Pacifica? The first thing they told us was that they pollinated or dispersed the seeds of a substantial number of plant species—forty-one in total—at this site. Most of the flowers visited by bats were produced by canopy trees whose principal blooming time was the dry season. Flowers of kapok and balsa, both of which are important economically, were particular favorites of our bats. In contrast, most of the fruits eaten by bats came from understory shrubs (for example, *Piper* and *Solanum*) and small, early successional trees (for example, *Cecropia* and *Muntingia*) whose principal fruiting time was the wet season. The blooming seasons of bat-pollinated plants were relatively short (just under four months, on average) and followed a rather striking seasonal sequence in which overlap in flowering times of different species was quite low. The fruiting seasons of bat-dispersed plants, in contrast, were longer (a bit over five months, on average) and overlapped extensively. No sequential pattern was evident in these species. Overall, at least two kinds of bat flowers and five kinds of bat fruits were available every month of the year at La Pacifica.

Plant-visiting bats responded to this seasonally shifting array of potential food resources in several different ways. Two species, the pale spear-nosed and common long-tongued bats, concentrated on floral resources and were primarily nectarivores throughout the year. The former species appeared to migrate to different feeding grounds when flowers became scarce at La Pacifica, whereas the latter species was more sedentary and supplemented its diet of nectar and pollen with fruits and insects when flowers became scarce. The little yellow-shouldered bat underwent a strong seasonal switch from feeding at flowers in the dry season to eating fruit in the wet season. And two species, the Jamaican fruit-eating bat and the short-tailed fruit bat, ate fruit throughout the year but visited flowers opportunistically during the dry season.

Our pollen and seed data revealed that degree of dietary similarity, or "food niche overlap," among the six or seven most common species of plant-visiting bats depended on resource type. Their diets overlapped extensively when bats were visiting flowers, perhaps because flowers were superabundant at times and their nectar and pollen was easily accessible, even to bats without the morphological specializations, such as elongated snouts and tongues, of professional flower visitors. Their diets were much more distinct when they ate fruit, perhaps because the abundance of some fruits (e.g., those produced by shrubs) was chronically low, and fruits were competed for as resources more strongly than flowers. Thus, the short-tailed fruit bat appeared to be a specialist on *Piper* fruit, the little yellow-shouldered fruit bat on *Solanum* fruit, the common long-tongued bat on *Muntingia* fruit, and the Jamaican fruit-eating bat on *Ficus* fruit. Studies at many other sites in Latin America have now revealed similar dietary affinities in these bats throughout the neotropics.

Our dietary data also gave us important new insights into the evolution of plant reproductive seasons as well as the evolution of bat community structure. These data, for example, provided a potential evolutionary explanation for differences in the blooming and fruiting sequences of our riparian- and dry-forest plants. If pollinating bats feed opportunistically at flowers and if bat abundances limit fruit and seed set of tropical trees, then selection should favor reduced temporal overlap in flowering times, within seasonal climatic constraints, to reduce interspecific competition. By being more selective harvesters of fruit, bats create less competition for seed dispersers among tropical shrubs and trees and allow them to have longer and more broadly overlapping fruiting seasons. Finally, lower dietary overlap on fruit resources than on flower resources suggested to us that competition for fruit has had a stronger influence on the number of coexisting plant-visiting bat species in this habitat than competition for floral resources has.

By banding and recapturing bats, we also gained new insights into their foraging behavior. Our plant-visiting bats, for example, fell into three groups based on individual recapture probabilities. One group contained the two dedicated flower visitors and two other species with low recapture probabilities of around 5 percent. Another group contained the short-tailed fruit bat and the little yellow-shouldered bat, whose individuals had relatively high recapture probabilities of about 22 percent. The Jamaican fruit-eating bat had an intermediate recapture probability of 14 percent. We interpreted these data as indicating that the two small frugivores were relatively sedentary bats that foraged primarily along the rivers where we were netting. In contrast, the two avid nectar feeders were not as tied to feeding in riparian habitat and probably had much larger foraging ranges than the small frugivores.

As a result of our studies along the Río Corobici, a new picture of the web of interactions between plant-visiting tropical bats and their food plants began to emerge. This web contains a diverse array of plant species, ranging in size from understory shrubs to towering canopy trees. A seasonally changing but year-round supply of flowers and fruit is important for maintaining local populations of bats, whose foraging behavior in turn influences the pollination and seed-dispersal success of their food plants. Finally, tropical bat communities contain at least three groups of plant-visiting species—professional flower visitors, understory frugivores, and canopy frugivores. Wetter, less intensely seasonal tropical forests, such as those on Barro Colorado Island, in Panama, and at La Selva, have more species in each of these feeding groups than we captured at La Pacifica, but these groups are fundamental features of the structure of bat communities throughout the New World tropics. As we will see, bat communities in the Old World tropics have independently evolved a similar structure in their plant-visiting bats.

While I was involved in my rodent and bat studies, Marcia was busy with her life in San José but was also concerned about her pregnancy. She continued her guitar and Spanish lessons but quit her teaching job at Country Day School in early September. She had no problems during the early months of her pregnancy, so she agreed to teach "new math" once a week to the Country Day teachers. In this way she met Ruth Stevens, who became one of our dearest friends in Costa Rica.

Ruth and her husband, Steve, were missionaries assigned to the Latin American Mission and had lived in Costa Rica for eighteen years. Steve's main job was managing a chicken hatchery, but he and Ruth were also heavily involved in running an orphanage just north of San José. Now in their midforties, they had raised six children of their own. Ruth was a perpetual student and loved to read and learn new things. She kept us well supplied with paperback books from the States. She would open their comfortable wood-paneled house in suburban San José to us on many occasions over the years.

By late October, Dr. Esquivel suspected that Marcia had an "incompetent" cervix, which would account for her inability to carry a baby to full term, and told her to begin reducing her activity. Nothing medically threatening happened, however, until late November. I had just driven out to La Pacifica and was having dinner with the Hagnauers when a Guardia Rural officer from Cañas drove up with a radio message for me. At the time La Pacifica had neither permanent electricity nor a telephone. The message said, "Su esposa esta muy inferma. Favor, llame la Clinica Biblica." The *muy* was crossed out, I suppose to eliminate a feeling of panic on my part. It didn't work. In a panic, I dropped my fork, drove quickly into Cañas, and called the hospital. Marcia was badly dilated and was going to have a "purse string" operation to close her cervix the next morning. Otherwise she would lose the baby. I caught a domestic flight from Cañas to San José early the next morning and arrived at Clinica Biblica, a small private hospital run by the Latin American Mission, just before Marcia went into the operating room. Everything went fine. I stayed with her in the hospital that night and reluctantly flew back to Cañas the next day. Marcia was bedridden for a couple of weeks and began taking anticontraction pills, readily available at our local *farmacia,* to lessen her chances of expelling the baby. Ruth Stevens, Lynne Hartshorn, and neighbors in our apartment complex took care of her when I was in the field.

To make life as easy as possible for Marcia, we hired a maid to clean the apartment and do laundry. María was a tiny mestizo who told us she had given birth to twenty-one children; fifteen were still alive. Her fecundity astounded us as being exceptional, even for Costa Rica, whose annual population growth rate in 1970 (3 percent) was one of the highest in the world. Indicative of the pronatalist philosophy that permeated the country, the only medicine you couldn't buy without a prescription was birth control pills. María was a pleasant *empleada* but must have dreaded the days that I returned home from La Selva. Along with all of my research gear and specimens, I always brought home a large plastic bag full of smelly, muddy clothes. Even though I tried to rinse them out before we left the field, these clothes were often still caked with mud. Washing them by hand (we had no washing machine) must have been miserable work.

Marcia's next medical emergency occurred on 26 December, exactly one year after she had lost our first baby. She went into labor that night, and we went to the hospital by taxi at midnight. She was given anticontraction medicine intravenously and was told to stay in bed for the duration of her pregnancy. Dr. Esquivel figured that the baby would have to be born one month early, in late January or early February. We made plans for me to be in the operating room when the sutures were cut and the baby was born.

January was a difficult month for Marcia. She didn't like being bedridden, spending her time knitting and reading, but was scared about losing the baby. A young woman who was a student at the language school moved into our spare bedroom to help María and to cook and care for Marcia when I was in the field. Then Marcia's mom arrived from Michigan on 26 January, two days before I again left for La Selva. On the morning of 31 January, Marcia began having contractions, and she and her mom taxied to Clinica Biblica. She was immediately put on an anticontraction drip to delay the inevitable birth temporarily, until I arrived back in town.

Our son, Michael, was born at 6:40 that evening. He weighed four pounds, ten ounces, and was so weak that he had to spend his first four days in an oxygen incubator before being moved to a regular incubator for another three weeks. During this time he was fed through a tube passing through his nose to his stomach. His weight dropped to four pounds, two ounces, before it began to increase. By the time he was released from the hospital, in early March, however, he weighed a whopping five pounds, seven ounces.

And where was Daddy on Mike's birth day? I was at La Selva, unaware of what was happening in San José. The field station had no short-wave radio or telephone, so Don Wilson and family volunteered to drive to Puerto Viejo to deliver a note describing Marcia's condition to Rafael when he made his afternoon boat trip to town. Marcia had expected me to leave La Selva as soon as I got the news, but I didn't. She has never forgiven me for not being with her for Mike's birth. In hindsight, I realize I should have left, but at the time I didn't really see how my presence in San José would have been beneficial. Marcia had family and friends with her, and I wanted to complete a trapping session for the first time in three months. Although on the 31st I knew the birth was imminent, I didn't know that I had a son for another two days. Jorge Campabadal brought me this news along with a big cigar when he arrived at La Selva on 2 February.

The month Michael was in the hospital was frustrating for us. We wanted to hold and examine him closely but had to be content with seeing him, full of tubes, through an incubator window. Marcia told friends that he looked like me—long legs and feet, a pug nose, and wavy brown hair. His hair color immediately set him apart from the *tico* babies in the nursery. They invariably came equipped with a thick mop of black hair. Our pediatrician, a wonderfully warm gentleman named Dr. Miguel Ortiz, was guardedly optimistic about Mike's condition. He said that we shouldn't worry about Mike's small size and his slow rate of weight gain. He would quickly catch up in size once he was out of the incubator.

Dr. Ortiz was right about Mike's rapid progress when he came home from the hospital. He increased steadily in size and did all the things babies are supposed to do, namely eat, poop, and sleep. We adored him, even during his 2 A.M. feedings, which were my assignment when I was home from the field. We bought a leather rocking chair and spent hours day and night rocking him. Eventually we started taking him, wrapped up in a blanket, with us on shopping trips and to restaurants. *Tico* women invariably made a fuss over our sandy-haired baby—our *rubio*—wherever we went. *Ticas* of all ages certainly loved little babies.

One of our first stops in San José when we returned for summer fieldwork in 1972 was to visit Dr. Ortiz and reintroduce him to Michael. He was thrilled to see our chattering, bright-eyed son with his long ringlets of curly blond hair. A cursory examination was enough to tell him that Mike's early struggles had caused no lasting harm. It was then that Dr. Ortiz revealed that he had had serious concerns about Mike's having permanent brain damage from prolonged exposure to oxygen in the incubator. He had kept these concerns to himself to avoid unduly alarming us. Since Mike eventually graduated from Carleton College magna cum laude with distinction in physics, it's clear that any brain damage he may have suffered from the oxygen incubator was indeed minimal.

Our departure from Costa Rica in 1971 was set for late August. By then, I had arranged for storage of much of my trapping equipment in the OTS warehouse for a year. We would be back next summer to finish up my rodent and other research projects. I also had to arrange for the shipment of nearly forty live individuals of *Liomys* and *Heteromys* back to St. Louis, where I planned to start colonies for lab studies. All that was left for us to take on the airplane was our personal luggage, three large souvenirs—a painting, our leather rocker, and a beautifully painted wooden ox cart—and our seven-month-old baby and all of his trappings. We had to change planes in Miami and could barely manage to get all of our gear onto a push cart and wheel it to the TWA check-in desk. One of my good friends from St. Louis, the ecologist Owen Sexton, happened to be returning home from Panama on the same day. He watched from a distance as a young couple with a baby and a tremendous load of baggage struggled through the Miami airport. When we met in the TWA departure lounge, Owen told me that he really felt sorry for that family. He was shocked when I said, "Guess what? That family was us!"

The following summer we returned to Costa Rica for two months of fieldwork. I wanted to retrap my grids to record annual survivorship and population densities of my two rodent species as well as to complete our two-year study of bats at La Pacifica. I brought a University of Missouri undergraduate along with me as my field assistant. We Flemings moved into a new *cabina* on newly acquired property at La Pacifica. Werner Hagnauer was now in the tourist business, in addition to all of his other activities, and had built a restaurant and several *cabinas* on the east side of the Corobici, across the river from the main property.

This was the first time Marcia and Mike had been outside San José. They thoroughly enjoyed living on a farm in rural Costa Rica, away from big cities. Marcia actually looked forward to driving into Cañas, a farm town of a few thousand people, each week to poke around and do the grocery shopping in the small *mercado central.* When I wasn't busy with fieldwork, we hiked

Figure 9. Marcia carrying Mike, 1972, La Pacifica. Photo by T. Fleming.

along the river and borrowed horses and rode around the *finca*. Mike traveled in a carrier on my back or Marcia's (figure 9). He became fascinated with howler monkeys and cottontail rabbits, calling them "mun-tees" and "rat-chees," respectively. After Michael was asleep in his bed at night, Marcia would sometimes come down to the Corobici to help net bats.

That summer La Pacifica had a steady stream of visiting biologists. Some of them joined us when we operated our nets. On one occasion a pair of young American newlyweds came down to the river. The husband was a biology graduate student, but his wife was not a biologist. We were all standing around talking between net checks, when the young woman suddenly began to scream and jump up and down. Her husband did not hesitate to pull down her slacks. Out jumped a thirty-centimeter ctenosaur—the local arboreal iguanid lizard, which attains a length of over one meter. It had climbed up her leg in the dark. This incident immediately reminded me of one of my biggest frights in Panama. I was netting bats alone at Rodman one night and was walking along the edge of a shallow stream between nets. Suddenly I felt something running up my leg toward my crotch. In the dark, all I could picture was a GIANT SPIDER in my pants. I frantically swatted at it and succeeded in killing a poor tree frog, which dropped lifeless out of my pants. It's amazing how the mind magnifies the size and threat of innocuous creatures when they are crawling up your leg in the dark. That woman along the Corobici must have thought she was being attacked by a dragon!

After two trapping sessions at La Pacifica, my assistant and I drove to Puerto Viejo for thirteen days of trapping at La Selva. As usual, we had to set out and check our traps in the pouring rain. We were the only researchers at the station and had to cook our own meals. Life had become spartan at this field station, and in the absence of refrigeration, our meat and fresh vegetables spoiled. We checked the traps at La Selva for the last time on 23 July, removed them from the grid, and packed them up for the trip back to San José.

I didn't know it then, but this would be the last time I ever set and checked a grid full of rodent traps in the tropics. Before leaving La Pacifica, I had decided that that site was becoming unsuitable for further rodent work. Werner Hagnauer told me of his plans to convert some of the largest remaining tracts of forest (but not Fleming's Woods) into pasture within the next year. If I was going to continue studying rodent populations in tropical dry forest, I needed to find a new study site. The La Selva site, of course, was still suitable for rodent work. In the end, though, the question of new study sites became irrelevant, because I was unable to obtain support from NSF to conduct experimental studies on the regulation of tropical rodent populations by social behavior. Instead, the officer in charge of the ecology program at NSF—the person with the final say on the disbursement of research funds—recommended that I seriously consider directing my research efforts toward full-time studies of tropical bats. He said the foundation would be more receptive to proposals dealing with tropical bat-plant interactions than to those dealing with tropical rodent populations. This was the final impetus I needed to become a full-time bat man. In this business, when someone from the major funding agency for basic research in the United States tells you that he's more likely to give you money to do X than he is for you to do Y, your research direction becomes clear. Naturally, you become an Xpert.

Ray Heithaus and I wanted to continue collaborating on bat studies, so we began to make plans to study in detail the foraging and food-choice behavior of two common phyllostomids, the short-tailed and Jamaican fruit-eating bats. For logistic reasons, we wanted to continue working in Guanacaste Province but felt that we needed to change our study site to a less-disturbed location than La Pacifica. We had heard that a recently established national park, Santa Rosa National Park, about seventy kilometers northwest of La Pacifica, near the Nicaraguan border, contained good stands of dry forest and had an accessible cave inhabited by *Carollia* and other bats. As our plans developed, we set 1974 as the year to begin our new studies.

Although Ray and I left La Pacifica for a new study site after 1972, other bat biologists continued to study bats there successfully through the 1970s and into the early 1980s. Dennis Turner, a doctoral student at Johns Hopkins University, studied the foraging behavior of vampire bats and discovered that they prefer to feed on cows that sleep on the periphery of herds. Revealing his applied bias, Werner Hagnauer half-jokingly indicated that he was pleased to finally see someone learning something useful about bats. Jack Bradbury, an assistant professor at Cornell University, and his wife, Sandra Vehrencamp, studied the social organization and foraging behavior of three species of sac-winged bats and the carnivorous false vampire bat that roosted on or in large hollow trees along the Corobici. And Jerry Wilkinson, one of Jack's first doctoral students when he moved to the University of California–San Diego, studied the social organization and foraging behavior of common vampire bats roosting in many of the same trees. We'll revisit some of these studies in the course of our work with the social behavior of short-tailed fruit bats.

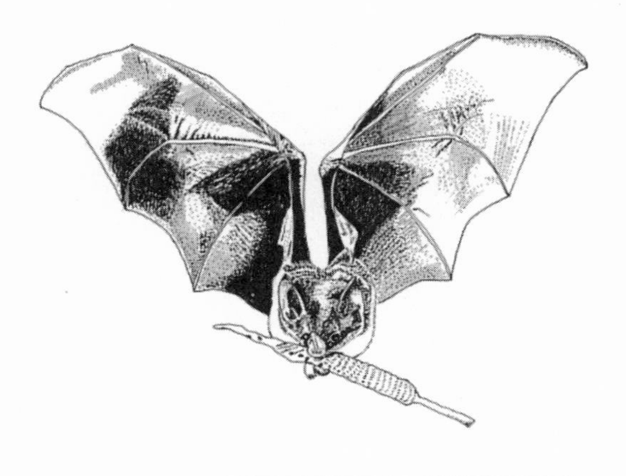

4 El Duende

I was recently flipping through the pages of a new general ecology textbook when a familiar picture caught my eye. In the section on animal foraging behavior was a photo of a short-tailed fruit bat, wings held aloft and a corn-cob-shaped *Piper* fruit in its mouth. The text accompanying the picture indicated that this bat is an optimal forager; that is, it maximizes its net rate of energy gain while searching for its favorite food. I had to smile at seeing some of the results of our long-term research on this bat in a textbook. Of all the examples of optimal foraging in animals, why had the author chosen to feature *Carollia perspicillata* in his book? I'll probably never know the answer to this question. Instead, I must be content knowing that at least one textbook writer has read my book on the behavior and ecology of one of the neotropics' most common bats. Over ten years of fieldwork went into that book. Several field seasons were devoted to discovering how *Carollia*—the name I will use for this bat—forages for food. It is humbling to realize that some of our major findings can be easily summarized in a few sentences in a textbook.

Short-tailed fruit bat carrying a *Piper* fruit. Redrawn by Ted Fleming, with permission, from a photo by Merlin D. Tuttle, Bat Conservation International.

Much of our work on the short-tailed fruit bat took place in and around a small creek bed called Quebrada El Duende, in Santa Rosa National Park in northwestern Costa Rica. The creek is located in a small valley a few hundred meters from Santa Rosa's main hacienda. The *duendes* living in this *quebrada* are two species of phyllostomid bats, the short-tailed fruit bat and the common long-tongued bat. Anyone sitting by the *quebrada* around sunset long ago would have seen many bats leaving a small tunnel-like cave. Silently flitting through the dappled moonlight, these bats might easily inspire a feeling of mystery and wonder. The allure of one of those *duendes* would keep me returning to this creek bed for over ten years.

By the end of 1972, my research had progressed from capturing and sacrificing tropical bats in Panama, mainly for reproductive studies, to marking and releasing them for behavioral and ecological studies in Costa Rica. This research had given us new information about the diversity and relative abundances of tropical bats as well as information about their reproductive cycles, diets, and movement patterns. But these studies told us relatively little about the lives of individual bats—how often they reproduced, how long they lived, where they foraged, and how they interacted with each other inside and away from their roosts. And for plant-visiting species, we had little information about their impact as pollinators and seed dispersers on plant populations. How important are phyllostomid bats in the overall economy of tropical forests? Are they merely interesting ornaments on the tree of life, or do they play an important functional role in these habitats? Our work at Santa Rosa aimed to answer these questions.

Back in 1966, when I began my Panamanian studies, a major source of information about the ecology and behavior of neotropical bats was a 1961 monograph on the bats of Trinidad and Tobago written by George Goodwin and Arthur Greenhall. As he had done in "The Mammals of Costa Rica," Goodwin focused on the taxonomy of Trinidadan bats. Ecological information, including the rabies status of different species, was provided by Greenhall, who summarized many years of observations made by members of the Trinidad Regional Virus Laboratory.

Bat-transmitted rabies was a major medical problem on Trinidad and was the main reason that so much attention had been focused on these mammals. Between 1925 and 1935, for example, eighty-nine people and thousands of cattle had died from this disease—"the highest mortality from rabies-infected bats thus far recorded anywhere," noted Greenhall (Goodwin and Greenhall 1961, 196). An antirabies program aimed at reducing vampire bat populations began in 1934. In addition to the control of the common vampire, which is a major reservoir for rabies among neotropical bats, control of Jamaican and short-tailed fruit bats was also recommended. Greenhall unsympathetically wrote: "These two groups are extremely abundant throughout Trinidad and Tobago and, owing to their wasteful feeding habits, damage a wide variety of wild and cultivated fruits. They also defile buildings, both inside and outside, by their droppings, in addition to their involvement with viral and fungal diseases" (198).

Greenhall's ecological notes indicate that the short-tailed fruit bat occupied a wide variety of roosts, including caves, hollow trees, tunnels, wells and mine shafts, and under leaves. It often lived in colonies of a hundred or more individuals of both sexes, but females sometimes roosted away from males. Pregnant females had been collected in February through October, and lactating females were noted in May through October. Individuals were thought to feed twice a night and carried at least twenty-three kinds of fruit to "digesting places"—temporary roosts away from fruiting plants—to eat. *Carollia* often roosted with vampire bats and was "one of the bats that is frequently found to be infected with rabies on Trinidad . . ." (251).

The Jamaican fruit-eating bat was also rabies-positive on Trinidad. Unlike *Carollia,* however, it seldom roosted with vampires. Instead, it lived in small colonies of up to twenty-five individuals under palm leaves and in the dark shade of foliage; it occasionally roosted in hollow trees and well-lit caves. Its reproductive records spanned the same months as those of *Carollia,* and it was known to eat at least fifty kinds of fruits. The mango, introduced into the New World early in the eighteenth century, was a favorite fruit, and bats sometimes completely stained their fur yellow-green with its syrupy juice. In flying to a digesting place, individuals of *Artibeus* occasionally dropped mangoes—THUMP!—on the tin roofs of houses, causing considerable consternation to people trying to sleep. To avoid this bombardment, people sometimes cut down the mango trees around their houses.

Prior to 1961, the best source of information about the lives of bats was *Bats,* a classic book written by Glover Morrill Allen, professor of biology at Harvard College. In his book, Allen summarized nearly all of the literature on bats and frequently pointed out areas of bat biology that were poorly known. He stated, for example, that "very little careful study has been made of the diet and food preferences or their seasonal variation in the American fruit-eating bats. Probably very few fruits are available at all times of the year, but bananas [introduced to the New World tropics in the mid-nineteenth century] may be one of these, and their pulp is greedily taken by certain species, particularly by the common *Carollia perspicillata,* of Central and South America" (Allen 1939, 102).

On the senses of bats, Allen wrote, "The question arises, how are bats able to live in dark places and to secure their insect prey by night without constantly dashing into obstacles in flight?" (134). After recounting experiments by Spallanzini in Italy in 1794 and by Hahn in the United States in 1908, Allen speculated that bats detect "minute echoes set up by air waves reflected from objects or from moving insects in their flight." Ironically, as Allen wrote this, a brilliant young Harvard undergraduate, Donald R. Griffin, had just discovered that vespertilionid bats produce ultrasonic sounds, that is, sounds above the range of human hearing (above about twenty kiloherz). He was now, in collaboration with Robert Galambos, conducting experiments that demonstrated that little brown bats could "see" in the dark using echolocation.

It wasn't until 1953 that Don Griffin turned his attention to studying the echolocation behavior of neotropical fruit bats. Using Panamanian individuals of the ubiquitous short-tailed fruit bat, Griffin obtained puzzling results when he tried to record their echolocation calls. Even when he held individuals of *Carollia* a few centimeters from his microphone, he could not detect high frequency sounds. Yet when he blindfolded these bats, they readily flew and were able to avoid thin wires, just as *Myotis* bats did in his early studies. Like *Myotis* bats, *Carollia* was reluctant to fly after its ears were temporarily plugged with collodion. Griffin concluded that *Carollia* must be an echolocating bat, even though he could produce no physical support for this hypothesis.

Griffin finally solved the "mystery of *Carollia*" by employing more sensitive microphones and amplifiers. It turns out that, unlike vespertilionid bats, which produce loud echolocation calls and are thus "shouting" bats, *Carollia* and other phyllostomid bats, including vampires, are "whispering" bats. Their echolocation calls contain only about one-thousandth as much energy as the calls of vesper bats. Thus, whereas the insectivorous vesper bats and many other kinds of microbats can detect prey items or other objects from a distance of several meters, phyllostomid bats need to be much closer to objects before detecting them. For finding their botanical "prey," plant-visiting bats often use their visual and olfactory senses to a much greater extent than their auditory senses.

Beginning in 1974, Ray Heithaus and I planned to learn more about the lives and ecological function of phyllostomid bats by focusing on Arthur Greenhall's two pest bats, *Carollia perspicillata* and *Artibeus jamaicensis*. We were anxious to use plant-visiting bats as a model system to test predictions of optimal foraging theory and to study plant-animal coevolution, two emerging hot topics in ecology in the mid-1970s. Interest in optimal foraging began in 1966 as a result of a paper written by the intellectually peripatetic Robert MacArthur and Eric Pianka. In their paper MacArthur and Pianka introduced "optimality thinking" into ecology by proposing that optimal diet breadth—that is, the number of different prey species that maximize an animal's net rate of energy gain—is strongly influenced by the energetic costs associated with the pursuit and handling of different prey. The diets of optimally foraging animals, they theorized, should include only those prey species yielding a net energy gain after the costs of pursuit and handling have been subtracted from their average energetic value. Were fruit-eating bats optimal foragers? we wondered. Did the fruits they selected to eat maximize their acquisition of energy and other essential nutrients?

At about the same time that optimal foraging became a hot ecological topic, plant-animal coevolution also emerged as an important research subject. The seminal paper in this area appeared in 1964 and was authored by a pair of precocious Stanford University biologists, Paul Ehrlich and Peter Raven. Ehrlich and Raven combined their encyclopedic knowledge of butterflies and plants, respectively, in a broad review of the effects that each group has on the other's evolution and diversity. They pointed out that the larvae of different groups of butterflies tend to feed on taxonomically restricted groups of plants and suggested that herbivore feeding pressure causes plants to evolve specific chemical defenses that reduce their palatability to generalist herbivores. These defenses, in turn, cause insects to become specialized herbivores. Through time, these reciprocal coevolutionary influences between plants and insects—influences that produce biochemically specialized groups of plants and feeding specialists among herbivorous insects—can increase the diversity of both groups. By implication, other kinds of animals that interact closely with their food plants, such as pollinators and frugivores, should also undergo coevolution. Ray and I wanted to know the extent to which plant-visiting phyllostomid bats are coevolved with their food plants. Like herbivorous insects, were these bats feeding specialists on fruits that provided them with an especially rich array of nutrients?

The first of my eleven research trips to Santa Rosa National Park began in late June 1974. On this two-month trip I was accompanied by Bill Sawyer, a master's student in anthropology at Washington University. Now an assistant professor of biology at Northwestern University, Ray remained in Evanston that summer, because his wife, Pat, was about to give birth to their first child.

Bill and I spent only one day in San José before heading west to Guanacaste Province and Santa Rosa National Park, one of the first parks in Costa

Rica's rapidly expanding national park system. While in San José we visited the office of the park service to talk with Mario Boza, who, along with Alvaro Ugalde, had helped launch Costa Rica's national park service in the early 1970s. Boza and Ugalde had convinced Karen Figueres, the wife of President Pepe Figueres, that conservation of Costa Rica's natural resources should be a high national priority. Young and idealistic, Mario and Alvaro were persuasive, and doña Karen worked hard to gain support from government officials, legislators, international financial institutions, and conservation groups for the fledgling park service. As a result, an executive decree issued in March 1972 led to the formation of the Servicio Nacional de Parques.

In our meeting, Mario indicated that we could expect full cooperation from the park service in our bat-plant studies. He felt that we would be gathering valuable information not only about the park's fauna and flora but also about how important parts of the tropical dry-forest ecosystem worked. This information was badly needed by the park service for drawing up management plans for Santa Rosa and other parks. Bats were particularly maligned in Latin America because of the rabies problem, especially in important cattle-raising regions, such as Guanacaste. Mario was anxious to present a more positive view of bats to park visitors. Our studies would supply him with the information he needed to help improve the public's image of bats.

Soon Bill and I were motoring in a VW Beetle out to western Costa Rica. Once we were past the large farms and ranches around Cañas and Liberia, the provincial capital, the land became wilder and less populated. After we crossed the Río Tempisque, the largest permanently flowing river in this part of Guanacaste, the land began to rise and was dissected by small basaltic canyons. When we could see intact forest south of the highway, we knew we were approaching the national park. Over the years, I would feel like I was coming home when I could finally see this forest. Attending to all of the details involved in each field trip often was a hassle. The final challenge always was getting all of my equipment and assistants jammed into or onto one vehicle for the drive from San José to Santa Rosa. But once the park was in sight, a great load fell off my shoulders, my heartbeat quickened, and a smile broke out on my face. I was almost home once again.

Although our first glimpse of Santa Rosa was of forest, it would be a mistake to think that the park was heavily forested. Owing to several centuries of human disturbance, first by the Nicaro Indians and then by Spanish cattle ranchers, its ten thousand hectares are a patchwork of forests of different successional ages interspersed with fire-maintained grasslands domi-

Figure 10. La Casona and Monument Hill, Santa Rosa National Park, 1977. Photo by T. Fleming.

nated by African elephant grass. Once inside the park, we drove south through an expanse of tall grass dotted with bat-pollinated calabash trees. After a few kilometers the park's gravel road dropped down from a plateau and entered a short stretch of tall evergreen forest, which gave way, in turn, to more grassland and patches of deciduous forest in a broad river valley. After a couple more kilometers we arrived at the park's administration area—a series of tin-roofed wooden or cinder-block buildings nestled in a grove of small trees.

As we pulled to a stop, we were greeted by don Miguel, the park's elderly administrator, who had been expecting us. After formal introductions, he took us to the small wooden house we would use as a field station for the next several years. The park did not have permanent electricity, but don Miguel told us that the park generator ran for a few hours each morning and evening. He also indicated that we were welcome to eat at the park's *comedor,* where meals were served at 6 A.M., 11 A.M., and 5 P.M.—perfect hours for our fieldwork.

Once we had settled into our little house, Bill and I began to explore the park by car, on foot, and on horseback. Our first stop was La Casona, a reconstruction of Santa Rosa's main hacienda when it had been a cattle ranch in the late nineteenth century (figure 10). The Casona was located about two kilometers from the administration area at the base of a small hill. Thick-walled and roofed with old barrel tiles, it was built around a central patio containing brilliant purple bougainvillea flowers. Its main veranda had a commanding view of southern parts of the park as well as a series of centuries-old stone corrals, shaded by ancient Guanacaste trees. In 1974, the Casona housed a modest historical museum commemorating an important date in Costa Rica's history—20 March 1856. It was on that day that a small band of Costa Ricans attacked and defeated a ragtag army of *filabusteros* led by the infamous southern American William Walker in "la Batalla de Santa Rosa." Anticipating the demise of slavery in the United States, Walker had dreamed of conquering all of Central America and setting up a new slave state. His defeat at Santa Rosa crushed this dream and sent him and his troops dashing back into Nicaragua, about thirty kilometers away, where he was elected president in July 1856. Walker was ousted from that country in 1857 and eventually was captured and executed in Honduras in 1860.

My goals in our first summer's fieldwork were relatively modest. First, I wanted to explore the park for potential places to capture bats and to study bat food plants. Second, I wanted to band as many individuals of *Carollia* and *Artibeus* and to gather as much information about their diets as possible. And, finally, I wanted to begin a series of experiments aimed at discovering how fruit-eating bats find their food.

Roads at Santa Rosa were limited in number, so most of our explorations were on foot. After a breakfast of *gallo pinto* (a tasty combination of fried rice, black beans, and onions, which I love but which seems to be reviled by most Latin Americans), scrambled eggs, fried plantain, and coffee, we would set off in the cool morning air, often wading through dew-soaked grasslands, looking for study sites. One of our first walks took us two kilometers west of the administration area to one of two dammed up "*lagunas*" that had provided water for cattle during the area's six-month dry season. A couple of hectares in extent, the *laguna* was surrounded by scrubby forest, in which we encountered our first troop of white-faced monkeys. We flushed a pair of Muscovy ducks and several fulvous tree-ducks from the pond.

We spent most of our time wandering through evergreen and deciduous forests along the park's main road. Evergreen forest dominated by tall *Hymenaea, Manilkara* (chicle), and *Mastichodendron* trees occurred in moist pockets along the southern edges of plateaus. This kind of forest was choked with a thick understory of shrubs and treelets, and we had to crawl through many spots on our hands and knees to make any progress. Deciduous forest, the dominant habitat over most of Santa Rosa, was a bit easier to walk through. In places that had escaped fire and tree removal, the understory was quite open. Here the forest contained large *Luehea, Bursera,* and *Calycophyllum* trees and felt "old." In yet another area where the plateau dropped down to the river valley, we encountered a thick grove of *Cecropia peltata* trees—an early successional species whose fruit is a favorite of many species of birds, monkeys, and bats. This site, which we dubbed Roadway, would certainly be an excellent place to net bats, I thought.

As we wandered about, we began to regularly encounter troops of white-faced and spider monkeys. Seeing capuchin monkeys had been an uncommon event in the Canal Zone; they were even scarcer at La Pacifica. Though three species of monkeys—capuchins, spiders, and howlers—occurred at La Selva, they were not common, and I seldom saw them there. Initially highly vocal and upset by our presence, the capuchins and spiders at Santa Rosa quickly adjusted to us and calmly went about their daily business in the forest canopy as we worked on plants and bat nets below them. Less conspicuous except when males were roaring, howler monkeys were also fairly common at Santa Rosa. It became a game for me to try to spot howlers stretched out and resting on branches before they spotted me. In or near fruiting fig trees was a good place to look for them. I eventually got to study the feeding behavior of all three species in some detail when I began watching frugivore traffic in fruiting *Cecropia* trees in the early 1980s.

The presence of monkeys in good numbers at Santa Rosa made me feel as though I was finally working in an area with a reasonably intact vertebrate fauna. This feeling was reinforced by occasional signs of large cats and Baird's tapir, animals that have been hunted out in many parts of Latin America. Nearly everyone who spent any time at Santa Rosa, except me unfortunately, saw a puma or a jaguar. My luck at spotting cats was so bad that I missed seeing a mountain lion bound across the main park road about fifteen meters behind me one morning. The only reason I know this is that the cat crossed right in front of a botanist, who was too surprised to tell me to turn around until it was too late. The largest cat that I saw at Santa Rosa was the jaguarundi, a small-headed, chocolate-brown, snake and bird eater about the size of a gray fox. An individual sometimes ran across the park road when I bicycled between patches of fruiting *Cecropia* trees at dawn in the summer of 1981.

My luck at spotting tapirs was a bit better. In some years, tapir sign, including muddy footprints on the main park road once it was paved, was fairly common in upland parts of the park. I saw my first tapir in June 1976. On horseback, I was exploring a small branch of the Río Cuajiniquil, the river that had carved the broad valley in which we did most of our work, when I flushed an adult from a shallow pool of water. I got a brief glimpse of it before it silently disappeared into the forest.

We all got a much better look at a tapir one afternoon in May 1981. Two park guards found an adult female resting in the slimy green water of the *laguna*, and they rounded up the biologists to see it. She was apparently in no mood to run away from the small crowd of people, though she displayed her nervousness by constantly flicking her white-tipped ears and extending her thick, flexible trunk and blowing air at the water's surface. Several slash marks on her neck indicated that she had recently been attacked by a large cat.

That summer a graduate student from Michigan State University was beginning a study of the park's tapir population. He was eager to radio-track them, and here was a golden opportunity to capture and collar one. Displaying their former cowboy skills, two park guards on horses quickly roped her around the neck and left rear foot. We then pulled her out of the water, and the guards proceeded to hog-tie her. Once she was sufficiently immobilized, the student slipped a leather collar bearing a radio transmitter, powered by a two-year battery, around her neck and cleaned her wounds with antibiotic cream. Upon release she retreated into the water in the gathering darkness. As soon as we left, she slipped away and kept moving until she was out of radio receiver range. Unfortunately, the student was unable to detect her for the next six months. Having survived attacks by a cat and people on horses in rapid succession, she undoubtedly was anxious to get as far away from the park as possible. We figured she headed straight for Nicaragua and returned to Santa Rosa only when she thought the coast was clear.

In addition to large charismatic mammals, the park contained substantial populations of coatis, collared peccaries, and nine-banded armadillos—animals that we saw almost daily. Bird life was also rich in the park. We often woke up to the churlish, antiphonal songs of a pair of rufous-naped wrens hopping around on our tin roof at sunrise. Raucous families of magpie jays flew Indian-file from tree to tree in open, parklike woods near the administration area. Noisy flocks of parrots and parakeets flew in and out of fruiting fig trees everywhere. In deep forest, I sometimes caught fleeting glimpses of crested curassows, the largest terrestrial birds in the park. A handsome black male, his head topped with a crest of forward-curling feathers, often accompanied a group of equally handsome rusty-brown females.

During our initial explorations at Santa Rosa, we found *Piper* shrubs to be common members of the understory in most forest habitats. This was in marked contrast to La Pacifica, where *Piper* plants occurred only along the Río Corobici; I never saw them in dry forest away from the river. After a couple of days, I had encountered four species of *Piper*s, easily recognizable because of differences in the size, shape, and aroma of their leaves, and began to take notes on their relative abundance at different sites. Up to three species co-occurred in some places, but each species seemed to have a different distribution among habitats. Since I already knew that *Piper* was a major fruit in the diet of *Carollia*, I began to plan my first serious study of plants. I would document the flowering and fruiting biology of the *Piper*s to see how their ecologies differed. As in our work with bats and plants at La Pacifica, my thinking was still dominated by questions about resource partitioning and how taxonomically and ecologically similar species managed to coexist—a major theme in community ecology in the 1970s. I quickly decided that *Piper*s would be a nice group of plants in which to study niche relationships.

Plants of the family Piperaceae occur in tropical forests throughout the world and are one of the dominant members of the forest understory in the neotropics. Costa Rica's flora contains over ninety species of *Piper*, an extraordinarily high diversity for a single plant genus. Most neotropical species are shrubs or small trees, and most species live in moist or wet lowland forests. In the Old World, however, most *Piper*s are vines. One of those vines, *Piper nigrum* from southern Asia, is important economically, because its seeds are the source of black pepper. Highly variable in leaf size and shape, all *Piper*s are nevertheless easily recognized by their spikelike flower stalks (inflorescences) and fruits. These stalks are eight to twenty centimeters in length and usually stick straight up (or down in a few cases) out of the foliage like thin candles. In fact, the Costa Rican name for *Piper*s is *candela* or *candelillo*. This morphology allows pollinators and frugivores easy access to flowers and fruit. Each light-colored inflorescence contains hundreds of tiny flowers, which develop into single-seeded fruits when pollinated. In 1974 little was known about the pollination biology of *Piper*s. My work in Panama and at La Pacifica indicated that bats were important dispersers of *Piper* seeds.

In addition to getting a feel for the distribution of plants providing fruits and flowers for bats, we used our explorations to locate a series of sites where we could net and band bats. One obvious netting site was the park's nature trail (Sendero Natural), which began near the Casona and continued north in a loop for about eight hundred meters. Passing through patches of relatively old forest as well as successionally young forest, the nature trail was well endowed with potential bat food. Of greatest interest to me in this area, however, was an eighteen-meter-long tunnel through which flowed the Quebrada El Duende. Over time this seasonal stream had carved a "natural bridge" out of the soft rock. The maximum width of the tunnel was about eleven meters, and it ranged from one to two and one-half meters in height. It served as a day roost for five species of bats, including *Carollia*, the common long-tongued bat, and the common vampire bat. The nature trail passed right over this roost. To avoid disturbing the bats, we didn't enter the Sendero roost in 1974. But by netting near the roost, we were able to band many of its *Carollia* inhabitants.

Our other netting sites included resource patches, such as patches of *Piper* and *Cecropia* plants, as well as natural flyways—old roads, trails, and streambeds routinely used by bats in traveling from roosts to feeding areas or between feeding areas. The one habitat from La Pacifica that was conspicuously missing at Santa Rosa was extensive riparian forest. Though we encountered many dry streambeds at Santa Rosa, they contained flowing water only at the height of the wet season and hence lacked well-developed riparian vegetation. A continuously flowing river, like the Río Corobici, where we could go swimming, was the only thing I missed from La Pacifica.

From early July until mid-August, we netted bats nearly every night and captured and released about twelve hundred individuals of twenty species. Most of our captures came from our two target species, the short-tailed and Jamaican fruit-eating bats, which we marked with numbered, aluminum forearm bands. Fecal samples that we collected from *Carollia* contained seeds from eleven species of fruit, of which three kinds of *Piper* seeds were most common. Fecal samples from *Artibeus* contained six species of fruit with seeds of three or four species of Moraceae (the fig family)—*Ficus* species, *Chlorophora tinctoria,* and *Cecropia peltata*—being most common. At Santa Rosa as at La Pacifica, these two bats had very different diets and hence interacted with different sets of plants. Most of *Carollia*'s food plants were either understory shrubs or early successional trees, whereas most of the food plants of the larger *Artibeus* were canopy trees.

By early July, our lives had settled into a rather full daily schedule. After breakfast, we usually moved our mist nets and poles from one netting site to another. Then we collected plant data of one kind or another until lunchtime. After a lunch break we returned to the field and did more plant work until dinnertime, after which we opened the mist nets for several hours. After closing the nets, we relaxed back at the house by drinking a warm rum and Coke and eating saltine crackers by candlelight—simple but satisfying pleasures. Throughout the years I worked at Santa Rosa, this became my usual daily routine, with many variations depending on what particular projects I was working on. For many years, sixteen- to eighteen-hour field days were my standard modus operandi.

Early in our netting program at Santa Rosa we discovered that short-tailed fruit bats come in two different color phases. *Carollia*'s usual color is either dark gray or brown, and the bats we found in the Sendero cave area were typical in color. However, at a site about two kilometers north, most

Carollia individuals were bright orange-red. It immediately struck me that bats with orange fur must be living in a different roost from those with brown fur—a roost containing strong ammonia fumes, which can bleach dark fur. If this was true, then we could begin to delimit the foraging ranges of at least two *Carollia* roosts by looking at the number of brown or orange bats captured at different netting sites. Fur color thus became an important piece of data that we recorded for each *Carollia.* Later in the summer, we found a site two kilometers north of our first "orange" site that contained only brown *Carollia*s. We guessed that these bats were unlikely to be coming from the Sendero roost, four kilometers away, because our longest recapture distance up to that point was only about one and one-half kilometers. Instead, they were probably coming from a second "brown" roost. We eventually used radiotelemetry to confirm my hypothesis that we were capturing bats from three different roosts. Bill and I also found another large concentration of orange bats in the coastal lowlands of Santa Rosa, over six kilometers south of the brown Sendero roost and about eight kilometers south of our known Red roost. Santa Rosa National Park thus contains at least two major *Carollia* roosts of the brown flavor and two major roosts of the orange flavor.

Soon after we had our netting program running smoothly, I began to plan my first behavioral experiment with bats. I wanted to know how bats search for ripe fruit and hypothesized that bats might use at least two different search strategies to do this efficiently (optimally?). The differences revolved around whether bats are constantly "on the alert" for ripe fruit whenever they are away from their day roost or whether they search for ripe fruit only in the vicinity of known fruit sources. Whether bats use the first or second strategy, I reasoned, might depend on the spatial and temporal distributions of their fruit resources. To maximize their encounter rate with ripe fruit (a reasonable goal for an optimally foraging frugivore), bats might be expected to be constantly alert for fruit whenever it is more or less uniformly distributed in space and time. Whenever fruit is patchily distributed in space and time, however, bats might not be constantly alert for fruit while commuting. Instead, they might confine their searching to areas where they had previously found a fruiting plant, relying on a well-developed locational memory to do so. In hindsight, these two strategies sound overly simplistic, but in 1974 they represented a reasonable starting point from which to explore some of the details of the foraging behavior of frugivorous bats.

Devising an experiment to discriminate between these two search strategies was seemingly straightforward. We would build artificial fruit "trees" or "bushes," move ripe fruits around the forest, and see whether or not bats found them. Our fruit bushes were simple cross-shaped structures about a meter and one-half tall, built of two saplings lashed together. Bill and I fastened six wire spikes to the horizontal bar on which we impaled ripe fruits. Each of our experiments involved a series of control poles placed under or near fruiting plants and a series of experimental poles placed in areas likely to be used by bats but not containing ripe fruit. If bats were constantly alert for fruit while commuting from their roost to a feeding area or between feeding areas, then we expected them to find the experimental fruits as readily as the control fruits. If bats were alert for certain fruits only in the vicinity of known fruits, we expected them to find the control fruits much more often than the experimental fruits.

The first fruits we tested were those of the currently fruiting *Piper*, *P. amalago*. To collect enough ripe fruits for a night's experiment (seventy-two or more fruits), we had to emulate *Carollia* bats and become proficient at distinguishing ripe from unripe fruit. One of the classic characteristics of bat fruits in general and *Piper* fruits in particular is that, unlike most bird fruits, they don't change color when they ripen. Thus, most ripe bat fruits are green, whereas most ripe bird fruits are red, black, or blue. To discriminate between ripe and unripe fruits, bats generally use olfaction and hence don't need color cues, which would be quite useless in the dark anyway (figure 11).

Neither Bill nor I had evolved to find ripe *Piper* fruits for a living, so we had trouble telling ripe from unripe fruits. We found we could not use visual cues for this, and we certainly couldn't use olfactory cues. We were therefore limited to using tactile cues for finding our quarry. To find enough fruits, we slowly walked through the forest feeling every ripe-sized *Piper* fruit we encountered. If the fruit was a bit soft (don't squeeze too hard—you'll damage the fruit!) and could be easily plucked off a branch (unripe fruits adhere tenaciously to branches), then it was ready for our use. We began to spend hours as fruit pickers because, as I was discovering by monitoring the maturation and disappearance rates of marked fruits, ripe *Piper* fruits were a scarce commodity, even on fairly large shrubs.

In our first experiment we wanted to see whether bats could discriminate between ripe and unripe *Piper* fruits, so we placed the two kinds of fruit alternately on our poles just before sunset. When we checked the poles during the evening and the next morning, the answer was clear: eighteen of twenty ripe fruits had disappeared, whereas only one of eighteen unripe fruits was gone. As expected, *Carollia* had no trouble in distinguishing ripe from unripe fruit.

Figure 11. Short-tailed fruit bat *(Carollia perspicillata)* approaching a *Piper* fruit. Photo © Merlin D. Tuttle, Bat Conservation International.

Once we knew that bats would pluck fruits from my artificial bushes, we went on to conduct a large number of fruit relocation experiments during the rest of the 1974 field season and in part of 1975's. In addition to two species of *Piper* shrubs, we tested the fruits of three species of trees: one early successional species, *Muntingia calabura,* and two canopy trees, *Chlorophora tinctoria* and *Ficus ovalis.* In conducting these experiments, we would rush out after supper, bags of ripe fruit in hand, and quickly set up our fruit-laden poles in control and experimental areas before driving to a third site to net bats.

As I've indicated, collecting enough ripe fruit for our experiments often took a fair amount of time, especially when we were working with *Piper* and *Chlorophora.* To obtain enough fruit of the latter species, which is a tall tree, we either had to find short plants with accessible branches or search the ground under fruiting individuals for freshly fallen fruit. Like other frugivores in the Santa Rosa forest, we had to learn where the best fruit sources were. Our previous exploratory wanderings came in handy for this.

On one occasion, we actually used white-faced monkeys to locate and supply us with ripe *Chlorophora* fruits. During one of our fruit searches, we heard a group of monkeys moving around in a tall tree some distance away. When we reached them, we were pleased to discover that the monkeys were in a tall *Chlorophora* tree, chowing down on the juicy, mulberry-like green fruits. But how were we going to collect enough ripe fruits for our experiment? The answer was delightfully simple: make faces, shout, and otherwise annoy the monkeys, so that they would throw things at us. Sure enough, when we did this, they began heaving fruits in our direction. Within fifteen minutes we had sixty ripe fruits—our quota for the night. We thanked the monkeys and left them to their feeding. It's a shame we couldn't hire that group to collect fruit for us every day.

Results of our experiments told us that bats apparently did use two different methods of searching for ripe fruit. Bats that ate *Piper*, primarily *Carollia*, were constantly alert for fruit, as I had expected. It didn't matter whether we placed fruit near *Piper* plants or in little islands of forest surrounded by grassland a kilometer from the nearest fruiting plant. The probability of bats taking these fruits was over 90 percent everywhere we put them. I never put ripe *Piper* fruits under my pillow at night, but I often predicted that if I had, a hungry *Carollia* would have found them.

Results for the other fruits were very different. Control fruits of *Muntingia*, *Chlorophora*, and *Ficus* had much higher disappearance probabilities than the relocated fruits, which were seldom taken. Bats (most likely *Artibeus* in the case of *Chlorophora* and *Ficus*) appeared not to be constantly alert for these patchily distributed fruits except in the immediate vicinity of fruiting plants. From these experiments, we learned that spatial distribution patterns of fruiting plants has an important effect on the food-searching behavior of frugivorous bats.

In addition to constantly searching for ripe fruits, I began to look for animal-dispersed seeds wherever I went. I had been aware of seeds squirting out the rear ends of bats for some time and could now easily identify, at least to genus, most of the seeds that bats excreted. All of the seeds that passed through bats were small, at most a few millimeters in length, but were morphologically distinct. Thus, *Piper* seeds looked like dark little bullets, *Cecropia* seeds were little Rice Krispies, *Chlorophora* seeds resembled flattened gray kidneys, *Muntingia* seeds were tiny white ovals, and *Ficus* seeds were round, tan, and embedded in a thin gelatinous coating.

As I walked through the forest in general and around fruiting plants in particular, I began casually to search the vegetation for seeds that had been dispersed by bats and other frugivores. Little clumps of seeds were not hard to find, especially under and around fruiting plants. A search of the lower leaves of a fig tree, for example, revealed seeds of *Muntingia* and *Cecropia* as well as those of *Ficus*. Similarly, I noted *Cecropia* and *Piper* seeds on the leaves of shrubs under a fruiting *Chlorophora* tree. Once I was attuned to looking for seeds, they seemed to be everywhere, sometimes in clumps containing two or more species.

These observations caused me to write in my notes of 25 July:

> These instances [of mixed-species associations of seeds], I'm certain, are of common occurrence so that, through time, one might expect the common occurrence of bat- (or bird-) dispersed plant species growing together. . . . This, in turn, could lead to the development of "mosaic associations" of tree or shrub species in tropical forests—patches of animal resources surrounded by plant species not dispersed by animals. Light gaps would facilitate the temporary formation of such mosaics because some of the animal-dispersed species (e.g., *Cecropia, Piper*) grow best in early successional situations. Resources that occur in discrete patches, in turn, could have a profound effect on the foraging behavior of fruit-eating animals such as bats. It seems likely, for example, that bats should *not* use a random search technique to locate ripe fruit but should learn the locations of such resources and fly swiftly between them, spending a little time searching for new resources to add to the route when old sources begin to "dry up." One might propose that fruit-eating bats locate fruit sources using a "location image" acquired by experience [gained] from living in the same area for long periods of time. . . . Once at a resource patch, the bats may switch from vision or audition as the primary sensory modality to olfaction, supplemented by vision and sound, in that order, to locate ripe fruit. Our observations that ripe *Ficus* are taken under fig trees but not away from them seems to be consistent with this scheme—bats are not looking for figs at Rockwood [one of our netting sites and experimental fruit patches] and hence pass them by, whereas they are "tuned into" figs under fig trees.

By the middle of my first field season at Santa Rosa, I was thinking about two major issues: how the defecatory behavior of bats (and other frugivores) might affect the distribution patterns of their food plants and how plant distributions might in turn affect the foraging behavior of bats. By the end of the field season I had come up with a new hypothesis that would guide much of our plant work during the next few years at Santa Rosa.

My hypothesis was that the foraging, feeding, and defecatory behavior of frugivorous birds and mammals causes their food species to occur in mixed-species aggregations. This hypothesis was based on the premise that bats and other vertebrates do not defecate seeds randomly in habitats. Instead, they tend to deposit seeds in heavy concentrations around their feeding trees and, in the case of bats, in such places as Greenhall's "digesting places," which we now call night roosts. One implication of this hypothesis is that plants with similar fruiting periods—seeds that are likely to end up in frugivore guts together and to be defecated together—are likely to occur together in space if they have similar germination and microhabitat requirements.

Early on I realized that the spatial co-occurrence of vertebrate-dispersed plants doesn't necessarily mean that frugivores have a major influence on the distribution patterns of their food plants. Bats may broadcast seeds widely and randomly, but mixed-species clumps of their food plants can still arise if different species have the same requirements for germination and seedling establishment. An obvious nonrandom spatial factor that is well-known for its influence on plant distributions is tree-fall gaps, where high light levels stimulate small seeds to germinate. If bat-dispersed seeds germinate only in gaps, then the distribution of gaps, not the foraging and defecatory behavior of bats, will ultimately determine plant distributions. Therefore, to test my hypothesis properly, I would have to control for the effect of habitat disturbance on plant distributions.

My "clumped within and between species" hypothesis was in contradistinction to a well-known hypothesis devised by Dan Janzen in 1970. Dan's hypothesis, which was meant to help explain how so many species of trees can coexist in tropical forests, was that species-specific insect seed predators cause trees to have "overdispersed" distribution patterns. That is, according to Janzen's hypothesis, adult trees of each species in tropical forests tend to be widely spaced rather than clumped together, because most of their seeds that land under or near their canopies suffer heavy mortality, either from larvae of insects such as bruchid beetles or from competition from their own kind. Only those seeds that are dispersed some distance from conspecific plants have a chance of escaping from their predators (or their competitors) and becoming established seedlings. More species of trees can be packed into a forest when individual species occur at low, uniform densities than when species occur in high, clumped densities. Seed predators thus help prevent any one species from becoming numerically dominant in tropical forests.

In addition to speculating about how frugivorous bats efficiently search for food and how they influence the distributions of their food plants, I was becoming fascinated by the fruiting and flowering behavior of the Santa Rosa *Pipers*. On 18 July I wrote:

> The *Piper* situation is becoming more interesting. There are far fewer fruits on *[P. amalago]* than when we began our censuses two weeks ago, but the fruits of *[P. pseudofuligineum]* are just "sitting" there in a quiescent state; none or few are being removed. *[P. marginatum]* is now starting to flower and should have some fruits soon, though they won't be common for awhile. We seem to be at the end of [*amalago*'s]

> fruiting period, though new flowers are still being produced. A major question is, when will [*pseudofuligineum*'s] fruits become mature and start to disappear? Is there little or no temporal overlap in fruiting between *[amalago]* and *[pseudofuligineum]*? Where does *[marginatum]* fit into the picture—is it bat- dispersed at all?

What I was beginning to observe was a temporal sequence of fruit maturation in the *Piper* species, a pattern similar to the one we had documented for bat-pollinated trees at La Pacifica. The three species that often co-occurred in dry forest—*amalago, pseudofuligineum,* and *marginatum*—did indeed have nonoverlapping periods of fruit maturation. The timing between *amalago* and *pseudofuligineum* was exquisitely close. Within a week of the end of *amalago*'s season, fruits on *pseudofuligineum* began ripening and, as in *amalago,* were being taken by bats the first night they were ripe. *Carollia* bats probably didn't miss a night of eating *Piper* fruit of one species or another during the wet season. The *Piper*-eating bats neatly switched from one species to another in lockstep. Different fruiting times allowed the *Piper* plants to share completely the same dispersal agents without competing. It was as if these plants had been reading the papers of Robert MacArthur! There was order and predictability in the world, at least in the small portion that I was starting to study intensely.

The nineteenth of August, our last full day at Santa Rosa in 1974, came all too soon. Our field season had been extremely successful. We had found Santa Rosa to be an excellent place to study bats and plants and had gathered an enormous amount of preliminary data about bat-plant interactions. Best of all, I was full of new ideas about how this system worked and was anxious to dig deeply into both bat and plant ecology by testing new hypotheses. It was easy to draw up a list of projects that would keep me busy here for the foreseeable future. This list set much of my research agenda for the next nine years.

Life at Santa Rosa had been spartan but comfortable. We had received little news from the outside world for most of the summer, because mail delivery to the park from San José was sporadic. We would go weeks without mail and then receive a big batch. My letters home reached Marcia and Mike a bit more regularly, because I mailed them from Liberia. The only world news to reach us, aside from letters from home, occurred on 9 August, when one of the park guards told us, "Nixon renunció anoche." News that Richard Nixon had finally been toppled by the Watergate fiasco was gleefully received. On the domestic front, Ray Heithaus wrote that he and Pat were proud parents of Michael Robert, born in late June.

Despite the obvious successes of the summer, I had two major concerns as Bill and I packed up our gear for our departure. My first concern was, where were two families, the Flemings and the Heithauses, and a couple of research assistants all going to live next summer? Our small house had been fine for Bill and me, but it obviously was not large enough for ten people. Other buildings in the administration area were fully occupied by park personnel, so alternative housing was not available. It looked like tents would probably be the answer to this problem. My second concern was, would we be successful in gaining financial support for additional field seasons? What was the likelihood of our gaining long-term support for all the research I wanted to do here? Actually, I wasn't overly worried about the financial side of things. As we said good-bye to our Costa Rican friends and to the plants and animals we had gotten to know so well this summer, I knew in my heart that I would be back to Santa Rosa in the coming years. I didn't know exactly how this would happen, but I was confident that it would. As we putputted along the park road toward the Inter-American Highway in our trusty VW Beetle, I knew I would be back. *El duende* was busy working its magic on me.

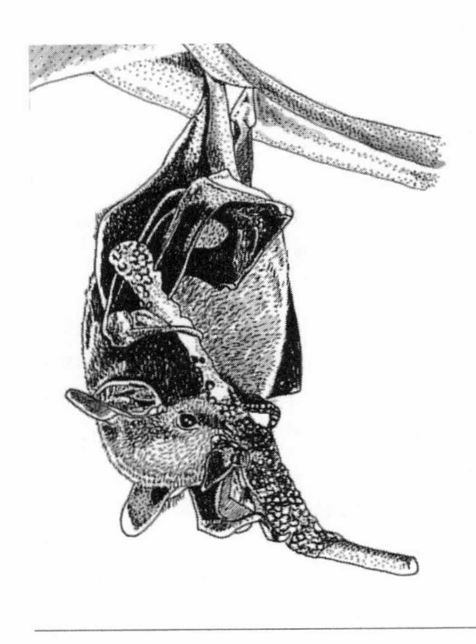

5 Three Hundred Nights of Solitude

Insectivory has always been the most common feeding mode in bats. Fully sixteen of the eighteen families and about three-quarters of all species of bats have evolved a diverse array of foraging styles for exploiting nocturnal insects. Among the insect eaters that occur in a tropical forest, such as Santa Rosa, are species that pursue insects far above the forest canopy (free-tailed bats), species that pursue insects in forest clearings and along streambeds (sac-winged bats and some vesper bats), species that hunt for their prey in cluttered vegetation (mustached, funnel-eared, and some vesper bats), and species that pluck insects off vegetation (certain phyllostomid bats). Given their vast numbers and protein-rich nutritional value, it is not surprising that insects are the food of choice for most kinds of bats.

Fruits, on the other hand, are routinely consumed by members of only two families of tropical bats—the New World leaf-nosed bats and the Old World pteropodids, or flying foxes. Worldwide, only about 250 species of bats are dedicated fruit eaters (or flower visitors). Why are there so few species of fruit-eating bats, and what are the consequences of being a fru-

Short-tailed fruit bat eating a *Piper* fruit. Redrawn by Ted Fleming, with permission, from a photo by Merlin D. Tuttle, Bat Conservation International.

givore rather than an insectivore? Over three decades ago the British ornithologist David Snow pointed out that there are both advantages and disadvantages for birds in being a frugivore. I'm sure the same advantages and disadvantages also hold for frugivorous bats.

One of the advantages is that fruits "want to be found" by their consumers. Whereas insects are continually trying to escape from their predators, the raison d'être of fruits is to be eaten so that their seeds will be dispersed, optimally to a good germination site. To maximize their chances of being found, most fruits advertise themselves with visual and/or olfactory cues. As we have already seen, bat fruits usually rely on olfactory, rather than visual, cues to attract bats. As a result, some bat fruits (and bat flowers) can be downright stinky. Another advantage to fruit eating is that fruits are relatively large compared with the average nocturnal insect and often occur in large patches. Patches of fruit may be some distance apart, but once a frugivore finds one, it often has a large mass of food at its disposal. Compared with insectivores, therefore, frugivores spend relatively little time searching for and pursuing their "prey." From his studies of fruit-eating manakins in Trinidad, Snow found that these colorful understory birds spend a small fraction of each day feeding and have plenty of time left over to concentrate on other biological imperatives, such as reproduction. As a result, lek mating systems, in which males spend long periods each day advertising for females with vocal or visual signals, are especially common in fruit-eating birds.

Countering these advantages are at least two major disadvantages to eating fruit. The first one is the low average nutritional value of fruit compared with insects. Gram for gram, fruits are poorer in proteins and fats than insects, although they are usually richer in water and carbohydrates. They can also be high in fiber, a nutritionally useless plant material. All in all, then, fruit eaters need to consume larger amounts of food to obtain a given amount of protein or fat than insectivores. A second disadvantage is that there are far fewer ways of pursuing fruit than there are of pursuing insects. As a result, there are fewer food niches for fruit eaters in a tropical forest than there are for insect eaters. This means that most tropical faunas and communities contain many fewer species of frugivores than insectivores. This generalization holds true for both birds and bats.

So how does a species such as the short-tailed fruit bat make its living? How much time does it spend searching for and consuming food each night? How far does it have to travel from its day roost to find food? How does it balance its need to feed with its need to mate and reproduce? And what are the botanical consequences of its food choices and foraging behavior? Are *Carollia* and its frugivorous relatives effective seed dispersers? Our fieldwork at Santa Rosa National Park was designed to answer all of these questions.

Our research with fruit-eating bats and their food plants at Santa Rosa can be roughly divided into two major phases. After the "introductory" field season of 1974, our work over the next four seasons (1975–78) concentrated on documenting *Carollia*'s foraging behavior in the wet and dry seasons, primarily through radiotelemetry, its food choice behavior through feeding experiments, and patterns of spatial and temporal variation in its food supply. After 1978, the last year Ray Heithaus worked on the project, I changed the focus to documenting *Carollia*'s social organization and mating system. On the plant side, our attention turned to detailed studies of the dispersal ecology of several of *Carollia*'s major food species. Although we had originally planned to study simultaneously both the short-tailed and Jamaican fruit-eating bats at Santa Rosa, limited grant funds forced us to concentrate on the more tractable of the two species. As we began our *Carollia* work, Douglas Morrison and Charles Handley began intensive studies of the Jamaican fruit-eating bat on Barro Colorado Island, Panama.

Radiotelemetry was the activity that dominated our field seasons during 1975–82. Being able to glue a tiny radio transmitter to the back of a nineteen-gram bat was absolutely essential if we were to document and understand the foraging behavior of short-tailed fruit bats. Fortunately, by 1975 the technology of miniaturizing radio transmitters, primarily through the creation of smaller and smaller transistors, had progressed far enough that 1.5-gram transmitters were commercially available. We ended up putting transmitters on 105 individuals of *Carollia*. I estimate that between 1975 and 1982, I spent about three hundred nights sitting alone on hills for up to four and a half hours at a time, listening to the *beep-beep-beep* of radio signals as bats foraged up to three kilometers away from me. Although the radio-tracking kept me very busy on many occasions, at other times I had relatively long periods of time between radio contacts to enjoy my solitude (and to struggle to stay awake). During those slow times, my mind would wander widely, allowing me to think about many persons, places, and things.

Since neither Ray nor I had ever radio-tracked an animal before, we and our two field assistants spent most of the 1975 field season learning basic telemetry techniques. Our first task was to accurately "track" a transmitter taped to our moving field vehicle from a nearby hill. Once we had learned the mechanics of pointing a seven-element antenna in the direction of the strongest radio signal and taking a compass bearing along the antenna arm, we graduated to tracking a person carrying a transmitter through the woods both day and night. It was relatively easy to line up the mirror of our Brunton compasses along the antenna arm during the day, but doing this at night was a bit trickier. The hard part was shining a headlamp along the antenna arm and lining up the arm's image on the compass mirror while holding the antenna steady in a blowing wind. Our tracking protocol called for us to take a radio fix on a different bat every three minutes. It took me a fair bit of time to learn to coordinate all the gyrations necessary for taking an accurate antenna fix at night in less than three minutes.

Finally, in late June 1975 we were ready to glue our first transmitter, complete with a thirty-centimeter antenna of stainless steel guitar string (a G string, naturally), on a bat. Early that evening, in a drizzly rain, we caught a male *Carollia* near the Sendero cave. I smeared a layer of Silastic medical adhesive on the bat's back while Ray held it and then settled the transmitter into the glue. We had planned to restrain the bat for about forty-five minutes to let the glue set before releasing it. However, the bat escaped after ten minutes and flew off strongly, stopping in the forest just north of the roost. Although we initially feared that the bat would quickly shed its transmitter, our fears were unwarranted. It began its normal foraging behavior later that night and retained its transmitter for many nights thereafter.

Although our first attempt at radio-tagging a bat had been rather clumsy, with a bit of practice, the tagging procedure became very smooth and routine. We quickly learned that *Carollia* is an ideal bat to radio-tag. It tolerated the transmitter well and didn't try to scratch or chew off its new "backpack." Knowing that these bats did not fuss with their transmitters, we gradually lightened each bat's burden by reducing the layer of dental acrylic we applied to the transmitter for protection against bat teeth. We were able to recapture enough of our tagged bats to verify that the transmitter and its glue caused them no harm. We were even able to recapture and retag a few bats in successive field seasons. We ended up recovering nearly 50 percent of our transmitters and placing them on new bats after replacing the battery. *Carollia* turned out to be a very cost-effective species to radio-track.

In addition to being tolerant of radio tags, *Carollia* turned out to be a relatively easy bat to follow because of its sedentary behavior. During the wet season, when we first began to track them, individuals seldom flew more than two and one-half kilometers from their day roost to their feeding grounds. Their relatively short flight distances, coupled with a very strong tendency to return to the same area to feed night after night, enormously simplified our task of quantifying their foraging behavior in considerable detail. But even with this "easy" bat, we often had our hands full when we tried to keep radio contact with up to six bats for as long as possible during each night's tracking session.

Our basic radio-tracking strategy involved taking simultaneous fixes on the same bat from three different tracking stations. In this way, we could pinpoint the locations of our bats through triangulation. Two of these stations involved "fixed" sites—usually hills or other promontories—overlooking forest patches. The locations of these fixed sites varied with the foraging locations of particular bats. We used some sites repeatedly, such as Monument Hill near the Sendero cave and Rattlesnake Hill, because they were within detection range of many of our tagged bats. Other sites were used only once or twice for bats foraging in out-of-the-way places. During the summers of 1975 and 1976, Ray and I spent many hours walking around the forests and fields at Santa Rosa, looking for suitable sites for "odd" base stations.

Our third tracking site was usually designated as a mobile unit. With this unit we hoped to get close to our tagged bats to pinpoint their exact locations. In particularly favorable situations, we were able to walk right up to bats in their night roosts. We seldom did this, however, because it turned out that *Carollia* bats are very sensitive to disturbance in their night roosts. Our close approaches usually caused them to fly off to a new location.

Instead of sitting on their duffs for several hours at a time, as the base station trackers did, people on mobile unit traveled by vehicle or horseback or on foot to get close to particular bats. I spent a lot of time as the mobile tracker over the years and probably traveled as many kilometers as the bats did in trying to keep up with them. The trouble was, my foot pursuit of the bats over hill and dale was slow, whereas the bats could fly between two feeding areas a couple of kilometers apart in a matter of a few minutes. In addition to keeping me slim, the mobile unit gave me a much better feel for where the bats were hanging out than if I had merely followed their movements at a long distance from the fixed base stations.

Horseback mobile units were particularly fun. In the dying sunlight of early evening the Santa Rosa ponies usually didn't mind carrying us and our equipment, which included a squawking walkie-talkie, a radio receiver and headphones, and a hand-held antenna and cables. But after dark, things were different. The horses became very skittish whenever the walkie-talkies started squawking or if they were touched by an antenna. It was almost impossible to take an accurate antenna fix on a jumpy horse, so we usually had to dismount to take our fixes. One night Ed Stashko, Ray's first graduate student, and I had been on mobile horseback along the southern boundary of the park. As we rode back toward the stable, our horses got jumpier and jumpier. The slightest touch by an antenna element made them jackrabbit ahead, causing the beams of our headlamps to shoot off wildly in all directions. Ed and I began to laugh hysterically, which caused the horses to become even more erratic, causing us to laugh even harder. After that episode, I usually elected to travel on foot rather than risk my neck on a nervous horse in the dark.

In addition to obtaining detailed information about the foraging movements of our bats, we used radiotelemetry to locate two major roosts north of the Sendero cave. In Tom Clancy fashion, our first objective was the so-called Red roost—one of the two roosts housing orange-furred *Carollias*—which we hunted in earnest during the summer of 1976. This roost turned out to be located on the top of a small wooded hill about five kilometers northwest of the Sendero cave. Its relatively remote location made it quite difficult to find. By radio-tagging orange bats that we caught at a feeding site, we were able to get a general bearing on the roost's position early in the field season. But locating the roost in a sea of abandoned pastures and forest patches was like trying to find the proverbial needle. We rode out on horseback in the appropriate general direction during the day on several occasions but could never detect a radio signal coming from the roost. Ray and I next camped overnight on a forested hilltop near where we thought the roost might be, only to discover that we were too far north when our tagged bat returned home just before dawn.

Ray and I made one last attempt to locate this roost in mid-August. That morning we rode our horses as close as we could to our proposed search area, which we had chosen with the aid of an aerial photograph of the park. After leaving the horses, we walked through forest to a dry streambed. I was constantly listening for bat #5 as we walked up the streambed. Suddenly, I heard a series of faint beeps. "CONTACT," I shouted to Ray. (Whenever one of us was wearing headphones, he or I always talked very loudly to the other, as though the person without headphones was equally deaf to outside sounds.) It was difficult to get a good bearing on the signal, and we quickly lost it a bit farther upstream and downstream. Finally, Ray suggested that we try walking up a small wash. When we did, the signal reappeared and became stronger with each step. In five minutes we had skirted a large patch of wild pineapple and were gleefully staring at a narrow underground opening at the base of a large dead tree. A booming radio signal was coming from inside a shallow cave. We didn't actually enter the cave at this time but were certain we had located the Red roost.

No one entered this roost until the 1980 field season, when we began to visit it on a regular basis. It turned out to consist of a narrow tunnel five meters long, through which we had to slither on our bellies; the tunnel opened up into a single semicircular chamber about thirty meters wide, twenty meters long, and two meters high. In addition to several hundred *Carollia,* it housed hundreds of individuals of two insectivorous species, Davy's naked-backed bat and the Mexican funnel-eared bat. All of the bats were bleached bright red or orange by strong ammonia fumes.

We didn't attempt to find the Cuajiniquil roost, a brown roost located about two and one-half kilometers north of the Red roost, until the summer of 1982. We used radio-tagged bats again to lead us to this roost over a period of several days. We tagged the bats at our Roadway netting site and refined our bearings on the roost's location over a series of three nights. Searches on foot during the day led us to a likely location for this roost early on. But we could not detect a radio signal coming from the targeted hilltop cave during the day, something we could easily do at the Sendero and Red roosts. We finally confirmed that our suspicion was correct by standing at the roost entrance at sunset and listening to a radio-tagged *Carollia* rocket past us, coming from deep underground.

In addition to searching for roosts on horseback and on foot, we spent a lot of time over the years wandering rather widely around Santa Rosa looking for the exact resources being eaten by our radio-tagged bats. As we had hoped, in the wet season many of our tagged bats fed within a couple of kilometers of their roosts in resource patches that were well known to us. Each field season we monitored rates of fruit production on marked individuals of *Piper, Muntingia,* and *Cecropia* in a series of these patches. We also systematically collected data on the amounts and kinds of seeds that bats and other frugivores defecated around fruiting trees and mapped the locations of many bat-visited plants in these areas. Other bats, however, went to patches of forest whose plant composition was unknown to us. By walking to these forest fragments, often through abandoned pastures full of head-high elephant grass, we discovered new resource patches. Sometimes these areas contained clumps of *Muntingia* trees or *Piper* shrubs. Others consisted of only a single fruiting plant, usually a *Cecropia* or a *Chlorophora* tree.

Through time, our knowledge of the locations of acceptable feeding areas became very extensive, but not nearly as extensive, of course, as the knowledge possessed by a roost of bats. When a bat left the roost shortly after sunset, usually at the same time each night, it seemed to know exactly where it was going. It flew directly to its current feeding area, which might be a patch of *Piper* plants or a single *Cecropia* tree, and sometimes stayed there all night. More commonly, however, a bat would feed in two or more different areas, sometimes separated by distances of up to three kilometers. We discovered that bats added new feeding areas and abandoned their old feeding sites gradually, over a period of several days. In making this switch, bats would briefly visit their new sites early in the evening before going to their usual feeding areas. They seemed to be comparing the food value of the new site with that of their current feeding area. Over the next few days they would spend more and more time in the new area as they phased out the old area.

By the time we had completed our radio-tracking studies in 1982, we came to view our nineteen-gram *Carollia* bats as extremely efficient foragers. They stayed fairly close to their day roosts to feed and minimized the amount of time they spent on the wing at night. During the wet season, they appeared to be highly selective in their diet, preferring fruit species with long fruiting periods, such as *Piper, Muntingia,* and *Cecropia,* rather than fruits whose trees produced enormous but short-lived crops, such as figs. By feeding on these fruits, *Carollia* could return night after night to the same feeding area to harvest fruit. But patches of these fruits usually contained relatively few ripe fruits each night, so bats often had to feed in two or more locations each night to harvest enough fruits to meet their energy needs.

By watching captive bats feed on a variety of fruits, two of my colleagues, Frank Bonaccorso and Tom Gush, discovered that *Carollia* bats also harvest fruits in very efficient fashion. Our bats rapidly ate several fruits, taking each one to their "night roost" in a corner of the flight cage, and then slept for nearly forty minutes each hour while they digested their food. As expected for a professional *Piper* eater, *Carollia* bats were extremely adept at handling their favorite fruit. They manipulated the *candelas* of *Piper* with their wrists. It took them less than two minutes to remove and ingest the pulp and seeds from each *Piper* fruit in corn-on-the-cob fashion. Then they would drop the stripped *Piper* stem and fly off to grab another fruit. These observations confirmed what our radio-tracking data told us: *Carollia* bats undergo bursts of activity followed by long periods of inactivity in the field. Their efficient food-harvesting behavior made these bats relatively easy but boring to radio track.

Studies of the Jamaican fruit-eating bat on Barro Colorado Island indicated that its foraging behavior was somewhat different from that of the short-tailed fruit bat. Roosting by day in canopy foliage or in tree holes, this bat fed nearly exclusively on figs. Because the density of fig trees is quite high on the island, these bats didn't have to travel very far from their day roosts—about eight hundred meters on average. After plucking a fruit from a tree, individuals of *Artibeus* flew two hundred meters or so to a night roost before beginning to eat. Unlike *Carollia,* which prefers to eat fruits that are low in fiber and easy to digest, the Jamaican fruit-eating bat prefers to eat fiber-rich figs, each one of which it chews very slowly and deliberately to squeeze out the nutritious juices, which it swallows. Then it spits out a wad of fiber containing many seeds before flying back to a tree to harvest another fruit. Depending on the size of the figs they are currently harvesting, individuals repeat this procedure five to ten times a night before returning to their day roosts.

While fieldwork dominated our lives both day and night at Santa Rosa during 1975–76, we also managed to have fun with our families and the park personnel. Marcia and the park cook, Reina, became fast friends during the two summers, primarily because Mike and Carito, Reina's young daughter, were constant playmates (figure 12). About the same age and size, Mike and Carito couldn't speak each other's language, but that didn't stop them from getting along terrifically. All day long they would run and jump and play imaginary games using scraps of wood and metal. When it rained, they wallowed in mud puddles and then washed off in our shower. When the sun shone, we would all go on hikes to the Casona or the nature trail, with Carito and Mike running far ahead of us. Sometimes we would climb to the top of Monument Hill near the Casona to read stories *(en inglés)* in the breezy shade of the large cement monument to the Batalla de Santa Rosa. Reina taught the kids how to make corn tortillas in her kitchen, and she showed Mike how to squelch the sharp pain of a chili pepper he had just tasted by putting sugar on his lips and tongue.

During Mike's first summer at Santa Rosa, his innate love of music, encouraged by his ex-musician father, began to blossom. He taught himself to whistle and then proceeded to skip around camp whistling all the time. He tried to teach Carito to whistle but became frustrated when she didn't pick it up easily. She would pucker her lips and, copying Mike, make the right facial expressions, but no sound would come out. Mike also began to pick up two sticks and pretend that he was playing the violin, using one stick as the violin and the other as the bow. When he wasn't whistling, he sometimes fiercely concentrated on playing the violin concerto he had heard on our tape player. Interestingly, he never picked up just one stick and imitated Miles Davis or John Coltrane, whose music he had also heard from birth.

When we returned to St. Louis that fall, we began to inquire into violin teachers for young children. It wasn't long before we discovered the Suzuki violin method and learned that Dr. John Kendall, who brought the Suzuki

Figure 12. Two Santa Rosa playmates—Mike Fleming and Carito Sauceda, 1975. Photo by T. Fleming.

method to America from Japan, taught at the Edwardsville campus of Southern Illinois University, just across the Mississippi River from St. Louis. We gave Mike Suzuki violin lessons for his fifth birthday and became deeply involved in this parent-child method of learning. Mike proudly carried his one-sixteenth-size violin with him when we flew to Costa Rica in 1976, and violin practice became a regular part of our after-lunch family routine. Mike was thoroughly enamored with the violin from the start, and it became an important part of his life from early 1976 on. By the time he reached high school, he was an accomplished violinist.

The summer of 1976 was notable for a number of reasons. It was a time of considerable scientific accomplishment through the hard work of our research team, which included two graduate assistants and an unofficial postdoc, as well as occasional input from Pat Heithaus and Marcia (figure 13). But it was also a time of considerable anguish for me as I watched my close friendship with Ray Heithaus disintegrate. One of our problems was that our families did not get along in the cramped living conditions. Our two families lived in tents, and the three students occupied the two small bedrooms in our wooden "field station." Seven adults and two children were using a single toilet and shower. Pat and Marcia occasionally had acrimonious run-ins over child care, laundry facilities, and other aspects of camp life. Another problem was my intense devotion to fieldwork. Though not a domineering taskmaster, I expected my colleagues to work as hard as I did.

Figure 13. The 1976 field crew: Ed Stashko, Bob Jamieson, Ted Fleming, Ray Heithaus, and Randy Lockwood (from left). Photo by M. Fleming.

Ray also believed in hard work, but he also tended to sympathize with the graduate students when they complained that they were working too hard. A field schedule that included work during much of the day and half the night turned out to be unappealing to some people. I had less patience with such complaints than Ray did, and the students resented me for this. My work ethic eventually turned Ray off, too.

Despite the sociological problems, the summer of 1976 ultimately was a time of considerable joy for Marcia and me because we finally adopted a baby girl. After Marcia's two traumatic pregnancies, we abandoned the idea that we would have another child ourselves. But we still wanted to raise two children, because Marcia didn't want Mike growing up an only child as she had. Beginning in 1974, we enlisted a lawyer in San José to help us find a baby girl for adoption in Costa Rica. We had decided to adopt a *tica,* because baby girls were more available in Costa Rica than in the United States.

Though our lawyer assured us that she was looking for a baby for us, she had no solid leads by midsummer of 1976. Thus, on 14 July we decided to return to San José to inquire at the Oficina de Patronato about the availability of older orphans. Mike, Marcia, and I took one of the buses that passed by the park entrance back into San José. After checking into a downtown pension, we went to the Patronato and received permission to visit the Hogar Infantil, located in a suburb of San José, to view potential adoptees the next morning.

Eager with anticipation, Marcia and I bussed to the Hogar and spent two hours meeting three young *tica* orphans. We felt strange looking at these lovely children and having to make a decision that would profoundly affect one of their lives. None of these kids seemed to be exactly right for our family situation, but the youngest one, a malnourished girl named Marta, appealed most strongly to us. Supposedly eighteen months old, Marta was very small and infantlike. But her dark curly hair, mahogany skin, and bright eyes affected us. We received permission to take Marta to our pediatrician, Dr. Ortiz, in San José for a checkup before making a decision about adopting her.

Marta passed Dr. Ortiz's physical examination. We had been told by a social worker at the Patronato that Marta's mother was mildly mentally handicapped. Dr. Ortiz assured us that, except for her easily correctable malnutrition, Marta was in good physical health. He was more circumspect about her mental health, saying that time would tell about this. Generally elated by this news, we walked downtown to the Patronato to indicate that we were ready to adopt Marta. To celebrate, we stopped at Pops, San José's largest chain of ice cream stores, and bought chocolate-covered ice cream cones. Marta giggled and smiled and thoroughly enjoyed eating ice cream. She was losing her initial reticence with us and was quickly winning her way into our hearts. But back at the pension, a message from our lawyer was waiting for us. It contained good news for us but bad news for Marta.

When we had arrived in San José in early June, our lawyer told us that an *empleada* of one of her friends was pregnant. The maid wanted to give her baby up for adoption if it was a girl; if it was a boy, a more valuable child in a male-dominated Latin society, she wanted to keep it. The baby, born the afternoon of 15 July, was a healthy girl. She was available for adoption. Our lawyer needed to know, did we want her? We told her that we would make our decision that night and let her know early the next day.

Two days before, we had wondered if we were ever going to have a daughter. Now we had to choose between two children—little Marta, who was winning more and more of our affection by the hour, and an unseen newborn. We agonized over the decision all night. While Marta fitfully slept in a makeshift crib, we wavered back and forth in our hotel room. Marta or

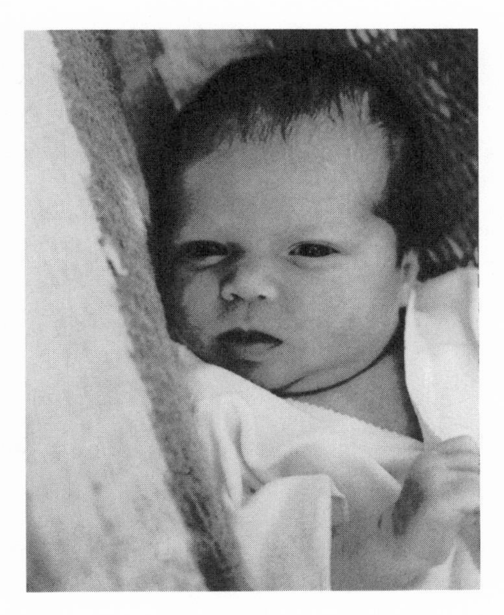

Figure 14. Cara Marie Fleming, July 1976, Santa Rosa. Photo by T. Fleming.

Cara Marie, our chosen name for the newborn? Cara Marie or Marta? By dawn we had reluctantly decided to return Marta to the Patronato and to accept the newborn *tica*. We didn't actually see Cara for two more days, but when we did, we knew we had made the right decision. Our new daughter, dressed in a tiny yellow jumper and brought to us from the hospital by our lawyer, was beautiful. She had fine brown hair, light skin, and delicate features. Dr. Ortiz took one look at her and said, "Adopt her, she's in much better health than the orphan." Our long wait was finally over (figure 14).

Two days later, Marcia, Mike, and I stepped off the bus in rural Guanacaste with our five-day-old daughter and a load of quickly purchased baby paraphernalia in our arms. It was nine o'clock at night, we were in the middle of nowhere, and it was about to rain. The bus driver must have really wondered what we were doing. Ray picked us up in our rented Landcruiser, and we ducked into our pitch-black tent just as it started to pour.

Cara had little trouble adjusting to camp life. After all, all she had to do was eat, sleep, and look pretty. Mike was thrilled to have a baby sister finally. The three of us took turns washing her dirty diapers in the *pila* (a cement wash basin). We kept her formula away from hungry ants by placing it in a shipping container in our tent. She sometimes fussed before going to sleep at night, but overall, she was very easy to care for.

Unfortunately, by the time we were scheduled to leave Costa Rica in late August, her adoption paperwork was not complete. Marcia and Cara stayed in Costa Rica, waiting for her release, when Mike and I returned to St. Louis. After waiting two weeks, Marcia reluctantly accepted that the legal red tape was going to take longer than we thought. So she asked Ruth and Steve Stevens, our friends who had raised six children of their own in Costa Rica, to take care of Cara so that she could come home. It wasn't until 11 December that Cara permanently joined us. Her release from Costa Rica was so sudden that Ruth did not have a chance to call us before getting on a plane for the States. Ruth called us from Lambert Field in St. Louis, saying, "I've got an early Christmas present for you. Would you mind coming to the airport to pick her up?"

Most of my field time at Santa Rosa occurred during the wet season, when the vegetation was lush and food levels for frugivores were high. But I also spent two months there in 1977 and shorter periods in 1980 and 1983 during Guanacaste's six-month dry season. With its brilliant blue sky, cool nighttime temperatures, and still-leafy vegetation, January became my favorite month at Santa Rosa. By early February most trees and shrubs had lost their leaves, and the forest floor was covered by a thick layer of crisp, dry leaves. Days became hotter, and strong northeast trade winds blew both day and night as the dry season intensified. The trade winds forced us to build rigid wooden frames to support our radio antennas. Even with the supports, on many nights it was a physically demanding job to hold the antenna steady enough to take radio fixes. By early March the ground was cracked and rock hard, which forced us to use a pickax to dig holes for our mist net poles. By mid-March, the air at Santa Rosa and elsewhere in Guanacaste was filled with smoke and debris from grassland fires.

During the decade I worked at Santa Rosa, extensive dry-season fires, often lit by poachers trying to hunt deer, raced through the park's grasslands every two years, on average. Fires raged over large areas of the park in late February and early March 1977. It was common to see several Swainson's hawks circling and many swallows flying along the advancing walls of fire, looking for easy prey to capture. By late February, the fires had overrun two of our radio-tracking base stations and nearly surrounded our living area. One night I went to bed early because I was scheduled for a 1 to 5 A.M. tracking session. At 9 P.M. I awoke from a deep sleep and immediately felt pure terror. Our research house was filled with choking smoke, and flames lit up its rooms. Still groggy from sleep, I initially thought the house was on fire but soon discovered that the fire was confined to the forested hillside just behind the house. Outside, sparks were flying everywhere in the soot-filled air. I grabbed a broom to beat out any sparks that landed near the house. Fortunately, the hillside fire soon died down, and none of the park's buildings was in danger of burning. My two field assistants, Pat Skerrett and Mike Zimmerman, both biology graduate students at Washington University, soon returned from their tracking stations to indicate that the fires had not harmed them. Over warm rum and Cokes, we spent the rest of the evening talking about the fire's effect on the park and its wildlife.

The dry season was a time of general food scarcity for Santa Rosa's frugivores. Most of the fruits that *Carollia* ate in the wet season were unavailable then. *Piper* fruits, for example, were restricted to widely scattered patches of evergreen forest and to a few plants growing along the moist banks of seasonal streams. *Karwinskia calderoni,* an uncommon canopy tree that produces small purple drupes, and the common second-growth ant-acacia, *Acacia collinsii,* whose seeds are embedded in a waxy yellow pulp, became mainstays in *Carollia*'s diet. To harvest these fruits, bats had to travel about twice as far from their day roosts as in the wet season. And instead of night-roosting close to their fruit sources, bats often returned to the Sendero cave between feeding bouts after most trees and shrubs had lost their leaves. All in all, *Carollia* bats flew much longer distances while foraging in the dry season than in the wet season.

Other animals had similar resource problems during the dry season. Water was in extremely limited supply, and terrestrial and arboreal mammals were forced to travel long distances every day to drink. Many mammals and other animals routinely visited a small shady pool of water in the Quebrada El Duende just below the Sendero cave. One afternoon I sat quietly behind rocks near the pool and watched a white-tailed deer cautiously look around for several minutes before lowering its head to drink. Its behavior suggested that ambushes by pumas and jaguars were not uncommon at dry-season water holes. After it left, a group of five collared peccaries, including three adults and two juveniles, entered the streambed and drank. The air became thick with their musky, onionlike smell. When they left, a group of four spider monkeys approached the stream through the trees. Unfortunately, they spotted me and refused to leave the trees for a drink as I had seen them do on other occasions. Finally, a male long-tailed manakin, resplendent with its bright red cap and light blue wing patches set against a solid black background, flew to the pool and drank. On another occasion, I found a similar male at the edge of another pool of water. This time, however, it was firmly in the jaws of a medium-sized boa constrictor that had been patiently waiting, well-camouflaged in deep leaves, for a thirsty animal to come by.

In addition to our radio-tracking studies, I spent many daytime hours during the 1977 field season mapping plants with my assistants, Pat and Mike. I was still keen to test my "mixed-species aggregation" hypothesis, and forest conditions during the dry season were perfect for rapidly mapping plants in a variety of different sites. We laid out temporary grids, counted the number of trees of all species, and mapped the locations of all bat-dispersed trees and shrubs on several plots up to two hectares in area.

As we worked in the field and back at the house, Pat, Mike, and I had a series of lively discussions about how best to determine the role played by bats and other frugivores in the creation of mixed-species clumps of plants. From our mapping, it was becoming obvious to us that bat-dispersed plants had clumped distributions and that mixed-species clumps were not uncommon. Such clumps often contained a mixture of one or more species of *Piper* and *Solanum* shrubs and juveniles of *Cecropia, Muntingia,* and *Chlorophora* trees—just the kind of association I expected on the basis of our knowledge of the wet-season diet of *Carollia perspicillata.* But we still didn't know whether the foraging behavior of *Carollia* and other bats was necessary *and* sufficient to account for these clumped distribution patterns. The effect of edaphic conditions, including soil types and disturbance history, on these patterns still needed to be determined.

In one of our discussions, we came up with the following idea. From Ray's and my 1975–76 careful mapping of seed clumps deposited by bats around fruiting trees, we knew that fig trees were especially good at "attracting" a high density and diversity of bat-dispersed seeds. Our transects around fruiting *Ficus* trees, for example, contained six to eight different kinds of seeds, compared with only two to four kinds around fruiting *Muntingia* trees. Knowing that fig trees were "hot spots" for the deposition of bat-dispersed seeds, we predicted that mixed-species clumps of bat plants should occur more frequently in the vicinity of these trees than around other kinds of trees growing under the same soil and disturbance conditions as the fig tree.

To test this prediction we set up small grids measuring forty meters per side, centered on five pairs of *Ficus* trees and *Calycophyllum candidissimum* trees. *Calycophyllum* produces wind-dispersed seeds and has no special affiliation with bats. We were careful to match each pair of trees in terms of site conditions. Then we counted the total number of trees and the number of bat-dispersed trees and shrubs present in each grid. Bat-dispersed plants turned out to represent about 10 percent of all of the trees and shrubs growing around both kinds of trees. Contrary to our hypothesis that, by attracting large numbers of frugivores, fig trees attract mixed-species clumps of bat plants, our data supported the null hypothesis that no difference in density and species richness of bat plants occurs around *Ficus* and *Calycophyllum* trees. The density of *Piper* shrubs was two to four times higher around fig trees, but this difference was not statistically significant, undoubtedly because of our small sample size of target trees. *Ficus* trees in disturbed sites had conspicuously higher densities of bat plants than trees in less-disturbed sites had, indicating that edaphic conditions clearly played an important role in determining the density and distribution patterns of bat plants at Santa Rosa.

Our overall conclusion from this little exercise was that, while bats and other vertebrates are necessary for moving fleshy-fruited seeds around tropical habitats, their spatially nonrandom foraging and defecatory behavior was not sufficient to account for the distribution patterns of their food plants. Mixed species clumps are the joint products of frugivore foraging behavior *and* habitat disturbances. Because of their abundance and mobility, fruit-eating bats clearly move many seeds around within and between habitats. A *Carollia* bat feeding on *Piper amalago* fruits, for example, will ingest and defecate about three thousand seeds in one night. When feeding on fruits of *Muntingia calabura*, it will defecate around sixty thousand seeds in one night. Multiply these numbers by the three hundred to four hundred or so bats in the Sendero cave, and you have a whole lot of seed dispersal going on. Because of the sheer numbers of seeds they handle, *Carollia* and its frugivorous relatives must play an important role in the regeneration of tropical forests. The rapid colonization of abandoned agricultural fields by bat-dispersed species of *Cecropia, Piper, Solanum,* and *Vismia* in many places in the neotropics provides strong support for this contention. But habitat conditions, particularly the availability of sunlight at ground level, are also important in this process. Soil and sunlight conditions ultimately determine where seeds of many bat plants will germinate and become established as seedlings.

As described in the next chapter, the next major phase of my *Carollia* work at Santa Rosa got off to a rather shaky start in the summer of 1979. It wasn't until January 1980 that we really began to study the social organization of *Carollia* in order to obtain a more complete picture of the lives of individual bats and how they deal with the potentially conflicting demands of feeding efficiently and maximizing their production of surviving, reproducing offspring.

The key to the success of our studies of *Carollia*'s social behavior was Charles F. "Rick" Williams. I met Rick when I went to the University of Oklahoma to give a research seminar in the fall of 1979. Rick was in his last undergraduate semester at Oklahoma, where he was completing a double major in botany and zoology. His speciality in zoology was animal behavior, and he had assisted Doug Mock, then an assistant professor of zoology, in his studies of sibling competition for food in nestling herons and egrets. Rangy in build and with a thick mop of curly brown hair, Rick displayed a quiet maturity and intelligence. He was keen to attend graduate school and jumped at a chance to become my research assistant for a year and a half in Costa Rica. He enrolled as a master's student in the spring semester of 1980 at the University of Miami, where my family and I had moved in 1978.

In early January 1980, Rick and I flew to Costa Rica to begin the next phase of my *Carollia* project. Among our pile of luggage and field equipment were two ten-speed bicycles. Now that the main park road at Santa Rosa was paved, we could comfortably travel by bike to many of our study sites, most of which were aligned north-south along the road. I didn't have enough money in my grant to pay for a rented jeep for the duration of Rick's stay in Costa Rica, so bikes became a cheap and healthy way to travel around the park. Bikes were our only means of mechanical transportation during our entire 1980 summer field season.

I stayed at Santa Rosa until the end of January to introduce Rick to the bats and plants we had been studying for six years. Rick was quick to learn everything, and we soon began to have detailed discussions about how best to study *Carollia*'s social behavior. The first problem we faced was giving our bats a distinctive mark so that they could be individually identified inside their day roosts. In 1979, I had begun marking bats permanently with a ball chain necklace bearing a numbered aluminum bird band—a marking system devised by Charles Handley in the mid-1970s during his study of the natural history of bats on Barro Colorado Island. Bats rarely removed or were injured by a necklace when it was properly applied. But it was obviously impossible to identify a bat with such a mark without capturing it, so we needed another marking system for behavioral studies. We eventually settled on a system based on two or three colored plastic forearm bands. Males were banded on the right forearm and females on the left. As he did in many aspects of our collaboration, Rick took the initiative in designing the color-code system and a bookkeeping system to keep all the data organized.

When we began our behavioral observations, relatively little was known about the social lives of phyllostomid bats (or most other bats, for that matter). Working at La Pacifica, Gary Stiles, Sandy Vehrencamp, and Jack Bradbury had recently reported that the carnivorous false vampire, the largest member of the Phyllostomidae, is monogamous and lives in extended family groups (see chapter 7). Jack, who had recently moved to the University of California–San Diego from Cornell, was the leading student of bat sociobiology. He and his postdoctoral student Gary McCracken were currently studying the polygynous (one adult male living and mating with several adult females) mating system of the greater spear-nosed bat, the second largest phyllostomid, on Trinidad. They would find that groups of females roost together in caves for years and that the single adult male that guards each group against the intrusions of other males has nearly exclusive mating rights with them. In the course of his study of the foraging ecology of the Jamaican fruit-eating bat on Barro Colorado Island, Jack's doctoral student Doug Morrison had just discovered that adult males defend tree holes that contain a "harem" of adult females. Finally, Jerry Wilkinson, Jack's newest doctoral student, was studying the social behavior of common vampires at La Pacifica and Santa Rosa when Rick and I commenced our study. In 1980–81, Jerry, Rick, and I often worked side by side at the Sendero cave.

Jerry's work revealed the first convincing case of altruistic behavior in mammals. Like greater spear-nosed bats, female vampires live together for years in groups that don't necessarily contain close relatives. It turns out that finding a blood meal can sometimes be difficult and that not all vampires feed successfully every night. If a vampire misses feeding for two nights, it is in danger of starving to death. This probably seldom happens, however, because females will share their blood meal with an unrelated adult as well as their own offspring when necessary. Jerry's calculations showed that this act of altruism can mean the difference between life and death for the other bat. By presenting starved adults from different roost groups to well-fed females in captivity, Jerry learned that females selectively fed only members of their own group. They did not feed strangers. By selectively feeding close "friends" and kin, female vampire bats maximize the chance that someone will help them in their time of need.

The social organization of *Carollia perspicillata* (figure 15) turned out to be rather difficult to study, because this bat is very easily disturbed inside its day roosts. Instead of calmly remaining in their roosting groups, as greater spear-nosed bats do, *Carollia* bats easily panicked whenever someone entered the Sendero cave. The common long-nosed bat, which also roosted in the Sendero cave by the hundreds, also quickly took flight, so that the cave

Figure 15. A cluster of *Carollia perspicillata* (short-tailed fruit bats) in the Sendero cave. Photo © Merlin D. Tuttle, Bat Conservation International.

became filled with hundreds of milling bats as soon as someone entered it. If we sat quietly on the cave floor without making a sound, however, the bats soon began to settle down, but in dark recesses of the cave rather than in their usual roosting locations. It would take a lot of patience on Rick's part to get the bats habituated to his quiet presence before he could make sense of who was regularly roosting with whom. In the end, Rick's patience won out, and he was able to discover the major features of *Carollia*'s social organization, something that Jack Bradbury had once predicted would be very difficult to do.

With Rick Williams on board the project, my enthusiasm for working at Santa Rosa, which had been thoroughly dampened by the events of 1979 (see chapter 6), began to soar again, and I left Costa Rica at the end of January eager to return for more fieldwork that summer. Although it still wavered from time to time, my enthusiasm for working at Santa Rosa remained high through the rest of the project, which officially ended in the summer of 1984, ten years after it had begun. This enthusiasm enabled me to weather a field season (summer 1980) in which the park lacked electricity of any kind, which made soldering batteries and antennas to radio transmitters very difficult and meant that all of the water used by the park had to be hauled in by truck. That summer my field assistants and I had to travel everywhere on foot or by bike. My enthusiasm also helped me to launch new studies of the reproductive biology of *Piper* plants and the dispersal ecology of *Piper, Muntingia,* and *Cecropia.* Because of his strong botanical interests, Rick played an integral part in these studies while he was documenting *Carollia's* social system. In fact, after he finished a master's degree with me, Rick entered the doctoral program in botany at the University of Wisconsin and studied the mating system of several species of temperate Umbelliferae for his Ph.D. degree.

Rick Williams wasn't the only person to make major contributions to my overall project in the early 1980s. Another of my graduate students, Larry Herbst, and two postdoctoral associates, Frank Bonaccorso and Don Thomas, also joined my research team for one or more field seasons. Larry did a master's project with me on the nutritional ecology of *Carollia.* When he wasn't busy assisting Rick and me on various projects, Larry preserved the fruits eaten by *Carollia* in the wet and dry seasons for nutritional analyses and carefully measured the digestive efficiency of captive bats fed known quantities of fruit. Larry proved to be a tenacious if somewhat unlucky fieldworker. He managed to survive various calamities at Santa Rosa, including burning up many of his 1980 fruit samples in a makeshift drying oven and getting thoroughly soaked by rain every time he climbed a *Cecropia* tree to collect ripe fruit in the extremely wet summer of 1981. To commemorate the latter situation, I devised the "Herbst law of field ecology." This law states that the amount of rain falling in the afternoon at Santa Rosa is directly proportional to Larry's distance up a *Cecropia* tree. Whenever we had crucial work that had to be done in the afternoon, we asked Larry *not* to climb a *Cecropia* tree. Larry took his calamities and our joshing in good spirits and never let his problems get him down. His affable personality, which was reflected by his bright red hair, kept him smiling in the face of impending disaster. His fine spirit later allowed him to weather a disastrous run-in with his doctoral adviser, Craig Packer, at the University of Minnesota. After spending a year and a half tranquilizing lions and taking blood samples in the Serengeti of Africa, he and Craig had a falling out, and Larry quit graduate school. He eventually earned a D.V.M. degree at the University of Florida.

Frank Bonaccorso and Don Thomas worked with me during the summer of 1982. I first met Frank in early 1971 at La Pacifica, when he was one of three enthusiastic neophyte bat researchers in an OTS course dealing with the ecology of tropical vertebrates. The other two students included Dennis Turner, who ended up studying the foraging behavior of vampire bats at La Pacifica, and Donna Howell, who studied the interaction between *Leptonycteris* bats and one of their major food plants, *Agave palmeri,* in southeastern Arizona. A graduate student at the University of Florida, Frank documented the structure and feeding ecology of the bat community on Barro Colorado Island in his doctoral research. By 1982, he was a veteran bat worker. At Santa Rosa, he conducted his own research on the feeding rhythms of captive phyllostomid bats and pitched in on other projects with bats and plants. After our field season together, I hired Frank as my laboratory assistant to complete a study of the genetic structure of *Carollia* populations based on the technique of starch gel electrophoresis.

I first met Don Thomas in the fall of 1976 when he was a master's student at Carleton University, in Ottawa, working with Brock Fenton, whom I consider to be my generation's most knowledgeable bat researcher. In 1982 Don had just finished his Ph.D. at the University of Aberdeen, where he worked on a two-year study of the foraging and seed dispersal ecology of three species of pteropodid fruit bats in Ivory Coast, Africa. Don was eager to gain some experience with New World fruit bats, so I invited him to join my group at Santa Rosa. Don brought a tremendous amount of energy, enthusiasm, and good cheer into the field. His studies of rates of seed rain, the composition of the soil seed bank, and the energetics of *Carollia* as determined by the new technique of doubly labeled water contributed much new information to our overall project. And he improved the *esprit* of our research group by insisting that we relax and drink a gin and tonic ("to ward off malaria") before supper every day.

When Larry Herbst and I arrived at Santa Rosa in early June 1980, Rick had his behavioral study well in hand. He had color-banded many individuals by capturing them in mist nets when they left the cave at sunset or entered the cave just before dawn and had habituated the Sendero bats to his presence. With the aid of a dim headlamp and close-focus binoculars, he was now regularly censusing the cave's *Carollia* population and watching behavioral interactions among known individuals. He discovered that bats were also roosting in the large horizontal trunk of a pochote tree about a hundred meters west of the Sendero cave. Rick sawed two repluggable viewing ports into this trunk, through which we were able to watch bats at close range without disturbing them.

By the end of the 1981 summer field season, we had a fairly complete picture of *Carollia*'s social organization inside and away from its roosts.

When I started sitting in the Sendero roost to watch bats in the summer of 1979, what had first appeared to be an anonymous collection of bats took on a nonrandom spatial and behavioral organization as soon as we began to study them as recognizable individuals. Females, for example, were not randomly dispersed over the one- to two-meter-high ceiling but were clustered into about fifteen groups near the center of this long, tunnel-like cave. As in *Artibeus* and *Phyllostomus* bats, each cluster of *Carollia* females was guarded by a single adult male bat, who defended his group and its roosting site against intruding males both day and night. We called these males harem males; and the groups they defended, harems. The bulk of the males in the cave did not roost with females but instead formed relatively large "bachelor" groups, which roosted around the periphery of the cave a few meters away from the central harem area.

One of Rick's most interesting findings was that the number of females roosting in the cave and *Bombacopsis* tree changed seasonally, but not the number of males. Females were scarce in the dry season and did not become common in the roosts until after the rainy season had started in mid-May or early June. During the long dry season most harem males remained in their territories in the harem area during the day, even in the absence of females. Males that guarded the largest harems, containing up to eighteen females in the wet season, were most tenacious about guarding their territories during the dry season. We were now aging all of the *Carollia* we captured by examining their upper cheek teeth with an otoscope, the device that doctors use to peer into our ears, and scoring their degree of wear. These males turned out to be the oldest, but not necessarily the largest, males in the cave.

We never did discover where females went during the dry season but hypothesized that they must have migrated to the moist slopes of the volcanic cordillera some eighty kilometers north of Santa Rosa. By regularly censusing all known roosts in the northern half of Santa Rosa throughout the year in 1983–84, we confirmed that females did not simply move from the Sendero roosts to other roosts in the park during the dry season. When females were missing from the Sendero roost, they probably had left the park, presumably moving to habitats where food levels in the dry season were higher than those in the parched and leafless lowlands of western Guanacaste Province.

Our discovery that females, but not males, underwent seasonal migrations has a number of important implications for the biology of this species. For example, from our recapture data at Santa Rosa and my previous studies of *C. perspicillata* in Panama and Costa Rica, we knew that nearly all adult females give birth to two babies (in two pregnancies) each year, one in the latter part of the dry season and, after a postpartum estrus, another in the middle of the wet season. When females returned to the Sendero cave in May or June, usually to the exact same roosting site and often to the same harem male they had left several months earlier, they were in advanced pregnancy. But unless females store sperm for many months, something for which we have no evidence in any phyllostomid, the babies they were carrying could not possibly have been fathered by any male in the Sendero cave. Instead, their babies must have been fathered by males living in their dry-season roosts. Similarly, females were already pregnant with babies sired by males in the Sendero and other roosts at Santa Rosa when they migrated to their dry-season roosts. The upshot of all of this was that babies sired in one roost ended up being born in another roost a fair distance away. *Carollia* bats were regularly exporting individuals and their genes from one population to another in western Costa Rica.

This sexually biased migration system poses at least two interesting evolutionary puzzles. First, why do harem males tolerate the presence of babies that they could not possibly have fathered in their territories? Although Rick's observations indicated that harem males rarely interacted with these babies, they didn't attempt to kill them either, something that occurs regularly in lions and certain monkeys when males oust competing males during territorial or harem takeovers. We don't yet know why male bats tolerate such nonkin, which represent potential competitors for food and mates. But I have speculated that this tolerance may be related to the long time between successive estrus periods in this bat and to the possibility that females might not choose to roost in harems guarded by infanticidal males. Unlike lions and monkeys, female bats do not undergo an estrus cycle shortly after they lose a nursing young. Thus, a male cannot expect to inseminate a female shortly after he has killed a nursing pup. And female *Carollia* appear to be choosy about their mates. Those that we observed often changed harems several times, sometimes moving between harems in the Sendero cave and the *Bombacopsis* tree, before settling into one harem for a lengthy stay just before they ovulated and mated. Thus, by playing it cool toward these babies, males might end up with larger harems and sire more babies than if they were aggressive toward them.

The second puzzle is that, unlike females, which have two mating seasons and hence two reproductive opportunities each year, harem males participate in only one mating season a year. That is, by staying in one roost year-round, Santa Rosa males missed a chance to mate with females immediately after they gave birth in the late dry season. The only time they mated was in the late wet season, just before females left the park for their dry-season roosts. Again, we don't yet know why males behave this way, but it must be related to the importance of staying in one roost year-round. By staying in one roost, males gain three possible advantages. First, they avoid any mortality risks associated with migration and becoming established in a new roost. Second, they can become very familiar with good feeding areas around their roost throughout the year. And third, if they are harem males they can provide year-round guard of their territories, which are prime pieces of real estate, because that's where mating takes place. If they are bachelor males, they can monitor the harem males and possibly acquire a territory should one of those males disappear.

These seasonal movements also have important conservation implications. Although it encompassed over ten thousand hectares at the time of our fieldwork, Santa Rosa National Park was too small to contain the annual resource needs of its population of short-tailed fruit bats. To maintain a viable population of these bats, an area at least an order of magnitude larger than this is needed. And this area needs to be oriented in a particular direction. East-west expansion of the park into the Guanacaste lowlands would not serve the needs of this bat. Instead, the expansion would have to be in a northerly direction, toward the Cordillera de Guanacaste. Without the protection of upland habitats from human modification, the park's *Carollia* population would have been vulnerable to local extinction. If a common bat such as *C. perspicillata* needs a large annual range, I wondered, how many other seasonal residents of Santa Rosa, insects as well as birds and mammals, also need protected areas in the Costa Rican uplands as well as in its lowlands? Conservation biologists are used to arguing that large protected areas are needed to meet the needs of species of top carnivores such as jaguars and wolves—species with large *daily* home ranges. They tend to forget that small species, including many nectar eaters and fruit eaters, undergo seasonal migrations and hence have large *annual* home ranges that need to be preserved. After I stopped working at Santa Rosa, Dan Janzen used this kind of information successfully to effect the expansion of nationally protected land in western Guanacaste to include the volcanic slopes north of Santa Rosa. William Allen describes Dan's conservation efforts in his book *Green Phoenix* (2001).

In addition to discovering that the two sexes of *Carollia* differed in migratory behavior, Rick and I found that the nightly foraging behavior of harem males differed from that of females and bachelor males. Doug Morrison found a similar sexual difference in the foraging behavior of the Jamaican fruit-eating bat in Panama. Prior to 1980, we had radio-tracked bats whose social status was unknown to us. In 1980–81, we concentrated on tracking socially known bats. We quickly discovered that harem males were stay-at-home bats and that females and bachelor males were the only individuals that foraged more than about a kilometer from the roost. We also found that females roosting together in their day roosts did not forage near each other at night. Daytime roosting associations broke down temporarily when the bats began to feed.

Again, these behavioral differences between males and females were associated with the need for harem males to guard their mating territories inside the roost day and night. Instead of leaving the roost around sunset and returning shortly before dawn—behavior that was typical of females and bachelor males—harem males spent most of the night inside the roost. To feed, they would quickly dart out of the roost, grab a fruit, and then return to the roost to eat it. Unlike females, whose diet was dominated by the relatively nutritious fruits of *Piper* during the wet season, harem males were more likely to eat chunks of fiber-rich *Cecropia* fruit—fruits that were common just outside the Sendero roost. In the terminology of Tom Schoener, whose theoretical papers on optimal foraging had helped inspire our work at Santa Rosa back in 1974, harem males were archetypical "time minimizers" when they foraged, whereas females, with their heavy energy demands associated with two annual pregnancies, were more likely to be "energy maximizers" while they foraged (Schoener 1971). Harem males minimized their feeding time, often by eating less nutritious fruit, to make time for territorial defense. In contrast, females (and bachelor males) spent all night searching for and consuming more nutritious fruit than that consumed by harem males, in part to support their considerable reproductive effort.

When they weren't feeding, harem males actively defended their territories at night against intruding males, which included bachelor males and other harem males hoping to gain a larger harem. This defense often involved short boxing matches, in which two males pummeled each other with their closed wings while hanging from the ceiling a few centimeters apart. Boxing matches also occurred occasionally among harem males during the daytime. These fights undoubtedly took a physical toll on males. Dark-blue #6 was our top-ranked male—the one controlling the largest harem—from April 1981 through at least mid-1985. When we examined him carefully in March 1983, he was blind in one eye, his ears were tattered, he was missing one upper canine tooth and the other upper canine was broken, and his cheek teeth were worn flat, an indication of his advanced age. He was a tough old bat and undoubtedly sired far more babies than the average harem male.

I always looked forward to seeing him whenever I conducted cave censuses after Rick left the project.

While Rick was observing bats in the Sendero cave, I was busy watching other sets of animals visiting *Piper* flowers and *Cecropia* fruits in the summers of 1980 and 1981. By 1980, I had amassed many data on the wet-season flowering and fruiting behavior of Santa Rosa's five species of *Piper.* These data indicated that each species flowers and fruits at slightly different times of the year. Data from our bat netting program told us which species of bats, primarily two species of *Carollia* and the common long-tongued bat, eat *Piper* fruits and disperse their seeds. These bats quickly shift from one species of *Piper* fruit to another when one fruiting season ends and another one begins. But in order to complete my studies of the reproductive biology of *Piper,* I wanted to know who pollinates *Piper*'s tiny flowers and whether the different kinds of *Piper* are likely to compete for pollinators. Perhaps differences in fruiting times—differences that allow the five *Piper*s to share dispersers without competing—are merely consequences of differences in flowering times they had evolved to reduce competition for pollinators.

To answer these questions, I spent a lot of time watching different species of bees and syrphid flies visit *Piper* inflorescences. Individual flowers are tiny, measuring only a millimeter or two in diameter, and offer pollen but no nectar rewards to their visitors. Stingless bees of the family Trigonidae were by far the most commonly observed *Piper* flower visitors. Several species, including the small orange *Trigona jati* and the larger, jet-black *T. dorsalis,* delicately hovered in front of an inflorescence before landing and slowly walked up and down the flower stalk, scraping pollen into the pollen basket on their hind legs. *Trigona* bees were relatively sedentary and collected most of their pollen from one or two inflorescences on the same plant before flying off to their nests.

From a *Piper* plant's point of view, trigonid bees are relatively poor pollinators, because they rarely fly from one plant to another during a foraging bout. If flowers of these plants are self-incompatible, as a series of pollination experiments I conducted in 1982 indicated for *P. amalago* and *P. jacquemontianum* (but not *P. pseudofuligineum*), then they need a more mobile visitor, one that frequently moves from one plant to another, in order to set fruit. A relatively uncommon bee in the Megachilidae (leaf-cutter bees), an undescribed species of *Megachile,* fits this bill perfectly. About twelve millimeters long, this hairy bee quickly runs up and down

Piper inflorescences while scraping pollen into her abdominal pollen basket. Rather than visiting many inflorescences on one plant, this bee—which the world authority on megachilid bees, Charles Michener, later told me belongs to a subgenus that specializes in visiting *Piper* flowers—quickly moved from one plant to another during a foraging bout.

As in the case of their seed dispersers, I found that *Piper* species extensively share the same species of insect pollinators. *Megachile,* it turns out, are the most effective pollinator of each species of *Piper.* Only by flowering at different times of the year do these plants manage to avoid competing among themselves for pollinators. My pollination studies indicated that *Carollia*'s favorite fruits are the products of the pollen-collecting behavior of trigonid and megachilid bees. And my seed-dispersal studies indicated that plants producing *Megachile*'s favorite flowers depend to a large extent on the foraging behavior of *Carollia* bats for their dispersal. The lives of *Megachile* bees, *Piper* shrubs, and *Carollia* bats thus are inextricably intertwined in this tropical dry forest.

When I wasn't watching the foraging behavior of *Piper*-visiting bees, I observed fruit- eating birds and mammals visiting *Cecropia peltata* trees. In 1979 I began a study of the population ecology of this plant, an important member of early successional habitats and habitat disturbances in tropical dry forest. Now I was anxious to quantify fruit and seed removal by diurnal and nocturnal vertebrates. Our bat studies clearly indicated that *Cecropia* fruits were important dietary items for most of Santa Rosa's frugivorous bats, but we also needed data on which other species of animals ate these fruits and dispersed their seeds. To obtain these data, Rick and I spent many hours in 1980 and 1981 "staking out" fruiting *Cecropia* trees and recording which species of vertebrates visited them and how much fruit they ate per visit.

We supplemented our animal observations with data on the nocturnal and diurnal disappearance rates of ripe fruit in a series of trees along the main park road. To determine how many fruits had been eaten at night, I woke up each morning at 5 A.M., hopped on my bike, and rode to each tree, where I counted the number of fruits on our census branches. Rick repeated this process late each afternoon to measure diurnal fruit consumption. Differences between our morning and afternoon counts gave us the numbers we were after.

The bike ride in the cool morning air was exhilarating, and, despite the early hour, I looked forward to it every day. My main motivation for taking the morning shift, though, was to see animals along the road. Collared peccaries, white-tailed deer, and monkeys were usually out and about at this time of day. I always hoped to see a big cat cross the road, but, as I have previously indicated, I saw only an occasional jaguarundi.

Sitting quietly on a hillside during the day and observing the canopy of one or more fruiting *Cecropia* trees proved to be an excellent opportunity to do some serious monkey watching. Monkeys were not the most common daytime visitors to *Cecropia* trees—a couple of species of small frugivorous birds were most common—but they certainly were the most interesting animals to watch. Spider and howler monkeys went about finding *Cecropia* fruits in entirely different fashions. These fruits (technically, infructescences) are finger-sized and occur in clusters of four that project away from *Cecropia's* umbrella-shaped leaves on a tough peduncle, or stalk. Like *Piper* fruits, they do not change color as they ripen but remain light green in color and thus give animals no color cue when they are ripe. Bats and other nocturnal mammals such as opossums and kinkajous undoubtedly use odor cues to detect ripe fruit, but spider monkeys are visual and tactile searchers. They spent a lot of time sitting quietly on a branch looking for ripe fruit. Once they spotted a likely candidate, they went over and touched it to determine if it was soft, just as Bill Sawyer and I had done when we were collecting ripe *Piper* fruits. In contrast, howler monkeys were totally unselective feeders. Whenever they entered a *Cecropia* tree, they went directly to the nearest cluster of fruit, quickly stuffed several of them, both ripe and unripe, into their mouths, and began chewing. They fed as herbivores rather than as frugivores. Because of these differences in feeding styles, howlers acted as predators on *Cecropia* seeds rather than as potential seed dispersers, like the spider monkeys.

Perhaps because of their different feeding styles, howlers were able to feed peacefully in the same *Cecropia* tree with spider or white-faced monkeys. The last two species, both selective consumers of ripe fruit, however, interacted strongly the few times I saw them together in a tree. Late one afternoon, for example, a female spider monkey, golden-colored in the dying sunlight, was busy eating fruit when a white-face entered the tree. The smaller black and white monkey initially stayed well away from its larger cousin. Soon it became bolder, quickly ran along branches up to the spider, and briefly chased it before retreating to a distant part of the canopy. After a few of these mild "attacks," the white-face launched an intense attack and began chasing the spider all over the tree. I stared in admiration at the agility of both monkeys as they rocketed from branch to branch, sometimes traveling quadrupedally, sometimes by arm swings, but always with perfect confidence and coordination. The spider monkey finally leaped out of the *Cecropia* into the branches of a dead tree, where she hung by her arms and tail and stared back at the white-face before moving off to the south. The white-face stayed in the *Cecropia* for a few minutes and ate two fruits before leaving the tree.

All in all, we recorded twenty-eight species of vertebrates—fifteen birds and thirteen mammals—eating *Cecropia* fruit. Data that we collected from our early morning and late afternoon fruit checks told us that half the ripe fruits disappeared during the day and half at night. Unlike *Piper* fruits, which are exclusively eaten by bats at Santa Rosa, *Cecropia* fruits and seeds end up in the guts of a variety of frugivorous vertebrates. Tanagers, woodpeckers, and crested guans as well as *Artibeus* and *Carollia* bats, among other mammals, are potential dispersers of this pioneer tree. With its diverse array of dispersers and a relatively long fruiting season, it is not surprising that *Cecropia* seeds turned out to be widely distributed and common (one hundred to two hundred per square meter) in the soil samples Don Thomas collected in 1982.

By the summer of 1983 I was running out of steam on the Santa Rosa *Carollia* project. I had accomplished my major goals and had completed most of the studies that I regularly listed in the back of my field journals, beginning in 1974. We now knew more about the behavior and ecology of the short-tailed fruit bat than almost any other species of neotropical mammal. It certainly was one of the best-studied bats in the world. We finally had enough quantitative data to indicate the specific importance of *Carollia* and the general importance of frugivorous bats in the economy of tropical dry forests. Our data strongly supported the view that these bats play an important role in the dispersal ecology of this habitat. Pioneer trees, understory shrubs, and canopy trees all depend on bats for their effective dispersal. Never again would we be able to view these bats as negatively as Arthur Greenhall did when he wrote in 1961 that they were messy, wasteful feeders (Goodwin and Greenhall 1961).

To round out the project, I hired Liz and Rich Chipman to conduct an extensive netting study that lasted from June 1983 through June 1984. Newly married and fresh out of their undergraduate studies in wildlife biology at the University of Maine, the Chipmans were protégés of Frank Bonaccorso, who recommended them highly to me. Larry Herbst and I spent the month of June 1983 training Liz and Rich in the details of their job, which involved regularly censusing the Sendero cave, capturing and banding *Carollia* bats at our three major roosts and at a series of sites located throughout the park, and recording the flowering and fruiting activity of *Carollia*'s food plants. For one last time, I wanted to document seasonal changes in the size of *Carollia*'s population and its distribution in different habitats around the park.

When I left Santa Rosa in early July, I was confident that Rich and Liz would be excellent field-workers and that the project was in good hands. My confidence turned out to be well-placed. The Chipmans did everything I asked of them and had many adventures along the way. They encountered curious tapirs, stubborn but unaggressive rattlesnakes, scorpions, and, of course, many bats. They even survived sharing field quarters with Dan Janzen and his girlfriend, Winnie Hallwuchs, who began living at Santa Rosa for long periods in 1979.

When June 1984 rolled around, I was scheduled to return to Santa Rosa to help the Chipmans wrap up the project. But I never made it to the field that year. Shortly before D-Day (departure day), I woke up early one morning with tremendous waves of pain rolling through my abdomen. Marcia called my internist, who told her to take me to the emergency room of Larkin Hospital, three kilometers from our house, as quickly as possible. Less than twenty-four hours later, I was in a recovery room after undergoing my first surgery. My large intestine had become badly twisted upon itself as a result, I'm sure, of the stress that I was experiencing in preparing for another trip to the tropics. The surgeons removed half a meter of my large intestine. A bit over one week later, they operated on me again to fix the joint between the small and large intestines, which had fused shut. I was hospitalized for nearly a month. By the time I was released, I had lost a lot of weight and felt like a feeble old man.

It took me a while before I could think seriously about science again. Shortly before I left the hospital, however, I knew that I was on the road to recovery—I dreamed for the first time in a long time about *Carollia.* In my dream I tried to explain to someone why it has a male-biased sex ratio at birth. Liz and Rich's return to Miami after successfully wrapping up the project helped me begin to feel normal again. But it took me all summer to complete the recovery process.

Although my Santa Rosa project officially ended in 1984, I returned to Santa Rosa for ten days in June 1985, primarily to recensus my three major roost populations. Dan Janzen seemed genuinely pleased to see me, a warmth he rarely displayed in earlier years, and offered to lend me his jeep whenever I needed it. My field assistant, Joe Maguire, and I hiked out to the Red and Cuajiniquil roosts and camped overnight there for the last time. Liz and Rich had camped at these roosts once a month during 1983–84.

I last visited the Sendero cave on the morning of 22 June. After crawling in through the upstream entrance, I sat quietly in the dark for a few minutes, thinking about times past, before turning on my headlamp to census the bats. In my mind's eye I pictured my first visit to this cave in 1974. Although many parts of the park had changed as a result of fires, building projects, and plant succession, the Sendero area was pretty much the same as it had been eleven years earlier. I could almost hear the voices of Ray Heithaus, Rick Williams, Jerry Wilkinson, and Don Thomas as we worked with bats in and around this cave over the years. Though the cave had not changed physically, the identity of its inhabitants certainly had. Nearly three generations of *Carollia* had been born, matured, and died since I had begun to work there.

Coming out of my reverie, I flicked on my headlamp and pointed it at the main harem area. Many *Glossophaga* and *Carollia* immediately took flight, and the air became thick with bats. Above my head a slender lyre snake extended its head and neck from a crevice in the ceiling and assumed a strike posture. This snake is one of two species of bat-eating snakes (the common boa is the other species) that regularly occupied bat roosts at Santa Rosa. As the harem area cleared out, I saw the bat I was looking for. Dark-blue #6 still hung from the ceiling, his tattered ears twitching as he looked right and left, surveying the melee. I looked away briefly, and in that instant he was gone. I was confident that he would be right back to his hallowed spot shortly after I climbed out of the cave. "May you live to be one hundred!" I thought as I emerged from darkness into the bright morning sunlight.

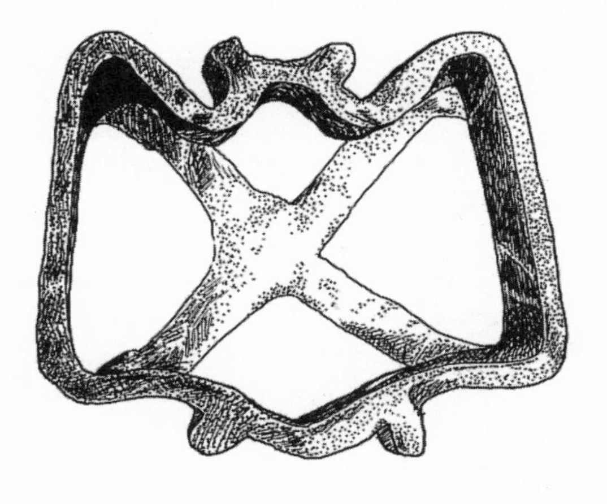

6 Anastasio's Last Stand

My office walls in the Cox Science Building at the University of Miami are decorated with a gorgeous red *mola* from Panama and a variety of framed photographs and colorful posters, mostly depicting bats and habitats where I have worked. In the midst of these is a rather small piece of scuffed leather, nicely matted and set in a dark wooden frame. This piece of leather is all that remains of a briefcase that I purchased in Costa Rica in 1971. Burnished into its tan leather is a dark cattle brand—a vespertilionid bat with rounded ears and broad tail membrane. The brand was applied to my briefcase in August 1979 by Santa Rosa park guards, who had found the branding iron on Finca Murciélago, located just west of the park. Until June 1979, Finca Murciélago (Bat Farm) belonged to Anastasio Somoza Debayle, Nicaragua's strongman president, who was driven from office on 17 July of that year. Just before Somoza was ousted, Costa Rica expropriated his ranch and turned it over to the National Park Service. Murciélago eventually was annexed to Santa Rosa.

The Murciélago brand in my office serves to remind me of a time when

Cattle brand from Finca Murciélago, Guanacaste Province, Costa Rica. Drawing by Ted Fleming.

political upheaval raged less than fifty kilometers from Santa Rosa and threatened to spill over from war-torn Nicaragua into peaceful Costa Rica. The summer of 1979 was a time when my research on bats and plants was nearly halted by larger forces affecting our activities in the park. That summer, a small army of Costa Rican soldiers practiced serious war games at Santa Rosa. Military officials were adamant that gringo scientists, namely Dan Janzen and I and our field crews, must leave the park. Only Dan's stubbornness in the face of constant pressure from the military enabled us to continue research that year.

I first became aware of the Sandinistas, Nicaragua's anti-Somoza guerrilla army, and the impending political turmoil in Costa Rica's northern neighbor in March 1977. On one of my days off at Santa Rosa, I read in the international edition of *Time* about the growing conflict between the FSLN (Frente Sandinista Liberación Nacional) and Somoza's Guardia Nacional. The article, entitled "Somoza's Reign of Terror," indicated that the Guardia Nacional was killing any peasants who were thought to be Sandinista sympathizers. In a footnote, *Time* explained that the Sandinistas were "named for Augusto Cesar Sandino, a guerrilla leader who fought against occupying U.S. Marines in the late 1920s and was executed in 1934 by the founder of Nicaragua's ruling dynasty, Anastasio ('Tacho') Somoza Garcia" (*Time* 1977, 29). Tacho's son "Tachito" was first elected president in 1967, twelve years after his father's assassination. By 1977, opposition to him was widespread in Nicaragua. Those opposing Somoza hoped that newly elected U.S. president, Jimmy Carter, with his strong human rights platform, would pressure Tachito into relaxing his dictatorial grip on the lives of his countrymen.

I filed this news report in the back of my mind and then forgot about it as my family and I prepared to move to England in early September 1977. We were excited about spending our first academic sabbatical at Oxford University, where I would be a visiting scientist at the Edward Grey Institute for Field Ornithology. At Oxford I temporarily left bats behind to study the communal roosting behavior of pied wagtails, lovely black and white birds about the size and shape of mockingbirds. Wagtails feed by picking tiny insects from the ground in meadows and pastures and along city sidewalks. The basic question I wanted to answer was, Do the communal roosts of pied wagtails serve as information centers about the locations of good feeding areas? The hypothesis that bird (and bat?) roosts serve as food information centers was a hot topic in behavioral ecology in 1977. It had been put forward as a general explanation for communal roosting in birds by Peter Ward and Amotz Zahavi in 1973.

To answer this question for wagtails, I had to get up before dawn in the cold, damp British winter and drive to a *Phragmites* reed bed northwest of Oxford. There I recorded the number of birds leaving the roost together and the direction flown by each group. I wanted to know if there was a consensus direction in the departure each day and whether it changed from time to time, presumably in response to changes in the locations of good feeding areas. At sunset, I counted the number of birds returning to this roost. I also carefully studied the feeding behavior of individually identifiable birds and set out experimental patches of fly larvae (fishing maggots) to see if birds that discovered rich food patches somehow communicated their locations to their roostmates. For three months I tromped around marshes, meadows, farmyards, and villages recording the feeding rates of my new avian friends. I became acutely attuned to the *che-bick, che-bick, che-bick* flight calls of pied wagtails and used these calls to locate birds for new observations. In the end, all of my data suggested that birds that found my experimental patches were tight-beaked about their locations. From these and other observations, I concluded that wagtail roosts were unlikely to function as food information centers.

It was a thrill for me to work at the Edward Grey, because its first director, David Lack, had been one of my intellectual heroes during my undergraduate and graduate school days. Reading David Lack's *Darwin's Finches* had been one of the highlights of my undergraduate studies. Lack died in 1972, and the Edward Grey was now run by Chris Perrins and young John Krebs, whose innovative research on the foraging behavior of great tits and the colonial roosting behavior of great blue herons had attracted me to Oxford. While I was at Oxford treading the halls where Charles Elton, Niko Tinbergen, and David Lack had worked, Santa Rosa, *Carollia* bats, and the Nicaraguan political situation were far from my thoughts.

By the time we left England in the summer of 1978, the political situation in Nicaragua had further deteriorated. On 10 January 1978, Pedro Joaquin Chamorro, the editor of Managua's influential *La Prensa* and popular leader of Somoza's growing opposition, was assassinated. Forty thousand mourners followed Chamorro's casket from the hospital to his newspaper office. Chamorro's death helped to increase Nicaragua's revolutionary fervor and signaled the beginning of the end of the two-generation Somoza dynasty. By mid-1978 the country's substantial middle class as well as the U.S. Embassy were openly critical of Somoza and began to call for his resignation. In August, a group of Sandinistas led by Eden Pastora stormed into the Nicaraguan Chamber of Deputies and took a large number of politicians hostage. After two days, Somoza capitulated to the guerrillas' demands for money and the release of political prisoners. In October and November a U.S. mediation team worked in Nicaragua, trying to get Somoza to resign, but this widespread dream was not realized until mid-July 1979.

In the fall of 1978, we moved into our new home in South Miami, a former farm town that was now surrounded by a rapidly expanding Miami metropolis. I had just joined the faculty of the University of Miami. Miami's biology department had attracted me because of its growing emphasis on tropical biology and a doctoral program not yet in existence in St. Louis. In addition, Miami was only about two and a half hours by air from San José, Costa Rica. Building on my exposure to behavioral ecology at Oxford, I submitted a new proposal to NSF to document *Carollia*'s social organization inside and away from its day roosts. After a break of nearly two years, I was anxious to begin a new phase of bat and plant research at Santa Rosa.

With a new grant in hand, I wrote to many of my colleagues in the spring of 1979, indicating that I wanted to hire a graduate student to conduct behavioral observations in the Sendero cave. Jack Bradbury recommended one of his senior undergraduates, "Jane" (a fictitious name). Jack indicated that Jane wasn't particularly interested in sociobiological theory but that she wouldn't hesitate to spend long hours in the Sendero cave watching bats. On the basis of Jack's recommendation, I offered the job to Jane, who enthusiastically accepted it. Unfortunately, once she got to Costa Rica, Jane rather quickly discovered that she was not comfortable living at Santa Rosa away from family and friends. For this and other reasons, she didn't stay with the project long.

Having hired a research assistant, I turned my attention to finding two more field assistants for the 1979 field season, which was scheduled to run from the end of May through mid-August. In the past I had always worked with male field assistants simply because the opportunity to hire a female had never occurred. In 1979, however, I hired two more women to complete my field crew. "Alice" (another fictitious name) was a graduate student in zoology at the University of Arizona and appeared to be well-suited to study the diurnal consumers of some of the fruits my bats were eating. My final assistant, "Rachel" (also a fictitious name), was enrolled in my undergraduate vertebrate ecology class at Miami. An excellent student, Rachel wanted to get some tropical field experience before she went to medical school.

My field crew was chosen by April 1979, but since the political situation in Nicaragua was getting hotter, I contacted my friend Alvaro Ugalde at Costa Rica's National Park Service to inquire whether the crisis was having any effect in western Costa Rica in general and at Santa Rosa, specifically. "Would it be unwise," I asked, "to bring three female assistants to

Figure 16. Dan Janzen at work, 1979. Photo by T. Fleming.

Santa Rosa with me?" In early May, Alvaro assured me that although Sandinistas were being harbored on the Costa Rican side of the border and were slipping into Nicaragua to fight, things were quiet at the park and were likely to continue that way for the foreseeable future. Relying on this information, I told Jane, Alice, and Rachel that our fieldwork would proceed as scheduled.

On 28 May, Alice, Rachel, and I flew to Costa Rica. Jane was still in school and would join us in late June. After my usual round of shopping and visiting friends in San José, the women and I packed our gear into an old Toyota jeep and headed off to Guanacaste. Foreshadowing a problem-filled field season, we got as far as the airport, twenty kilometers west of San José, before the jeep's transmission began to shake and grind. Our rental agency provided us with a new Toyota, and we continued to La Pacifica our first day on the road. The next day, we drove into Santa Rosa along its newly paved main road and began to set up camp in my usual field house. The new park director was Franklin Chaves, a young ornithologist and protégé of Dr. Gary Stiles, who taught ornithology and ecology at the University of Costa Rica. Franklin greeted us warmly and indicated that he was anxious to help us in our studies of avian frugivores in the park.

Dan Janzen (figure 16), dressed in gray work clothes with rolled-up long sleeves and partly laced-up leather boots, was also on hand to welcome us. He and his new girlfriend, Winnie Hallwuchs, a doctoral student at Cornell

University, and a crew of three male assistants would be working in the park for most of the summer. Although Dan had occasionally visited the park when I was there prior to 1979, he and I had never been at Santa Rosa together for long stretches of time. By 1979, he had decided to spend as much time as possible at Santa Rosa each year, and from then until the end of my *Carollia* project, my assistants and I always worked in the shadow of Dan.

From my viewpoint, Dan's presence in the park was a mixed blessing. On the positive side, he was an enormous source of information and inspiration about the natural history of the tropics in general and tropical dry forest in particular. Whenever I had questions about tropical plants and animals, I always went to Dan for answers. His knowledge (and opinions) about tropical biology seemed to be inexhaustible. In addition to being an unlimited source of information, Dan was an excellent sounding board for research ideas. Whenever I had a new question I wanted to investigate, I would turn to Dan for advice and suggestions. Finally, Dan was supportive of my research. Whenever we needed some logistical help, particularly during the summer of 1980 when my field crew and I worked without a field vehicle, Dan was generous in loaning us his jeep.

But I also paid a price, probably more imagined than real, in working close to Dan Janzen. I was sensitive to the scorn that Dan heaped on people who didn't subscribe to his particular brand of workaholism. As is widely known, Dan Janzen thrives on driving himself and those around him to work up to twenty hours a day, for days on end. Early in his career, one of the forces driving him to work this hard was burning ambition, ultimately rewarded by his election to the National Academy of Sciences. Another driving force was his overwhelming curiosity about nature. He wanted answers to seemingly a million questions—many of them posed in the form of provocative hypotheses. "Why does shit stink?" was one of his most famous questions in the early 1970s. It was easy to fill a twenty-hour day with work when you were simultaneously studying seed predation by rodents and bruchid beetles, seed dispersal by horses and cows, consumption of tropical foliage by moth and butterfly larvae, rodent population dynamics, and moth diversity and life histories.

I thought I was working on many projects and for long hours during my Santa Rosa years (and I was), but my work schedule paled in comparison to Dan's. He was usually up earlier and went to bed later than I did every day. Beginning in 1981, I usually took a short break after lunch and read or napped in a hammock slung between two trees. Dan often marched past me as I rested and commented on more than one occasion, "There's Fleming on vacation at Santa Rosa." He may simply have been joking with this, but I suspect he wasn't. I suspect that one of Dan's greatest fears was to be called a nine-to-five biologist.

Except for when he was sick with Ménière's disease, which forced him into bed with periodic bouts of nausea, I don't know if Dan ever took a break from work while I was at Santa Rosa. He and I never kicked up our heels and drank a cold beer together, something I did many times with other colleagues and park visitors. The one time I rode with him into Liberia, he let Winnie drive so that he could work on a manuscript. He never ate in any restaurant in town, saying he couldn't take time away from his work for this. He also frequently skipped meals in the park because he was too busy to eat.

In addition to (potentially) experiencing Dan's scorn, one had to accept that he was the alpha male at Santa Rosa. Although the National Park Service owned and operated Santa Rosa, Dan was the de facto person in charge of the park's management and research decisions. If our research impinged in any way on what Dan was doing, we were quick to hear about it. An extreme example of this occurred in July 1980 when he criticized two of my assistants for trimming a few saplings that blocked one of our main trails through the forest. He said he had been watching the growth of one of those saplings (without marking it!) for some time. I was astounded by Dan's attitude. His criticism, I felt, was mainly a territorial display in which he was proclaiming, admittedly in a relatively unthreatening way, that he was king of this jungle.

In Dan's defense, years later (in 1988) he did apologize to me for his "heavy-handed battering" of my life at Santa Rosa. He explained, "I obviously am never really sure about how much to interfere with [the] lives of others, but I feel relatively more secure about what is best for the park." As usual, Dan felt he knew what was best for Santa Rosa.

Dan's clout with the National Park Service and beyond was to become most obvious during the summer of 1979, when the Costa Rican "army," consisting of the Guardia Nacional and the Guardia Rural, set up camp in Santa Rosa. Unfortunately, Alvaro Ugalde was not clairvoyant and therefore had not anticipated an army takeover of the park when I spoke with him in May. But on 2 June, a day after we had arrived at Santa Rosa, at least two hundred troops pitched their two-man pup tents in the park's main camping area, about a kilometer from our living quarters. Eventually, about four hundred soldiers were camped in the park. If I had known this would happen, I definitely would not have hired three women to work with me that summer. I'm not sure even I would have gone to Costa Rica that year.

The growing military presence didn't immediately affect our lives to any great extent. We were too busy getting settled and beginning new studies of bats and plants. Our first task was to clear net lanes and trails that hadn't been used in some time. Shortly after we began this work, a period of heavy rains, called *temporales,* set in. Occurring early in the wet season about every other year, *temporales* dump relatively large amounts of rain on the Pacific coast of Central America for up to a week before atmospheric conditions change. During our period of *temporales,* waves of rain blew in from the Pacific and inundated low-lying areas and streambeds at Santa Rosa. Quebrada El Duende quickly reached flood stage and completely filled the Sendero cave with swirling torrents of muddy brown water, forcing its chiropteran inhabitants to seek temporary shelter elsewhere. Cattle tanks such as the Laguna swelled in size and quickly attracted dozens of male frogs, whose choruses became deafening at night. From a distance, the frog ponds buzzed like giant beehives day and night during *temporales.*

Despite the wet weather, Alice, Rachel, and I worked with high spirits. Except for making our machete handles slippery, the rain didn't impede our vegetation clearing. In fact, it was pleasantly cool swinging a machete in the rain. After a couple of days of this, however, Rachel began to get a bit disillusioned with life at Santa Rosa. She grumbled, "I thought there would be sunshine and red and purple birds here, but all there is is soldiers and rain."

As the Costa Rican army began turning Santa Rosa into a military training camp, we had to adjust to the sight of soldiers carrying rifles day and night and, until they came to know us better, to their routinely stopping us with weapons drawn. Most of them showed their Nicaraguan political sympathies by wearing red and black armbands—the colors of the FSLN. But other than occasional catcalls directed at the women, they never seriously hassled us. In fact, two soldiers—Julio and Marcel—became our unofficial field assistants during some of their free time. The two *ticos* were particularly fascinated by our bat-netting operation and often helped us when we were catching bats at the Sendero cave. Julio was especially friendly and came by our house almost daily to keep us informed about the military situation. He warned us, for example, that the army expected Somoza to send his planes to bomb the park on the night of 22 June. Fortunately, that night came and went without an air raid. Unfortunately, Julio had no clout with his superiors, who became increasingly anxious to oust all gringos from the park in early July.

By the middle of June, the army had erected barriers across the park road where it joined the Pan American Highway and near our living quarters, effectively cutting the park off from contact with civilians. Small cannons were set up near the army headquarters, just down the road from our house. Beginning on 22 June, helicopter gunships firing round after round of blanks at invisible enemies began circling low over the park. The gunships swept low into a field, and groups of soldiers practiced entering and leaving the helicopters while they briefly hovered just above the ground. The booms of larger guns echoed through the woods, undoubtedly upsetting monkeys and other wildlife as well as some of us researchers. A round of live ammunition ricocheted off Dan Janzen's house near where two of his assistants were working. This understandably upset them, but Dan wasn't perturbed. "It's obvious you guys have never been in the army," said Dan. "If you had, you'd be used to this kind of mayhem."

The presence of soldiers in the park was to have its greatest effect on Alice. Basically shy and unused to group living, Alice was thoroughly flustered by their presence. Any attention the soldiers paid her, no matter how casual or innocent, made her feel very uncomfortable. I had hoped that Alice would be in charge of making quantitative observations of the birds that ate *Cecropia* and *Muntingia* fruits, but it soon became apparent that she could not work alone. She was fine when we netted bats as a group at night, but solitary diurnal work was not for her. Nor was she able to radio-track alone at night like the rest of us. By early July, Alice's contribution to fieldwork ground to a halt, and she left the project.

Field biologists working in foreign countries, of course, always face the risk of running into physically or psychologically uncomfortable situations stemming from political events beyond their control. A good example of this is the physical danger that Edward Goldman and Edward Nelson occasionally faced while working in Mexico. The thirty-six-year-old Nelson and his eighteen-year-old assistant, Edward Goldman, had been hired by C. Hart Merriam to conduct extensive faunal studies in Mexico, beginning in 1892. Although not as lawless as it had been in the mid-nineteenth century, Mexico was seething with prerevolutionary tensions in the years 1892 to 1906, when Goldman and Nelson worked there. Those years marked the latter portion of the reign of Porfirio Díaz, who gained power in 1876 and was finally overthrown in 1911. Díaz had done a great deal of good for Mexico by guiding it through an intense period of modernization, but this progress had occurred at the price of individual freedom. By the mid-1890s, Díaz's policy of *"pan y palo"* (bread and club) for quieting civil disobedience was beginning to breed deep resentment in the general populace. By 1895, Díaz's popularity was on the decline, and he was forced to increase levels of repression against his people to retain power. Lawlessness again began to rear its ugly head throughout Mexico.

Edward Goldman's field books, which now reside in the Smithsonian Archives, reveal two examples of this lawlessness. On 9 November 1892, at Lerma in the state of Mexico, he wrote:

> I was held up by 5 men on [the] road to Salazar just above Japala & robbed of gun, watch, . . . and other small items. 3 men attracted my attention in front by coming along [the] road in front & toward me but made as though to pass & when about 20 ft in front of me closed in & then I saw one man had a knife in one hand & a stone in the other. The others all had stones only. My gun was all ready off my shoulder as I never pass anyone on the road at night without taking it down. When I saw their intentions I started to shoot but at that instant 2 men from behind reached me; one grabbed the butt of [the] gun & the other struck me in [the] left temple with a stone, cutting a gash to [my] skull. Then all of them were on top of me & took everything I carried but did me no bodily harm. (Goldman, fieldbook 2)

After the bandits ran off, Goldman walked to the nearest town and reported the incident to the police, who spread the word of the robbery to other pueblos. The next day the five robbers were apprehended nineteen kilometers from the crime scene, and Edward was able to recover nearly all of his stolen possessions. Goldman, now nineteen, calmly noted, "If the 2 behind [me] had been 2 seconds later, I would undoubtedly [have] killed one of those in front."

The second incident occurred in May 1893 on Sierra (Cerro) Malinche, in the state of Veracruz. The setting here was pine forest at an elevation of about thirty-six hundred meters on the slope of the old volcano. Goldman was staying at a *rancho* consisting of several adobe huts with straw roofs facing a square; each building had only one opening, a door. There also were sheds for horses, burros, and poultry, but "the huts occupied by the people [were] used almost as much by the animals as the sheds." Edward rode to a higher elevation to trap rodents. He sat resting on a pine log one day with an old Mexican, who warned him, "Tenga mucha cuidada porque hay gente mala en el cerro." He suggested that if Edward was confronted by suspicious people, he should tell them that he and the *jefe politico* of the nearby town of Huamantla were members of a party of fourteen people climbing the mountain.

Goldman's notes of 15 May begin:

> I am in a hole in the ground, as I write a band of robbers, about a dozen they tell me, are watching for me to come down from the woods & are

> also watching the rancho so that I cannot make my escape but will try to do so this night if I am not found before. If they find me, they may or may not kill me. The people at the rancho here are doing all they can to prevent my being found. My hole is behind the rancho from the road under some bushes. I write at 10 min to 2 P.M. [in a hand that does not reveal his fear]. (Goldman, fieldbook 3)

Goldman remained in the hole until 4 P.M., when it began to rain. Some of the robbers returned to Huamantla, but others remained near the rancho, waiting for Goldman to appear. During the rain, Edward slipped into a hut until dark and then moved to another hut until 2 A.M. the next morning. Then an Indian took him by a circuitous route down to Huamantla. Goldman had been dissuaded by the ranchers from using the main road to Huamantla, but he wrote, "I think I could have dodged them [the robbers] without difficulty, however." Such is the confidence of youth!

Over the years in Mexico, Nelson and Goldman must have had many encounters with *"gente mala"* and developed methods for minimizing the risk that bad guys presented. Goldman's son Luther told me of one trick his father had taught him. He and his dad were camping in a canyon in Mexico, and Edward told Luther that they should put their bedrolls out in plain view before sundown. If there were any bandits in the area watching them, he said, they likely would secure their rifles in the rocks so that they were aimed at the center of each bedroll. After sundown, when the bandits believed their victims were asleep, all they had to do was pull the trigger to shoot them. Of course, the savvy "victims" moved their bedrolls to safer places as soon as the sun went down.

Edward Nelson also alluded to the tensions that he and Goldman faced when he compared the character of the people they encountered in Baja California with that of the people on mainland Mexico:

> We found them [the Baja Californians] everywhere honest and friendly. There was a general absence of revolvers and machetes among the people of the peninsula, in marked contrast to the general custom of carrying one or the other of these weapons throughout most . . . parts of Mexico. During this trip [through Baja in 1905–6], owing to the obviously peaceable and friendly character of the people, even in the remotest sections, for the first time in Mexican territory we felt no need of weapons for self-protection. (Nelson 1916, 11)

More recently, biologists George Schaller and Craig Packer have described in their popular books *The Year of the Gorilla* and *Into Africa,* respectively, how political tensions in Africa disrupted their field studies. Schaller was forced to leave his study area in the Congo when revolutionary fervor swept through that country in 1960. Four of Packer's student associates at Tanzania's Gombe National Park were kidnaped by Zairan rebels in May 1975 and held captive for up to two months before being released. After this incident, foreign students were banned from working at Jane Goodall's research station. Closer to home, two of the graduate students in my department, Annette Olson and Mohammed Bakkar, had to abandon their field projects when Liberian rebels overran Sierra Leone in the early 1990s.

On the first of July, the army told us that all U.S. civilians had to leave Santa Rosa. The army viewed us as "security risks." In typical military fashion, they worried that whenever we went to Liberia to buy gas and supplies, we would tell the enemy (whoever that might be—how were we supposed to distinguish friend from foe?) about the military situation in the park. This infuriated Dan, who decided to call Costa Rican president Carazo and argue that Costa Rican national parks must be free from army control during peacetime. While Dan fumed and put pressure on the Costa Rican government through the U.S. Embassy, a battle for control over the park raged in San José. Alvaro Ugalde was eventually successful in convincing someone in power that the army must not be allowed to wrest control of Santa Rosa from the National Park Service. By 5 July, the army had reduced its pressure on us. We were told we could continue to work in the park, but, for our own safety, we were not to wander around the park at night. Neither Dan nor I took this restriction seriously. There was no way that either of us could continue our research programs without night work.

Although gringo researchers were allowed to live and work at Santa Rosa, the park was still closed to other civilians in late June and early July. This restriction put a rather strong damper on my family plans. Marcia and the kids were scheduled to come to Santa Rosa to spend three weeks with me, beginning in mid-July. This was to be their first trip to Santa Rosa in three years, and we planned to live in a roomy tent, as we had in 1976. Given the realities of life at Santa Rosa, we changed our plans and spent a few days together at Monteverde, in the highlands of Guanacaste, and at La Pacifica before Marcia and the kids returned home. Our celebration of Cara's third birthday, complete with a chocolate cake and candles, at Monteverde's popular Pensión Quetzal and horseback rides around La Pacifica—reminiscent of the summer of 1972, except that now Mike could ride his own horse and Cara was the babe in arms—were the highlights of their brief visit. Marcia was anxious about my safety at Santa Rosa because of the daily accounts of the Nicaraguan military situation she was reading in the *Miami Herald.* She and my colleagues at La Pacifica almost convinced me to abandon the field season. But in the end, I decided to continue our work in the park. If Dan Janzen was going to stick it out, I thought, so would I. Professional peer pressure won out over my wife's concerns.

Fortunately for everyone, Somoza's reign ended in mid-July. On 26 June the U.S. ambassador to Nicaragua had met with Tachito in Managua with a message from our government telling him to resign immediately. Somoza's response was twofold. Don't let communists take over Nicaragua, he requested, and when I resign, treat me as a friend by granting me asylum in the United States. On 11 July the Sandinista junta, safely housed in San José, Costa Rica, demanded that Somoza hand power over to them and disband the Guardia Nacional. Somoza finally submitted his resignation to the Nicaraguan congress early on the morning of 17 July. He and his family then fled to—where else?—Miami, my new hometown. After some last-minute political machinations, the Sandinistas rolled into Managua as the new rulers of Nicaragua on 20 July. Fifty thousand cheering Nicas filled the central plaza to greet their new leaders enthusiastically.

As the Nicaraguans were celebrating their liberation—at least from one form of tyranny—the Costa Rican Guardia began to pack up and pull out of Santa Rosa. By 21 July there were fewer than fifty soldiers left in the park. Many soldiers, including Julio and Marcel, came by our house to say "Adios!" before leaving. We wished them the best of luck as their lives returned to normal. Their departure had an uplifting effect on everyone left in the park. We all felt as though a tremendous psychological burden had been lifted from us. To celebrate the park's return to normalcy, Franklin Chaves and assistant park director Alberto Salas cooked a special meal for all the researchers. We drank beer, ate *papas fritas*, cole slaw, and grilled fish, and generally had a good time. Somoza could have bombed us that night, and we wouldn't have felt a thing.

After the army left the park, I still had trouble focusing my mental energies on science. Though we had not been in great physical danger, the military events had indirectly taken a substantial toll on my psyche. Two of my three assistants—Alice and Jane—had effectively been more of a burden than a help to me, and I ended up doing a larger than normal amount of fieldwork to compensate for their limited assistance. Unlike Alice, Jane was not bothered by the presence of soldiers, but she showed little enthusiasm for any of my research and wasn't comfortable living at Santa Rosa. There was no way that I could trust her to run my research program alone when I returned to Miami in mid-August. So I worried about finding a good research assistant to replace her. I desperately needed someone to help rekindle my enthusiasm for the *Carollia* project. Fortunately, I did find a talented assistant in the person of Rick Williams. But, of course, I didn't know about Rick's availability when I left Santa Rosa on 13 August.

Despite all the problems that occurred during the summer of 1979, the field season was far from being a bust. Some aspects of our work, especially radio-tracking bats of known social position within the Sendero cave, failed to yield many useful data. But other projects went well. Rachel was able to complete an experimental study of the importance of ants and rodents as predators of small, bat-dispersed seeds. She discovered that both groups of animals were heavy consumers of these seeds. Jane and I were able to gather preliminary data addressing the question, Why do bats take fruit to night roosts to eat rather than eating them in the fruiting plant? Our observations with a night vision scope indicated that the bats we temporarily restrained in fruiting plants were more likely to be "hassled" by other bats than those we restrained in night roosts. Finally, for the first time, I had begun to observe the behavior of *Carollia* bats inside the Sendero cave.

Until 1979, we had spent little time inside this cave to avoid unduly disturbing its chiropteran inhabitants. But if we were going to learn as much as possible about *Carollia*'s social life, we would have to break this moratorium. Thus, in mid-June I sat just inside the cave's upstream entrance and began watching bats for the first time. I didn't learn much that first session, because it took nearly an hour for the several hundred bats to calm down after I had entered the cave. But I did begin to get a feel for the general distribution of *Carollia* in the cave. I saw one group of seven bats that looked like a harem near the center of the cave, but none of those bats was banded, so I didn't know their sexes. Banding as many bats as possible with recognizable marks became a top priority with me.

Subsequent visits to the cave were more profitable, because the bats were less disturbed by our presence. Jane and I were able to watch a harem containing one male and three females for forty-five minutes in late July. These and other *Carollia* rested quietly, groomed themselves with their tongues and hind feet, and stretched their wings while we watched. Near the upstream entrance we watched other bats, most likely common long-tongued bats, undergoing what we called hover chases. One bat would alight on the cave wall, and another would hover with rapid wingbeats about half a meter in front of it for a few seconds. Then the stationary bat would take wing, and the hoverer would briefly chase it. After the first bat landed again, it was again approached by the hoverer, and the sequence would begin all over. I watched in fascination while one pair of bats underwent these chases for more than ten minutes.

On 12 August, my last day at Santa Rosa in 1979, I sat in the cave for an hour and a half and watched bats. Again the bats were little disturbed by my presence, and they remained quietly in place, mostly resting. One bat landed near me and was immediately visited by two or three other bats, each of which hovered briefly near it before landing for half a minute about fifteen centimeters away. The first bat, a rare red-colored individual (probably recently immigrated from the Red roost), seemed totally oblivious to me. I left the cave that day feeling confident that with enough patience, we were going to learn the secrets of *Carollia*'s social system. Now that the political revolution in Nicaragua was over, I could concentrate on a new phase of research at Santa Rosa.

7 *Vampyrum*

When he's photographing bats, Merlin Tuttle leaves nothing to chance. Everything about his photographic setup—camera, lenses, flashes, infrared beams, and background material—must be in perfect working order and perfectly placed before he makes his first exposure. His excruciating attention to detail certainly pays off. The portraits and action shots of bats that he has been taking since the late 1970s make Merlin the premier photographer of bats in the world. And his artistry with these normally shy and elusive subjects has been the key to the success of Bat Conservation International, the organization he founded in 1982. Merlin's beautiful portraits of a wide variety of interesting species have made bats much more sellable to the general public as animals with critical conservation needs. With a current membership of about fourteen thousand, BCI is the world's most important agency for disseminating information about our bat fauna and helping to protect it.

I have known Merlin for many years, having first heard about his ex-

---

False vampire bat *(Vampyrum spectrum)*. Redrawn by Ted Fleming from an illustration (© John D. Altringham, 1996) in *Bats, biology and behaviour,* by John D. Altringham (1996), by permission of Oxford University Press.

traordinary collecting abilities when I was working in Panama in 1966. But I never had the chance to work with him in the field until the summer of 1987, when he and I went on a two-week bat photography expedition to Costa Rica. He wanted to collect pictures of tropical bats for a long-term book project on bats of the world, and I wanted first-class photographs of bats for a book I had just written, entitled *The Short-tailed Fruit Bat.* It turned out to be relatively easy to get the photographs I needed, including a dramatic cover photo of a *Carollia* with up-stretched wings, carrying a ripe *Piper* fruit in its mouth. But *Carollia* wasn't the only bat on Merlin's wish list. Much more coveted was a series of pictures of the false vampire bat, *Vampyrum spectrum,* a member of the Phyllostomidae. Weighing 150 to 190 grams as an adult, *Vampyrum* is the New World's largest bat. We were in a position to realize Merlin's wish, since we were working with *Carollia* and *Piper* in western Costa Rica at Finca La Pacifica, where I knew the location of a *Vampyrum* roost.

The roost was located in a large hollow espavé tree along the Río Corobici, near where Ray Heithaus, Paul Opler, and I had spent many nights catching bats fifteen or so years ago. To get to this tree, Merlin, our volunteer assistant Pat Morton, and I carried two large loads of photographic gear a kilometer through a dusty pasture into a narrow strip of riparian forest several hours before sunset. Pat was a master's student in the Department of Zoology at the University of Wisconsin. She had been working in Costa Rica for over a year on an educational program aimed at increasing public awareness of the beneficial nature of Latin American bats. To the average Latin American, all bats are *"vampiros,"* warranting destruction wherever possible. Until recently, little public understanding or appreciation has existed concerning the beneficial roles that neotropical bats play by eating insects, pollinating flowers, and dispersing seeds. Pat's book *Murciélagos tropicales americanos* has helped to educate a Hispanic public about their bat fauna.

We worked quietly setting up the equipment to avoid disturbing the three bats roosting nine meters above us inside the tree. Merlin first placed an infrared beam and its detector precisely where we thought a bat would emerge from the triangular hole at the tree's base. Next we placed four flashheads on pointed poles beside and in front of the expected flight path. With a wooden ruler, Merlin carefully measured the distances between the flashes and our intended target and made small adjustments until he had just the right exposure, as determined by a flashmeter, on all sides of the imaginary subject (figure 17). Then he set up his camera with its zoom lens so that it was four meters away from the exit hole and critically focused on the plane

Figure 17. Merlin Tuttle with his photographic setup at La Pacifica, 1987. Photo by T. Fleming.

of the infrared beam. Next came numerous tests of the beam and flash system to make sure that everything was ready for a perfect exposure. Finally, because we wanted to capture the first exiting bat for detailed pictures of its feeding behavior under more controlled conditions, we placed two short mist nets across its most likely departure routes.

Our work was completed an hour before sunset, which gave us time to relax and watch the swift water of the Corobici as it flowed past the giant roost tree. As the forest gradually darkened, each of us took our assigned places around the roost tree. Pat and I sat next to the nets to make sure that our quarry did not escape. Merlin sat behind his motorized camera and held its shutter open with a cable release. In the increasing darkness, he could open the shutter for at least three minutes without exposing the film, although occasional flashes of distant lightning forced him to advance the film more frequently. As we sat in the dark, each of us could mentally see the picture we were about to get. When the flashes went off, they would illuminate for $^1/_{10,000}$ of a second a large bat with a wingspan of nearly one meter. Perfectly highlighted by the flashes would be the bat's reddish-brown fur, its large light-colored ears, its large fleshy nose leaf, and its bright black eyes.

Scientific knowledge of this magnificent member of the Phyllostomidae dates from Linneaus's tenth edition of *Systemae Naturae,* published in 1758. At that time this bat was called *Vespertilio spectrum.* In 1815, the French systematist Constantine Rafinesque formally named the genus *Vampyrum,* which contains a single species distributed from Veracruz, Mexico, to central Brazil and Peru. Four years later he became professor of botany and natural history at Transylvania College in Lexington, Kentucky. *Vampyrum*'s generic and common names undoubtedly reflect the fact that early European naturalist-explorers of the New World associated this large, powerful bat with reports that certain neotropical bats fed on the blood of mammals, including humans. Its large, robust canine teeth certainly give it the appearance of a Dracula-like vampire.

Probably because their skulls also contain large, daggerlike canines, several additional phyllostomids were given vampirelike generic names by nineteenth- and early twentieth-century systematists. Thus, certain stenodermatine bats were named *Vampyrodes, Vampyressa, Vampyrops,* and *Vampyriscus.* Later natural history observations would reveal that, instead of being used to desanguinate other animals, these canines are used to pierce the thick skins of fig fruits, which the bats then carry off to a night roost to eat. Until the feeding behavior of true vampires, classified in the genera *Desmodus* (figure 18), *Diaemus,* and *Diphylla,* became known, the largest members of the Phyllostomidae—species of *Phylloderma, Phyllostomus,* and *Chrotopterus* as well as *Vampyrum*—were considered to be blood-feeding bats well into the twentieth century.

Edward Goldman was normally a careful natural historian. However, in his "Mammals of Panama," he perpetuated a serious mistake about the behavior and food habits of *V. spectrum* into the early part of the twentieth century. Because he did not collect any specimens or make any observations of *Vampyrum* when he worked in Panama, he had to rely on observations made by Henry Bates in the mid-nineteenth century in Amazonian Brazil for his species account of this bat. According to Bates,

> The vampire was here by far the most abundant of the family of leaf-nosed bats. . . . Nothing in animal physiognomy can be more hideous than the countenance of this creature when viewed from the front; the large leathery ears standing out from the sides and top of the head, the

Figure 18. A common vampire bat *(Desmodus rotundus)* on the ground. Photo © Merlin D. Tuttle, Bat Conservation International.

> erect spear-shaped appendage on the tip of the nose, the grin and glistening black eye all combining to make up a figure that reminds one of some mocking imp of fable. No wonder that imaginative people have inferred diabolical instincts on the part of so ugly an animal. The vampire, however, is the most harmless of all bats, and its inoffensive character is well known to residents on the banks of the Amazon. (Bates 1863, 332)

This description, while not terribly sympathetic to its subject, does not provide grossly misleading information about *Vampyrum.* But Bates's account of its behavior and diet was way off the mark and indicates that he was confusing *Vampyrum* with another large phyllostomid bat. Bates continued:

> I used to see them, as I sat at my door during the short evening twilights, trooping forth by scores from a large open window at the back of the altar, twittering cheerfully as they sped off to the borders of the forest. They sometimes enter houses: the first time I saw one in my chamber, wheeling heavily round and round, I mistook it for a pigeon, thinking that a tame one had escaped from the premises of one of my neighbors. I opened the stomachs of several of these bats, and found them to contain a mass of pulp and seeds of fruits, mingled with a few remains of insects [Coleoptera]. (333)

This description cannot possibly refer to *Vampyrum,* which roosts in small groups and does not eat fruit and beetles. Bates clearly was describing the appearance and behavior of the greater spear-nosed bat, the New World's second largest bat, which roosts gregariously and is an omnivore. But Goldman's species account, erroneous as it was, was about all we knew about the false vampire in 1920.

Observations indicating that *V. spectrum* is a carnivore, not a sanguinivore or an omnivore, were first published in 1936. Working on Trinidad, Raymond Ditmars, whose books on reptiles I had devoured as a snake-chasing youngster, reported that bird feathers, bat wings, and rodent tails could be found on the floor of hollow trees occupied by this bat (Ditmars 1936). This was later confirmed by George Goodwin and Arthur Greenhall in their monograph on Trinidadan bats, and several workers have since reported finding the remains of bats, passerine birds, and arboreal rodents in the stomachs of false vampires that they killed and dissected. Still others have noted that captive *Vampyrum* "relish" eating birds and bats. Arthur Greenhall, for example, kept a breeding pair in captivity for over five years on a diet of white mice (two per bat per day), young chicks, small pigeons, and half-grown white rats. His bats refused to eat bananas, mangoes, papayas, and citrus that he offered them. Ironically, he did not offer them defibrinated cattle blood, the standard laboratory diet of true vampires, as a direct test of the vampire hypothesis.

Interestingly, *Vampyrum* is not the only bat in the world to be called a false vampire. Members of the Old World tropical and subtropical family Megadermatidae, which contains five species, also have this common name, presumably because they were originally thought to be blood feeders. Megadermatids physically resemble *Vampyrum* in having large ears, large eyes, and a fleshy nose leaf as a result of having the same hunting style. Africa has two species, *Cardioderma cor* and *Lavia frons*, both relatively small "sit and wait" insectivores that hunt from perches and capture their prey either on the ground or in the air. *Megaderma lyra* of southern Asia and the Malay Peninsula weighs about 60 grams and eats frogs, rodents, and other bats. The largest member of this family is the Australian ghost bat, *Macroderma gigas*, which weighs up to 140 grams. This handsome, intelligent-looking bat eats lizards, birds weighing up to 120 grams, small marsupials, bats, and rodents.

Most of what we currently know about the feeding ecology and social behavior of *Vampyrum* in the wild comes from a study conducted in 1973 and 1974 by Sandra Vehrencamp, Gary Stiles, and Jack Bradbury at the roost we were staking out at La Pacifica. They studied the diet of a group of five individuals (an adult pair, two subadults, and one suckling baby) by periodically collecting the vertebrate remains that accumulated on the floor of the roost for one year. They reported that these bats ate birds of eighteen species spanning a size range of 20 to 150 grams. Major prey species included the groove-billed ani, two species of parakeets, a motmot, a small dove, two species of cuckoos, and a wren. Each of these species is a common inhabitant of deciduous forest and second growth in Guanacaste Province. They are notable for roosting at night in groups (anis, doves, parakeets, and the wren) and/or having strong body odors (anis, cuckoos, motmot). Ornithologists have suggested that group roosting and strong body odors are adaptations for reducing predation by terrestrial or arboreal mammalian carnivores and snakes. *Vampyrum,* however, has capitalized on these traits to increase its hunting success. What apparently works to deter one group of enemies can be a strong invitation to dinner for another predator with different sensory and locomotory adaptations.

How does a bat as large as *Vampyrum* capture its prey? Greenhall suggested that it captures birds and rodents either by stalking them (from arboreal perches?) or on the wing. He reported seeing a *Vampyrum* agilely pursue a fruit bat through an open house in Venezuela. As an indication of its aerial agility, it can deftly eat bats captured in mist nets without getting caught itself. On several occasions in Panama, I encountered pairs of severed wings hanging in the upper shelves of mist nets, a clear indication that the false vampire was opportunistically raiding my larder. Its agility results from its broad but relatively short wings that allow it to fly slowly and to maneuver in cluttered places. Its broad wings also allow it to take off from the ground or from a perch while carrying a large prey item in its mouth.

Despite its obviously maneuverable flight, it still is not clear how *Vampyrum* captures groove-billed anis, its most common prey at La Pacifica. Anis sleep in groups in the middle of thorny shrubs in an obvious attempt to be as inaccessible to nocturnal predators as possible. It seems hard to believe that *Vampyrum* crawls into shrubs to capture anis, but the only other alternative—flushing a sleepy bird out of the bush—seems just as implausible. We obviously need some solid observations on how it manages to be so successful in catching these seemingly inaccessible prey.

Sandy Vehrencamp and Jack Bradbury tried to study the foraging behavior of the La Pacifica *Vampyrum* by radio-tagging three individuals. Only one transmitter, however, functioned for more than one night. The other transmitters apparently were destroyed by the bats when they groomed each other. On two of the four nights it was tracked, the tagged male foraged within 250 meters of the roost, to which it returned between foraging bouts lasting up to 130 minutes. On the other two nights, it never left the roost.

From these and other observations, Sandy and Jack concluded that each volant individual foraged solitarily and that "the time spent foraging varies greatly from night to night, and some individuals may not forage at all on certain nights."

All reports in the literature indicate that *Vampyrum* roosts in hollow trees, either solitarily or in small groups of up to five individuals. Groups probably represent monogamous families, a social system that is rare in bats as well as in mammals generally. Vehrencamp and her collaborators suggested that young false vampires stay with their parents for extended periods and are provisioned by them while learning how to hunt for themselves. Although Vehrencamp and her colleagues made no direct observations of food sharing, they concluded that the bird remains that accumulate in the roost represent food that one bat has brought back to share with the other bats each night. If food sharing is common in *Vampyrum,* then its social system resembles that of terrestrial carnivores such as foxes, wolves, and jackals. As we have seen, females of the true vampire bat *Desmodus rotundus* also share blood meals with their young and with other adult roostmates. In general, however, food sharing is rare in bats.

Now, fourteen years after the Vehrencamp study, we were sitting at the same roost tree waiting for one or more of the descendants of those bats to emerge and have their picture taken. As time passed with no sign of the bats, I became restless and finally suggested to Merlin that we cautiously check the roost to see if any bats were there. At 8 P.M., an hour and a half after we had begun our stakeout, we discovered that the roost was empty! The three *Vampyrum*—two adults and one juvenile—that had been present in the morning were gone. In our caution to avoid disturbing the bats that afternoon, we had deliberately not peered into the roost and had assumed the bats were still there. But what had become of them? A quick check of the flash system indicated that it was working perfectly and would have been set off if a large bat had left the roost after sunset. Because a close examination of the roost tree the next morning failed to reveal an alternate exit, we were forced to conclude that the bats had left the roost sometime between midmorning and early afternoon the previous day. Diurnal flight would be unusual in these bats, but the location of this roost is well enough known by local people and foreign biologists that the bats may have become very sensitive to disturbance.

We left La Pacifica the next day, disappointed that we had failed to get photos of one of the New World's most spectacular bats. Twenty-four hours after our stakeout, I could still see in my imagination the dramatic flight picture that we had missed. Merlin figured that to obtain pictures of this species, he would have to make a special trip to Trinidad, where the locations of *Vampyrum* roosts are well-known.

Three nights later Pat and I were netting bats in tropical wet forest at Finca La Selva while Merlin was setting up his photographic equipment in the three-meter square nylon screen cage that serves as his portable photographic studio. The cage, all of his equipment, and our personal effects were crowded into one of La Selva's comfortable new bunkhouse rooms, where we would work for the next five days.

By 1987, La Selva had changed considerably from the primitive field station it had been in 1970, when I first worked there. The field station now had a spacious two-story, air-conditioned laboratory building, and more labs were on the drawing board. The original field station building, now quaintly called the River House, still housed a few researchers, but most long-term scientists lived in a series of wooden duplexes near the lab buildings. Short-term visitors stayed in one-story cinder block bunkhouses across the Río Puerto Viejo near a new cafeteria-office-shop complex. A long suspension bridge allowed people to cross the river without getting their feet wet, even when it was swollen with torrential rains. Finally, visitors no longer arrived at the field station by boat. Instead, it was possible to drive a car on paved road from San Jose to the cafeteria-office complex and unload gear for transportation to other parts of the station by wheelbarrow or cart. In a period of about fifteen years, La Selva had evolved from a relatively remote and rustic field station full of tropical charm to a modern center where sophisticated tropical research could be done.

From the minute Pat and I unfurled our ten mist nets, we were busy handling bats of ten species. We saved a few unusual bats for Merlin and released the common frugivorous species that I had been studying for years. At 7:30 P.M., Pat began to remove a small species of *Artibeus* (now sometimes classified in the genus *Dermanura*), a fig-eating bat that is notorious for emitting loud, harsh screams when handled. These screams, whose adaptive significance is still poorly understood, often attract other bats of the same or different species into the area. The screams sometimes also attract bat predators, as I learned at Santa Rosa. I had been removing a screaming Jamaican fruit-eating bat from a net one night when—THUD!—a three-hundred-gram woolly opossum jumped out of a tree behind me and landed on my bare head. My first thought was that a *Vampyrum* had just flown into me. Before I could regain my composure, I saw the opossum leap from my head into the net, still intent on getting that bat.

Tonight, Pat's bat did not attract an opossum. Instead, it called in a

Vampyrum! I was just beginning to remove another bat from the net when a large animal flopped into the net between Pat and me. Having captured two *Vampyrum* at La Selva the previous summer, I immediately guessed what we had snagged. I dove for the struggling bat to prevent it from quickly chewing its way out of the net. In short order I removed the bat from the net and held it firmly but gently in my gloved hand for Pat to examine. We were both ecstatic at our blind luck. Merlin would get his flight and feeding pictures of *Vampyrum* after all.

Merlin indeed made the most of this opportunity to photograph our 160-gram adult female. He spent the better part of the next three days and nights photographing the bat eating birds that he had captured in second-growth forest away from the biological station. The bat proved to be a most cooperative subject. Art Greenhall had reported that his captive *Vampyrum* were tame and gentle. Our bat behaved similarly and quickly adjusted to captivity as soon as we began to feed it. It could be handled without gloves, for example, on the night we caught it. It was completely tolerant of Merlin's moving it around on branches for close-up pictures of its feeding. Only occasionally did it become churlish and only when we took away its meal (a bird) to prevent it from becoming satiated before the end of a photo session. Instead of venting its anger by biting, the bat merely made a rapid clicking noise with its large teeth. Finally, when we finished photographing the bat each night, it did not object to being placed in a large cloth "sleeping bag" so that it would not disturb the small fruit bats that we also kept in the cage. The one night we did not bag the bat, she constantly flew around the cage, landing noisily on its walls, apparently in an attempt to capture the more agile fruit bats. I finally got up at 4:30 A.M. and bagged the persistent *Vampyrum* with my bare hands so that I could get some sleep.

In three days, Merlin was able to take a large series of photos of *Vampyrum* eating a variety of colorful tropical birds. As he took a picture of the bat chewing on a yellow tanager (figure 19), I suggested to him that perhaps Bat Conservation International could use this photo for its annual Christmas card. BCI could send a card to the National Audubon Society with a warm season's greetings message on it. After briefly considering this, Merlin nixed it on the grounds that Audubon could easily dredge up a photo of a barn owl eating a bat in retaliation. And a photo, no matter how artistic, of a large bat merrily chomping on a poor little bird wouldn't sit well with many of BCI's members. Hence BCI will continue to let photogenic but non-threatening species of bats grace the cover of its Christmas cards.

Because its carnivorous food habits place it relatively high on tropical food chains, the false vampire is an uncommon bat. As a result, relatively

Figure 19. A false vampire bat *(Vampyrum spectrum)* eating a tanager. Photo © Merlin D. Tuttle, Bat Conservation International.

few specimens of it exist in museum collections, and there has been a strong tendency for mammalogists to preserve every individual they capture (and to publish a note on its occurrence in a scientific journal). It is therefore somewhat surprising to find that the La Pacifica *Vampyrum* have managed to survive since 1977, the year the location of their roost was published in a scientific journal.

Even if they manage to avoid being captured by museum collectors, the La Pacifica false vampires face another threat to their survival. Riparian forest along the Corobici is currently the only habitat in this cattle-raising region that contains trees large enough to serve as roosts. Most other potential roost trees have been carbonized during the conversion of forests into pastures and farmlands. The number of such trees at La Pacifica has decreased in recent years as a result of the completion of the Arenal hydroelectric project in the late 1970s. In this project, the headwaters of the Corobici were dammed, and the flow rate of the river was changed by the periodic discharge of large volumes of water from the newly created impoundment. At times the Corobici now becomes a raging river that has severely undercut its stream banks at La Pacifica, causing many large trees, some of which housed colonies of bats, to topple into the river. Because of continued habitat destruction and persecution by man, this bat, along with its much larger mammalian ecological counterparts such as spotted cats (margays, ocelots, and jaguars), is clearly an endangered species in Latin America.

We released our *Vampyrum* on its fourth night in captivity. It circled the lighted area in front of our bunkhouse once before disappearing into the darkness with slow, steady wingbeats. Its large size and owl-like flight style make *Vampyrum* easy to distinguish from all other neotropical bats in the forest at night. On several occasions at Santa Rosa and La Selva, I have heard or seen *Vampyrum* flying overhead, possibly hunting other bats. Except in protected areas such as Santa Rosa and La Selva, however, I fear that such sights will become increasingly rare in Latin America. If *Vampyrum* disappears from the neotropical mammal fauna, the world will have lost one of its most interesting bats. Let's hope that the final repositories of this beautiful animal are not restricted to dusty museum cases, tissue collections in ultracold freezers, and Merlin Tuttle's vast slide collection.

8 Fooling Around with Flying Foxes

For my forty-sixth birthday, Marcia gave me a didgeridoo ("sound stick"), a musical instrument made by north Australian Aborigines from a tree branch that has been hollowed out by termites. One to two meters in length, this instrument is played much like a large brass instrument. By blowing air into one opening while vibrating their lips, musicians produce a low-pitched, droning sound. Using "circular breathing," a technique also employed by brass and woodwind players, didgeridoo players can produce a continuous sound for over ten minutes. In addition to droning sounds, skilled players can imitate all kinds of animal sounds—bird calls, dog barks and growls, frog croaks, and kangaroo hops, to name a few.

My didgeridoo is a relatively straight, meter-long branch on which are painted a series of roosting and flying black flying foxes, the common species along Australia's northeast coast. Above and below the bats are ochre-colored bands decorated with black cross-hatching, separated by white lines and dots. A large black python—a species that often feeds on flying foxes in their day roosts—with featherlike white scales winds around the branch

Tube-nosed bat. Redrawn by Ted Fleming, from a photo in A. Mitchell, *The enchanted canopy,* Fontana / Collins.

near each end. Marcia had this instrument made specially for me by an Aborigine from Kuranda in far northern Queensland.

We were living in Townsville, Queensland, at the time of my birthday. I had been awarded a Fulbright Senior Fellowship for the 1987–88 academic year to teach and conduct research at James Cook University, a small school of about four thousand undergraduate and graduate students. The zoology department wanted me to teach a course in plant-herbivore interactions in their new master's program in tropical biology. The assignment appealed to me because it represented a once-in-a lifetime opportunity to work in one of the world's most interesting floras and faunas. Ask any group of North America field biologists where they would most like to do research sometime and Australia would be close to the top of nearly everyone's list. Along with South Africa, it certainly was at the top of my mine. So when the Fulbright opportunity came along, I jumped at it, even though it meant once again disrupting the lives of my family members (we had returned from a year's sabbatical in North Carolina in the fall of 1985). Fortunately, Marcia was all for another move away from Miami. I only had to convince my two teenaged kids that a year "down under" would be worth the disruption to their school and social lives. Although Cara and Mike weren't exactly thrilled to be moving halfway around the world, living in Australia was such a positive experience for them that they were unhappy about returning to Miami after our year was over.

So in mid-August 1987 we rented our house (complete with our dog and cat) to two friends, packed thirteen trunks and duffel bags of personal gear, research equipment, and books, and headed off to Queensland, with stops in Michigan to visit our parents and Honolulu to play tourist for a few days. From Hawaii, we flew the nine-hour leg to Cairns in northern Queensland, where we waited five hours for a flight to Townsville, another 350 kilometers south (see map 3). By the time we arrived at the partly furnished house we were renting on the outskirts of Townsville, we were dead tired. That evening Mike and I fell asleep in the middle of dinner at the home of my academic host, Helene Marsh.

After a long night's sleep, we began to get our bearings in our new setting. A city of about a hundred thousand people, Townsville is located on the Coral Sea in a region of relatively low and very seasonal rainfall. In fact, at the time of our arrival, Townsville was suffering from a six-year drought. Everything looked parched and dry and reminded us of the American desert Southwest. The color of our front lawn pretty much matched our dull yellow, two-story, cinder block house. But our backyard was a lovely oasis featuring a ten-meter-long pool. The owner of our house was in the swimming pool business and spared no expense when he built his own pool. The Flemings would spend a lot of time playing and relaxing in this pool over the next year. When our mango trees were laden with fruit, I would float on my back at night and watch the dark shadows of large flying foxes passing overhead on their way to feed.

Across the road from our house was a park that ran along the Ross River. The park was full of beautiful, cream-barked eucalyptus trees. Sulphur-crested cockatoos screeched from these trees' airy crowns, and a flock of pink and gray galahs (ground-feeding parrots) foraged for seeds in the dried grass of a playing field as Marcia and I began to explore the park on our first full day in Townsville. Shortly after sunset, we spotted our first Australian flying foxes landing in flowering eucalypts along the river. These were individuals of *Pteropus alecto,* the black flying foxes on my didgeridoo. With a wingspan of one and one-half meters, these bats, members of family Pteropodidae, are among the largest in the world. On slow wingbeats they glided silently into the crowns of eucalypts and began to scramble along branches, looking for all the world like ungainly lemurs. In the increasing darkness we could hear bats clambering around in several trees. They were especially conspicuous whenever two individuals began to fight over a patch of flowers. During these short-lived squabbles, the bats made odd barks and gurgles—noises that I had never heard coming from New World bats. These odd noises would become familiar to us when we adopted three baby *alecto*s, beginning in mid-October.

That evening in August 1987 was not the first time I had seen flying foxes in the wild. My first encounter with them had occurred in October 1985 in Bombay, India. I was visiting Renee Borges, one of my doctoral students, who was studying the feeding ecology of giant Indian squirrels at two sites south and east of Bombay. Before taking me to her field sites, Renee gave me a brief tour of her hometown. One of our first stops was the Victoria Gardens and Zoo, primarily so that I could see a wild colony of giant flying foxes. About four hundred of these megabats, which weigh about one kilogram each, were roosting in the crowns of a couple of trees high above the zoo visitors. In midafternoon they were flapping their wings to cool themselves and occasionally uttering their characteristic harsh barks.

Although it was old and decrepit, the Victoria zoo and its shady walkways offered pleasant relief from bustling—no, call it chaotic—Bombay. Renee loves the city and can't imagine living anywhere else in the world. My opinion of Bombay at that moment was just the opposite. I couldn't imagine why anyone would want to live there. My colleague Steve Green, who knows Bombay and India well, had warned me that my first exposure to India on the drive from the airport into town would be an assault on my olfactory, auditory, and visual senses. He was dead right. The humid early morning air was permeated with the sickish sweet smell of vegetarian human excrement. The dilapidated condition of many buildings was depressing, but worse yet were the myriad of people sleeping in the street, on sidewalks, and everywhere you looked. Many of the sleepers were wrapped in shroudlike blankets and looked like corpses, just as Paul Theroux described in one of my favorite travel books, *The Great Railway Bazaar.* As we approached the center of the city, streets became absolutely choked with taxis, busses, motor rickshaws, bicyclists, and hordes of pedestrians. With car horns blaring and a kaleidoscope of shapes and colors flashing past my jet-lagged eyes, I thought I was in hell. I couldn't wait to get out into the Indian countryside, which Dan Janzen had told me was "thoroughly trashed."

Two days later I was in that Indian countryside, having driven with Renee, her fiancé, Ulhas Rane, and his jeep driver Ganesh eighteen hours south along the Bangalore road to the town of Magod in Karnataka State. Contrary to Dan's pronouncement, the land around Magod was lovely. Much of it was still heavily forested and consisted of a series of gentle hills and valleys that reminded me of the Blue Ridge Mountains in the eastern United States. Magod itself was a partly occupied "company town" that had been built in preparation for the construction of a major dam and hydroelectric plant. The hydro project had been blocked by an effective environmental coalition, and the forest around Magod was now slated to become a wildlife reserve.

For several days, Renee and Ulhas introduced me to the natural history of this area. Both excellent naturalists, they were able to name nearly all of the birds, butterflies, mammals, and plants that we encountered. The informal bird list that I always keep whenever I visit a new area quickly began to swell with names such as tree pies, jungle crows, racquet-tailed drongos, scarlet minivets, heartspotted woodpeckers, chestnut-headed bee-eaters, purple-rumped sunbirds, and red-whiskered bulbuls. It was exciting to see such a colorful bird fauna, but birds were just one of the animal treats in this forest. "Flying" lizards of the genus *Draco* glided from tree to tree by spreading their elongated ribs into a sailplane. Not much larger than the *Anolis* lizards that abound in my yard in Miami, *Draco* looked like a bright orange leaf as it drifted between trees.

An even more impressive glider was the Indian giant flying squirrel. One moonlit evening we watched one of these beautiful mammals emerge from its sleeping nest in a tree hole, groom itself for a few minutes, and then make a series of twenty-five-meter glides from the forest's edge deep into its interior. About one meter in total length, this squirrel spread its loose skin, ivory white on its belly, into a large, nearly square gliding surface, from which its feet and head barely protruded. From the tops of trees it silently launched itself into the moonlight, defying gravity for what seemed like an eternity. Upon landing on a tree trunk, it quickly hopped to the top and prepared for another glide. Although it disappeared rather quickly into the dark forest, I will carry the image of this beautiful squirrel, silhouetted against a full moon, with me for the rest of my life.

Renee's study animal, *Ratufa indica,* was also impressive in size. *Ratufa* is a diurnal tree squirrel—the largest in the world—and like the flying squirrel is about a meter long. Its dorsal color is a striking russet, whereas its head, tail, and belly are cream-colored, as are its ghostly eyes. Solitary (except for mother-young pairs) and territorial, these giant arboreal herbivores are leaf-and-twig eaters rather than seed eaters like our North American tree squirrels. When I was at Magod, they were eating the succulent twigs of a common canopy tree. After cutting a twig into short pieces, a squirrel would feed by sitting on its haunches on a large branch with its head and long tail hanging over the branch, forming an inverted V.

While in Magod, we took our meals at Yousef's teahouse, the only restaurant in town. A slim, middle-aged Muslim, Yousef took great pride in serving his wife's vegetarian cooking to a Westerner and took special pains to make sure my food was not too spicy. During the clear, cloudless days we ate outside under a bougainvillea-draped arbor while listening to exotic-sounding music from Sri Lanka on Yousef's portable radio. After dark we ate inside the tiny restaurant under the dim yellow light of incandescent bulbs. For breakfast, I usually had *uppma,* a yellow-colored ground wheat "fluff" mixed with coconut meat, and rich tea laced with water buffalo milk and a bit of sugar. Dinner usually included *poppadems, chupatis,* a veggie mixture of some kind, an omelet, and tea. Like my traveling companions, I began to eat my food using my right hand and, with Ulhas's help, even learned to roll rice into a sticky ball before popping it into my mouth.

One afternoon we set four mist nets across the clear stream running by Yousef's restaurant to sample the local bat fauna. We opened the nets at sunset and immediately caught a little blue kingfisher. A few minutes later we caught our first bats, three individuals of a species of *Rhinolophus.* The adult male horseshoe bat was very vocal as I removed it. About the size of *Carollia,* this bright orange bat has large mobile ears, tiny eyes, and a nose leaf pressed flat against its face, not free-standing as in the phyllostomids. As we photographed our bats, a crowd of men and boys (no girls) gathered around us to look at our captures. Most people were afraid to touch the bats, which they called *bawli* or *bawli-bawli,* but an old man screwed up his courage and touched a bat's wing. Afterward, he gave me a big toothy grin before melting back into the crowd.

At the next net check, I spotted a large bat with bright orange eyeshine in the first net. I excitedly removed my first pteropodid, the short-faced fruit bat, *Cynopterus sphinx,* from the net. Dull yellow in color, *Cynopterus* was about the size of a large *Artibeus* (about sixty-five grams) but had much larger brown eyes. It bit my gloved finger with considerable force. Most impressive, however, were its vocalizations. It uttered a loud and resonant series of *laddle-laddle-laddles* to let us know that it was upset. We photographed and released this bat, an immature female, and eventually caught two more of these *laddling* bats before closing the nets. Our final bats that evening were a series of tiny pipistrelles—all males—which must have been roosting close to our net site. These bats were molting from rusty-red to dark brown.

After four days at Magod, we drove another eighteen hours to Renee's second study site at Bhimishankar, east of Bombay. En route we stopped near the town of Manchar to see a colony of about 150 *Pteropus giganteus* roosting in a grove of neem trees. The colony contained adults, both males and females, none of which was nursing young. Most bats were hanging quietly in the shade with their wings tightly wrapped around them in capelike fashion. A few groomed themselves with their pink tongues and stretched their wings. One male took flight on a massive expanse of soft, black wing. It looked huge to me.

Pteropodid bats—commonly called flying foxes or megabats (figures 20–22)—are classified in a separate suborder (Megachiroptera) from all other bats, which are classified in suborder Microchiroptera (the microbats). Found only in the Old World tropics and subtropics, megabats have had a very different evolutionary history from that of microbats. Except for members of the genus *Rousettus,* which produce nonultrasonic clicking sounds with their tongues, none of the 160 or so species of megabats uses sound for orientation in the dark. Fossil pteropodids also show no anatomical evidence of ever having been echolocators. Instead of using ultrasound for nocturnal orientation, megabats rely on their extremely sensitive retinas for navigating at night. As a result, megabats have more "normal"-looking mammalian faces and heads. They have large eyes, a doglike snout (which gives them their common name, flying foxes), and rather small ears. Instead of hunting insects early in their evolutionary history, these bats have apparently always been plant visitors. Nectar and fruit (and leaves in larger

Figure 20. Indian giant flying fox *(Pteropus giganteus)*. Photo © Merlin D. Tuttle, Bat Conservation International.

species) are their only sustenance. They are the Old World ecological counterparts of plant-visiting phyllostomids of the New World tropics.

Although most people would naturally assume that megabats and microbats share a common ancestor, if for no other reason than they both have membranous wings, the veracity of this has been vigorously debated in the past couple of decades. In 1986, the Australian neurobiologist Jack Pettigrew challenged this idea by describing in a paper in *Science* a unique neural connection between the eyes and midbrain found only in megabats and primates. To explain this similarity, he proposed that megabats are flying primates, that the order Chiroptera is not a natural evolutionary group, and that flight has evolved twice in mammals. As might be imagined, Pettigrew's "flying primate" hypothesis caused a major stir among bat biologists and,

Figure 21. Wahlberg's epauletted bat *(Epomophorus wahlbergi)*. Photo © Merlin D. Tuttle, Bat Conservation International.

more broadly, among evolutionary biologists. Pettigrew and his collaborators began to look hard for more evidence to bolster his hypothesis, whereas other workers, primarily in the United States, began to muster evidence against it. U.S. workers muttered under their breath that this upstart Aussie couldn't be right and that Pettigrew must be suffering from a bad case of primate envy (Australia, of course, has no native primates).

Between 1986 and 1993 a large series of papers was published offering new evidence for and against the flying primate hypothesis. By late 1993, when a group of bat biologists gathered in London for a symposium sponsored by the Zoological Society of London, however, the debate had pretty much died down. Jack was there to present his viewpoint, but he was definitely on the defensive. The offensive was taken by Nancy Simmons, of

Figure 22. Dawn bat *(Eonycteris spelea)*. Photo © Merlin D. Tuttle, Bat Conservation International.

the American Museum of Natural History. Nancy had spent much time carefully reviewing all of the morphological, physiological, and biochemical evidence bearing on the evolutionary relationship between megabats and microbats. Her analyses strongly supported the hypothesis of bat monophyly (that is, their having a shared common ancestor). Megabats and microbats were sister groups, and flight had apparently evolved only once in mammals. For the time being, conventional wisdom seems to prevail among bat biologists, though Jack Pettigrew, for one, remains unconvinced.

Once we were settled in Townsville, I began to give serious thought to my Australian research. What I wanted to do was study the role of megabats in the dispersal ecology of one or more rain-forest trees. I also wanted to study the foraging behavior of an Australian fruit bat for comparison with neotropical fruit bats. However, what I wanted to do and what I ended up doing were two quite different things, mainly because Australia lacked baseline ecological data on its tropical plants and animals as well as established field stations in its rain forests. In the mid-1980s, research on the fruiting biology and dispersal ecology of Australian tropical trees was in its infancy. Most of this kind of work was being done at a government research station on the Atherton Plateau, about three hundred kilometers north of Townsville. Frank Crome was the only vertebrate ecologist among the ten research scientists at Atherton. For his doctoral thesis in the early 1970s, he had studied the feeding ecology of eight species of fruit pigeons and the fruiting biology of their food plants at Lacey Creek, near Mission Beach, located about midway between Townsville and Cairns on the Queensland coast. Frank was my closest colleague in Australia in terms of research interests, but the physical distance that separated us made day-to-day contact and close collaboration out of the question

In 1987, studies of the behavior and ecology of Australia's thirteen species of pteropodid bats were also in their infancy. Francis Ratcliffe, when a newly minted zoology student from Oxford University, had begun these studies in the late 1920s and early 1930s. Ratcliffe had been hired by the Australian government to investigate the "flying fox problem." In his delightful memoir *Flying Fox and Drifting Sand,* Ratcliffe described this problem succinctly: "Although inoffensive in the eyes of their fellow-beasts, however, their frugivorous habits have earned for them the enmity of mankind, and it was the unwelcome attentions of flying foxes to the orchards and gardens of the eastern seaboard that brought me to Australia" (1947, 4). In his travels up and down Australia's east coast, Ratcliffe discovered that flying foxes much preferred to feed on native flowers and fruit and that their overall impact on commercial fruit was insignificant. Sometimes occurring in vast numbers in their daytime "camps," four of Australia's *Pteropus* flying foxes were highly mobile and migrated long distances following the erratic blooms of eucalypt blossoms and figs. Ratcliffe also reported that flying fox numbers appeared to be declining as a result of habitat destruction and persecution by man.

In the fifty or so years since Ratcliffe's pioneering studies, not much has changed regarding Australian fruit growers' attitudes toward flying foxes, which were still officially viewed as "vermin" by the Queensland government when we lived in Townsville. Deforestation and development continue to chip away at their traditional roosting and feeding sites in New South Wales and Queensland, and farmers still regularly shoot up bat roosts near their orchards. In the early 1990s Australian bat biologists developed a continent-wide bat conservation action plan, but implementation of that plan has been and will continue to be difficult, according to long-time bat researchers Greg Richards and Les Hall. Two issues are particularly contentious for the preservation of flying foxes: forest conservation and excessive culling of bats living near fruit orchards. In less than two hundred years, European colonists have reduced Australia's forests to 50 percent of their former area, and rates of deforestation are still high. Farmers can still purchase licenses to kill flying foxes, and with only modest culling rates, they can quickly reduce population sizes of these bats because of their low reproductive rates (adult females produce a maximum of only one young per year).

Early in our stay in Australia, I began to explore different areas north of Townsville, looking for a suitable study site. On my first field trip, my son, Mike, and I accompanied George Heinsohn and his undergraduate class in wildlife ecology and conservation on a week-long camping trip. Our first stop was Kirrama State Forest, a mixture of open eucalypt forest bordered by moist tropical forest along streams, where George had been livetrapping small mammals and mist-netting birds for a number of years. On our first night at Kirrama, several students helped Mike and me to set up a series of mist nets along and across a shallow creek down the road from camp. Cockatoos and two members of the birds-of-paradise family—magnificent riflebirds and green catbirds—loudly proclaimed their presence in the canopy above us by screaming, yowling, or whistling sharply as we worked. Our catch that night was disappointing—a single male greater broad-nosed bat, a rather large vespertilionid. Back at camp, however, little red flying foxes were audible as they fed in flowering eucalyptus trees above us all night.

Over the next several days, I began to learn the local flora and fauna. George introduced me to three species of murid rodents, including a mean-tempered, arboreal white-tailed rat that weighs up to seven hundred grams and the northern brown bandicoot. Male whipbirds—ground-feeding, thrush-sized birds with a dark crest and long, dark-olive tail—gave their loud *thwacking* calls throughout the woods. Memorable plants included flowering *Syzygium* trees with red, powder-puff flowers scattered along their trunks; cauliferous figs, which also produced their large, green fruits on their trunks; massive elkhorn ferns stuck high up on the trunks of tall trees that served as nest sites for tree kangaroos and other animals; wait-a-minute or lawyer ("once they get their hooks into you . . . ") palms with their long, heavily barbed climbing leaves; and stinging trees, a member of the nettle family whose leaves and petioles are covered with extremely irritating hairs.

While George worked with some of his students doing his field exercises, I decided to conduct my own OTS-style field problem, assisted by other students. I proposed that we describe in detail the comings and goings of birds and bats in a flowering blue gum tree, located on a ridgetop just east of camp. The plan was for two or three people to watch this twenty-meter tree continuously from sunrise until about 8 P.M. and record the time and identity of each nectar-feeding bird or bat that entered the tree. Mike and I took the first watch and saw six little red flying foxes silently leave the tree just before dawn. Within two minutes of the last bat's departure, a scarlet honeyeater entered the tree to begin the day shift of nectar feeding. From then on, the tree was alive all day with the comings and goings of several species of colorful honeyeaters and noisy lorikeets. Interspecific interactions were very uncommon, and most birds seemed to enter and leave the tree independently of each other. The last bird left the tree at sunset, and within half an hour the little reds—up to seven at a time—began to feed in it. Once in the tree, the bats were sedentary and silent and only issued their harsh barks when another bat got too close. At wide intervals individual bats entered or left the tree.

The overall picture that emerged from our little study was that this tree served as a round-the-clock feeding station for nectar-eating birds and flying foxes. In our hikes around Kirrama, Mike and I noted that other blue gums were also full of lorikeets and honeyeaters by day and little reds at night. Our observation tree was not exceptional. I began to wonder how many other Australian flower and fruit sources were twenty-four-hour feeding spots. Were Australian tropical plants less specialized in their pollination and dispersal ecology than their New World counterparts, which tend to be pollinated either by bats or birds (or insects) but not by both? Did both birds and bats play equally important roles in the reproductive success of many Australian trees?

Before finally settling on a research project, I visited a number of other sites looking for potential study plants and animals. I recruited a James Cook master's student, Murray Hassler, to help me with my explorations. Murray was a distinctive-looking Aussie. He was tall and thin and had an enormous head and beard of long, curly brown hair. His field garb consisted of a pair of shorts, a holey T-shirt, and bare feet. Actually, bare feet were rather common among Australian biology students. I definitely felt overdressed in my nylon soccer shoes. The absence of dangerous snakes and land leeches, at least in lowland rain forest, allowed Australian field biologists to wear much more casual footwear than their neotropical counterparts.

One of the sites we visited was Lacey Creek–Mission Beach, where Frank Crome had done his doctoral work. Unfortunately, this area had been hit hard by a cyclone in February 1986, and its forest was still in bad shape. Many trees had been toppled, and much of the canopy was still missing a year and a half later. Lawyer vines and stinging trees were everywhere. In many places the sunny forest understory was an impenetrable mass of hostile vegetation armed with impressive thorns and irritating hairs. As a result, our exploratory hikes were confined to dirt jeep tracks, streambeds, and a couple of narrow trails that climbed up and down the local hills. Murray and I had a bit of a scare on one of those trails. A long elapid snake, possibly a taipan (one of the most venomous snakes in Australia), darted across the trail between Murray and me. Bare-footed Murray leaped forward and nylon-footed Ted leaped backward to give the snake plenty of room.

Before the cyclone struck, cassowaries, the largest frugivorous birds in the world, were fairly common in the Mission Beach area. Murray and I encountered two of these birds, a meter and one-half tall, on our hikes. These dark-bodied forest giants have naked blue necks and heads, crimson wattles, and a black horny casque atop their heads. We got only a brief glimpse of each bird as it walked silently through the forest. Much more common than the birds themselves were piles of seeds they defecated along car tracks and trails. About twenty centimeters in diameter, these piles often contained an amazing variety of seeds. One pile contained at least ten species ranging in size from tiny fig seeds to large *Eleocarpus* seeds measuring nearly four centimeters in diameter.

According to Barbara Jorissen, the well-known "cassowary lady" of Garner's Beach whom we visited one morning, the cassowaries were hard-hit by the cyclone. Over tea and cookies, this delightfully spry old German lady told us about the genealogy of birds in her area, most of which were the progeny of Barbara (her namesake), the cassowary matriarch. Prior to the storm, about sixty-eight birds lived around Garner's Beach, a subdivision of Mission Beach. In their desperate search for fruit after the storm, twelve birds had been killed by automobiles in the past year. Normally shy around humans, five individuals were still regularly visiting Mrs. Jorissen's house for food.

I didn't seriously consider studying seed dispersal by cassowaries (my student Andy Mack would do this a couple of years later in Papua New Guinea), but the easily found piles were just waiting to be studied. However,

Figure 23. Queensland blossom bat *(Syconycteris australis)*. Photo © Merlin D. Tuttle, Bat Conservation International.

instead of switching to cassowary dung, Murray and I stuck to our original plan and tried to see what fruit bats were in the area by setting mist nets across the narrow Lacey Creek and a forested trail. In three nights, we caught only four bats, all pteropodids—three Queensland blossom bats and one Queensland tube-nosed bat—a disappointing catch indeed by neotropical standards. Weighing only about fifteen grams, the blossom bat resembled a much scaled-down *Pteropus* in most respects (figure 23). It had woolly, fawn- colored fur, beautiful brown eyes, and a slightly elongated snout. Like *Cynopterus* in India, blossom bats were feisty and vocal. Each of my captives strongly objected to being handled and struggled violently to get free. My eventual collaborator Hugh Spencer calls these *Syconycteris* bats *Psychonycteris* because of their crazed behavior in the hand.

The tube-nosed bat *(Nyctimene robinsoni)* was the only frugivorous Australian bat that I figured I had much of a chance to study because of its abundance and because, unlike *Pteropus* bats, it often flew far below the forest canopy. Therefore, I was excited to see my first tube-nosed bat in a net.

I stopped to inspect it for a minute, and it escaped before I could grab it. Before it escaped, however, I noted its bulky build and large head and large eyes. It appeared to be much more powerfully built than the Jamaican fruit-eating bat, its closest ecological analogue among phyllostomid bats. Despite its reportedly being common in some areas of tropical Queensland, little was known about its behavior and ecology.

All in all, our reconnaissance trip to Mission Beach was disappointing. The forest was too highly disturbed to work in easily, and bat population levels seemed to be low. Murray and I reluctantly concluded that we would have to look farther to find a suitable field site. I began to worry whether I was going to be able to complete any research in the land down under.

Fortunately, a short lecture trip to Sydney and Brisbane in October got me out of my blue funk. My Australian colleagues were eager to hear about my research with phyllostomid bats. In both cities, they took me to visit flying fox day roosts, or "camps." These roosts, located in the tops of trees, are notable for a couple of reasons. First, their intense smell, produced by a combination of urine, feces, and musky body secretions, can be detected from a substantial distance downwind. Flying fox camps definitely are not beds of roses. And second, once the bats become aware of your approach, they erupt in a cacophony of harsh barks and beating wings as individuals begin to mill around inside the roost. Chaos momentarily reigns where all was relatively calm a minute ago.

At sunset in suburban Sydney and downtown Brisbane, I watched hundreds of grey-headed flying foxes or black flying foxes leave their camps on slow, steady wingbeats, heading for patches of flowering eucalypts many kilometers away (figure 24). At the University of Queensland, I visited a captive colony of three species of *Pteropus*, whose reproductive biology was being studied by Len Martin and his graduate students. I also visited Jack Pettigrew, who asked me an odd question. "Does *Carollia* defecate and urinate while hanging upside down or while hanging rightside up by its thumb claws?" he wanted to know. I assured him that, unlike pteropodid bats, which hang rightside up to relieve themselves, *Carollia* pooped and peed upside down, in typical microbat fashion. This is one of the many behavioral differences between mega- and microbats that Jack was cataloging.

Les Hall, one of Australia's most knowledgeable bat men, was my host in Brisbane, and he and his wife took Marcia and me on a tour of that lovely city, awash in lavender jacaranda blossoms during the height of spring. Les also introduced us to one of Australia's famous "bat mums," Helen Luckhoff. Helen is a pioneer in successfully rearing and releasing back into the wild orphaned baby flying foxes. During years in which eucalypt flower crops

Figure 24. Grey-headed flying fox *(Pteropus poliocephalus)* and eucalyptus flowers. Photo © Merlin D. Tuttle, Bat Conservation International.

fail, hundreds of female flying foxes die or abandon their pups in and around camps in Sydney and Brisbane. Helen and her crew of women volunteers jump to the task of trying to save as many of the babies as possible. Several times a day they gather to bottle feed their babies. Their success rate is remarkably high, and Helen told me that some of "her" youngsters sometimes return to her backyard, carrying their own babies.

After my talk at the New South Wales Zoological Society, Hugh Spencer, a neurophysiologist from the University of Wollongong, south of Sydney, suggested that we collaborate on a study of the tube-nosed bat in far north Queensland. Hyperactive and sporting a head of thinning gray hair and a scraggly gray beard, Hugh studied the electrical activity of neurons for a living. His real passion, however, was the ecology and behavior of pteropodid bats. Skilled at building electronic devices, he designed his own transmitters and receivers and was beginning to radio-track flying foxes for the first time in New South Wales. He was also contemplating leaving academia to build a rain forest field station at Cape Tribulation, at the base of the Cape York Peninsula, about 560 kilometers north of Townsville. There he had befriended Colin and Dawn Gray, the owners of Mermaleca Tropical Fruit Farm. Tube-nosed bats were common around the thirty-six-hectare farm, and Hugh said that they might be feasible to radio-track. In Sydney we made plans to work with *Nyctimene* for a couple of weeks each in November and December.

In mid-November, Murray Hassler and I drove up the paved coast road to Cairns in a rented university ute (utility vehicle), passing many kilometers of sugar cane plantations along the way. After picking up Hugh, we continued north to the Daintree River, which we crossed on a small ferry. Signs at the river's edge warned people to watch out for saltwater crocodiles, the most dangerous beast in this part of Australia. We drove the final thirty kilometers to Cape Tribulation along a muddy dirt road, nearly colliding with a large tourist bus coming around a tight corner. It seemed incongruous to see tourists being hauled in a big comfy bus to view some of Australia's last remaining rain forest. I had expected most people in this part of Australia to be rugged outback types rather than pampered tourists.

Cape Tribulation was named by Captain James Cook in 1770, during his first Pacific expedition. After the *Endeavour* ran aground on coral, Cook wrote: "Hitherto none of the names which distinguish the several parts of the country that we saw are memorials of distress; but here we became acquainted with misfortune, and we therefore called the point which we had just seen farthest to the northward, Cape Tribulation" (quoted in Russell 1985, 57).

Despite its melancholy name, Cape Trib is a place of unsurpassed beauty—a place that's often described as where the rain forest meets the sea. Cape Trib is dominated by turquoise water and deep green forest, by coral reefs and forests containing ancient, big-seeded trees and elegant *Licuala* fan palms. Part of its forest had recently been nominated for World Heritage status by the Australian federal government. The Queensland state government, however, vehemently opposed this designation and wanted timber cutters to continue logging it. Hugh Spencer and his environmental activist friends were currently in the midst of a campaign to stop the logging. By the late 1980s this area was also undergoing rapid development as a mecca for backpackers and the youth hostel crowd. When we worked there, one hostel was in operation, another was being built, and land was being cleared just behind the beach for additional tourist accommodation. Fortunately, the Japanese-financed megaresorts a bit farther south along the coast had not yet crept into the Cape Trib neighborhood.

There was no field station yet, so we took most of our meals with Dawn and Colin Gray and camped out in a rustic house owned by Hans Nieuwenhuizen, a stocky, red-bearded Belgian expatriate who earned his living as a handyman and nature guide. Before sunset on our first night at Trib, Hugh gave Murray and me a quick tour of the fruit farm, whose crops included jackfruit, lychee nut, soursop, star fruit, mango, rambutan, durian, banana, and papaya. Unlike most Australian fruit growers with their negative attitudes toward flying foxes, the Grays were enlightened farmers who did not resent or prevent fruit bats from taking an occasional meal at their expense. They even sent off to market soursop fruits that were missing a small "divot" of rind and flesh that had been scooped out by tube-nosed bats. After supper, Hans took us on a night walk through tall forest that paralleled the beach. We saw no mammals on this walk. Retracing the same path the next morning, I was struck by the large quantity of fruit that littered the ground, including species of fan and black palms as well as the large, prickly fruits of *Davidsonia* and the nutmeglike fruits of *Myristica.* As at other sites I had visited, the density of fruit-eating birds and other animals seemed to be very low. Only shiny starlings and Torres Strait pigeons were present in any numbers.

That night we began our work with *Nyctimene.* Dawn reported that this bat is an early flier, so at sunset we opened six nets that we had placed around a group of soursop trees. We quickly learned, however, that this was a big mistake. As soon as the nets were open, they began to catch scarab beetles. By 9 P.M. our poor nets contained at least a thousand of these big black creatures, which quickly became entangled in gobs of net. A large diadem horseshoe bat slowly drifted among the fruit trees on broad wings, hunting these beetles. It took us more than an hour to clear the nets of beetles, after which we finally caught four females of *Nyctimene.* Two of our bats were in late-term pregnancy.

Before outfitting three individuals with transmitters, we carefully examined this odd bat. Weighing about fifty grams, it was less feisty in the hand than *Syconycteris* and less vocal than *Cynopterus.* Notable anatomical features included beautiful yellow or chartreuse green spotting on its ears and wings; tubular nostrils (the source of its common name) that constantly twitched; a short snout and mouth containing robust teeth; large tan-colored eyes; a noticeably rounded cranium; a short thumb (in contrast with the markedly elongated thumbs of *Pteropus*); a short tail that projected freely from the tail membrane; and lax tan fur. The yellow spotting obviously serves to camouflage this bat in its diurnal foliage roosts. But why does it have flexible, tubular nostrils? It has been suggested that this bat eats soft fruit and that the tubes serve as "snorkels." But our observations at Cape Trib in November and December indicated that it was eating figs and the flesh of soursop, neither of which are mushy. We currently don't know the functional significance of these unusual nostrils.

Compared with my experience with the short-tailed fruit bat in Costa Rica, *Nyctimene* turned out to be a much more frustrating bat to radio-track. Individuals of *Carollia* seemingly never fussed with their transmitters, and we were able to track them for up to three weeks before their radios fell off. Hugh's transmitters lasted for far shorter periods of time on *Nyctimene,* because bats apparently fussed with them all the time. Our first three bats managed to pry their transmitters off in less than three days, which forced us to use a stronger glue to attach a radio to their backs. Even with stronger glue, none of our transmitters stayed on a bat for more than eight days. So we didn't gather nearly as much data per individual of *Nyctimene* as we had for *Carollia.*

Perhaps it's just as well that the transmitters had a short residence time, because the tubed-nosed bat turned out to be boring to radio-track. Most of the fourteen individuals that we tagged fed close to their day roosts. And since they were eating two species of fifteen-gram figs when they weren't snacking on soursop, they didn't need to eat many fruits each night to satisfy their energy requirements. We estimated that their nightly harvest was five to seven figs, or about one fig every two hours in a ten-hour night. A short flight from their day roost to grab a fruit every two hours or so makes for a pretty slow night of data gathering. The longest foraging movement of our most active individual was a round trip of about two kilometers from its day roost to a fig tree at the edge of an abandoned field.

Most of our nights at Cape Trib during both field trips were spent either capturing or radio-tracking *Nyctimene.* During the day we searched for our radio-tagged bats in their leafy roosts. We were able to get quite close to most bats but couldn't actually see them when they were roosting high in the canopy. A few bats, however, roosted within six meters of the forest floor, and we were able to spot each one of them. These bats were roosting solitarily with their wings wrapped tightly around their bodies. Backlighted against the pale green foliage, they resembled dead leaves as they slept. In December, when Mike served as our field assistant, we tagged three bats with transmitters equipped with a flashing LED light. We found the day roosts of two of these bats by spotting their blinking red lights among the leaves. Finding the day roosts helped us interpret our radio-tracking data, because most roosts were located within thirty meters of a fruiting fig tree. Two bats were actually roosting in fig trees. Like howler monkeys that camp out in fruiting fig trees for days, these bats didn't have to go anywhere to find food.

Although tube-nosed bats roosted alone during the day, they were not necessarily solitary feeders at night. Our field observations and netting results indicated that several individuals simultaneously harvested figs from the same tree. On 16 December, for example, we netted around two fruiting fig trees and caught nine individuals at one tree and thirteen at the other. Several of these bats were carrying figs when they hit the net. Thus, unlike *Pteropus* bats, which tend to stake out feeding territories in the canopies of flowering or fruiting trees, tube-nosed bats behave more like *Artibeus* bats by removing a fruit from a tree and flying to another tree before eating. From a plant's point of view, *Nyctimene* is a better potential seed disperser than *Pteropus.* Les Hall and Greg Richards have pointed out that socially subordinate flying foxes—bats that fail to establish feeding territories in fruiting trees—do most of the seed dispersal in their species. These bats feed by quickly flying into a tree, grabbing a fruit, and then carrying it off to a feeding roost before being harassed by a territory-holder. Greg calls this the "resident and raiders" foraging strategy (Hall and Richards 2000, 81–82). To my knowledge, no New World frugivorous bat exhibits this kind of foraging behavior.

Tube-nosed bats also differ from both *Pteropus* and *Artibeus* bats in their use of audible sounds for long-distance communication. Both when they are flying and when they are stationary in a tree, these bats whistle. Actually, the whistle is more like a high-pitched rodent squeal, but it is audible (to humans) from a fair distance. Who gives these whistles (only one or both sexes?) and for what purpose is currently unknown. But once you become attuned to the sound, vocal cues alone make it easy to assess the general activity levels of tube-nosed bats in an area. I'm sure these bats are also keeping tabs on each other with their whistles.

In addition to vocal communication, this bat likely also communicates with olfactory cues. When we began to catch males (away from the fruit farm, which seemed to attract mostly females), we noticed that the sexes differed strongly in smell. Males issue a strong citronella-like odor (the same smell I detected in male yellow-shouldered bats in Panama) from a chest gland. Females apparently lack this gland and do not produce a detectable scent. As in the case of audible whistles, we do not yet know the function of this scent. Do males mark their day roosts or feeding trees with scent? Do they spread scent on each other? Bats generally are richly endowed with scent glands, but their olfactory worlds have scarcely been explored by scientists.

In our two field trips to Cape Tribulation, we barely scratched the surface of the behavioral ecology of *Nyctimene robinsoni.* But that is generally the state of field research on bats in Australia. Except for the work of Francis Ratcliffe and John Nelson, who studied seasonal fluctuations in the size of camps of grey-headed flying foxes in New South Wales in the late 1950s, Australian pteropodid bats were little studied until the 1980s. This is not surprising given the size and mobility of *Pteropus.* Except for the Queensland blossom bat (and the closely related northern blossom bat) and *Nyctimene,* Australian pteropodids seldom forage within mist-net range, so capturing individuals for population and foraging studies can be a major operation. Chris Tideman and his associates at Australian National University have recently devised a giant harp trap to capture flying foxes in their camps. This trap consists of a series of vertical wires strung on a frame supported by two yacht masts (!) that force bats down into a collecting bag. Using this capture method, Peggy Eby, of the University of New England, was able to radio-tag grey-headed flying foxes and, with a small airplane, follow their seasonal movements, which sometimes included distances as great as eight hundred kilometers. The Queensland blossom bat has proved to be less challenging to study. Brad Law, a doctoral student at Sydney University, was able to conduct elegant field experiments in which he supplemented with sugar water the nectar produced by *Banksia* shrubs to demonstrate that food levels control the local density of this species in the coastal heathlands of northern New South Wales.

When Mike and I left Cape Trib and returned to Townsville just before Christmas, my official fieldwork was over. I had pretty much expended my limited research funds, and my teaching stint was coming up. Hugh had other obligations after Christmas. But our contact with flying foxes was far from over, because my family and I had been hand-rearing three orphan black flying foxes *(Pteropus alecto),* since mid-October. We became aware of the orphan *Pteropus* programs in Sydney and Brisbane shortly after we arrived in Townsville. Taking our cue from the "bat mums," we decided that we would adopt up to three flying foxes orphaned during the October birth season. So we put out the word to local wildlife authorities: "Please let us know if you need a home for orphan flying foxes."

The spring of 1987 was a good year for eucalyptus blossoms, so bat mothers were in good shape going into the birth season. Few babies were abandoned to die because of low flower levels. But flying foxes can be rather clumsy fliers, especially when females are carrying their young babies while searching for food. Mothers sometimes fly into obstacles or, worse yet, into power lines and die. If the babies survive these accidents, they can be quite easily adopted and reared until they are old enough to fend for themselves. This is how we acquired three black flying foxes, which we named Annie, Alexis, and Sam.

Annie was the first to join our family. Her mother had been electrocuted (as had another adult *Pteropus,* whose body hung on a power line near our house for nearly a week), but Annie survived. The tip of her right wing had been severely burned, however, and we eventually had to amputate it. Annie would never be able to fly. We estimated that she was about two weeks old when we took her in, and she already weighed eighty grams (about the same weight as an adult greater spear-nosed bat, the second largest phyllostomid!). We took turns feeding her infant formula from a pet baby bottle several times a day. Marcia even brought Annie and her paraphernalia in a Styrofoam six-pack cooler to Brisbane aboard Qantas during my October lecture tour. When she wasn't being fed, we wrapped Annie in a fuzzy washcloth and kept her in a blanket-lined cardboard box with a mesh top. Ginger, our orange cat, was fascinated to have a new animal around the house. Ozzie, our peach-faced lovebird, couldn't have cared less about the baby bat.

Shortly after we returned from Brisbane, Alexis and Sam showed up. Now we really had our hands full, because our cute little babies expected to be fed five times a day. For a while, our domestic lives seemed to revolve around making formula, washing baby bottles, and feeding our little charges. At this stage of their development, our babies, though cute, didn't have distinctive personalities. They were all wings, big brown eyes, and short black fur; they spent most of their time sleeping. Their most notable anatomical feature was their long, sturdy thumbs, which were tipped with a sharp claw. When the bats became active crawlers and fliers, their claws became inadvertent weapons that we constantly had to watch out for.

Gradually our babies grew and became more active. Between feedings, they hung upside down in the box, still resting on (and sometimes sucking) their "security blankets." By mid-December they each weighed about two hundred grams and needed only two bottle feedings a day. They began to nibble on pieces of papaya and mango. When they outgrew their original cage, we moved them to two metal cages. Annie and Alexis had been ganging up on their somewhat smaller "brother," Sam, so we kept them in separate "boy" and "girl" cages. While hanging in their cages, they would fre-

Figure 25. Ted with three-month-old Sam, a black flying fox *(Pteropus alecto)*, Townsville, 1988. Photo by M. Fleming.

quently stretch and flap their growing wings in an obvious attempt to fly. Their wing flapping was especially common around sunset. In addition to wing flapping, Sam showed us how (male?) flying foxes keep cool during the day. On hot summer afternoons, he would pee into a cupped wing membrane and then rub the urine over his head and chest. He may have felt better as the urine evaporated, but this habit certainly gave him a strong body odor.

Beginning in December, we let our flying foxes out of their cages for a while each night to crawl, climb, and eventually fly around our living room (figure 25). Before they could fly, they traveled around on our living room carpet doing what I called the "batterfly"—they swung their partly opened wings forward in unison and used their backward-facing hind legs to push off. When they reached our floor-length curtains, our furniture, or us, they would climb with alternate swings of each closed wing, firmly anchoring themselves at the end of each swing with their sharp thumb claws. To protect our skin from these claws, we began wearing long pants and long-sleeved shirts whenever we played with the bats. We also had to be careful to keep their claws away from our eyes. These baby bats were really gentle with us and never attempted to bite or scratch. But in their enthusiasm to get close to us, they could be careless with their claws.

Once they had some freedom of movement, our babies began to show us their personalities. From the very beginning, they were curious about their surroundings and loved to explore our house. Before they could fly, they climbed all over our furniture, into our laundry hamper, and onto Ozzie's cage. With a few sharp pecks, Ozzie quickly showed the bats who was the alpha male in our menagerie. They climbed on us and sniffed our ears and mouths while hanging upside down from our hair or glasses. Annie turned out to be the most people-oriented of the three bats and loved to cling to us for long periods. Sam, on the other hand, was a natural-born explorer. Once he learned where our garbage box was in the kitchen, he would make a beeline for it every evening to check out what we had just eaten. He appeared to have a real sweet tooth and loved to lick the frosting off cardboard cake boxes and the insides of waxed ice cream containers.

By January, our bats were weaned from milk, and we fed them their weight in fresh fruit every day. We tried to make a deal with our local grocery store to give us leftover fruit that they couldn't sell. When that supply became unreliable, we broke down and bought at least one kilo of fruit—papayas, mangoes, bananas, apples, and grapes—every day. By now the bats were flying, and we sometimes played "three corner catch" with them by letting them fly between Marcia, Mike, and me. Cara was quite intimidated by their size, now nearly three hundred grams, so she avoided the bats when they were not caged. Whenever the bats flew around our house—swoosh, swoosh, swoosh—on their velvety black wings, Cara retreated to her bedroom. Except for their nightly flight time, the bats now resided in a large mesh cage that I placed in an aluminum utility shed in our backyard. Cleaning up the mess that was left over from last night's fruit salad became my first chore every morning.

As the time of our departure from Australia drew closer, we had to decide what to do with our adolescent bats. Unlike Helen Luckhoff and her "bat mums," we were in no position to gradually introduce Alexis and Sam to wild flying foxes so that they could become incorporated into the local population. Annie, of course, would always be a captive. We briefly considered applying for permission to import them into the United States but rejected that idea because of the expense and work involved in keeping them well-fed when they reached their adult weight of eight hundred to one thousand grams. Besides, the United States has strict laws against the importation of *Pteropus* bats. California fruit growers are no fonder of large fruit-eating bats than their Australian counterparts are, though in reality flying foxes pose no serious threat to fruit crops in either country

In the end, we gave our "babies" to Hugh Spencer, who was well on his way to building a modest research station adjacent to the Mermaleca Tropical Fruit Farm at Cape Trib. Hugh would use the bats as wildlife ambassadors in his proposed rain-forest interpretative center. The Grays would provide a steady supply of fruit to our captives as well as to a few more that Hugh eventually incorporated into his flock. Hugh kept us informed about Annie, Alexis, and Sam in his Christmas letters. Sam turned out to be the alpha male in the captive group, and he mated with Annie (but spurned Alexis) during the annual breeding season. Annie proved to be an excellent mother, despite her handicap, and produced a baby every year. Unfortunately, after a few years in captivity Sam escaped from the flight cage during a storm and was never seen again. Hugh thinks Sam perished in the storm.

The last sight I had of Annie and Alexis was in August 1995, at the Tenth International Bat Research Conference in Boston. Hugh attended the meeting and brought a videotape of Annie giving birth to one of her babies. What was notable about this event was that Alexis, who was still childless, served as a midwife during the birth. While Annie struggled to give birth, Alexis supported her with her wings and licked her; she eventually helped the baby emerge from Annie's vagina and licked it clean. My bat colleagues and I watched in fascination at this rare example of intimate cooperation between two unrelated wild animals. But the idea of Alexis's aiding Annie in her time of need shouldn't be so strange. After all, Annie and Alexis had been companions their entire lives. They were family. Watching the video, I thought to myself, "Not only are they family, but they were once members of our family."

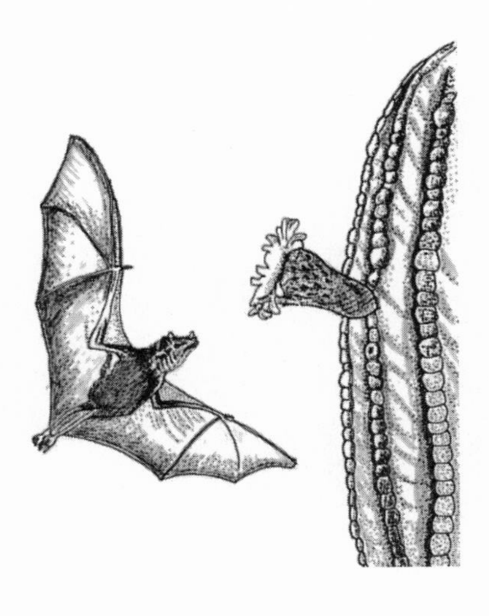

9 Tracy's Hypothesis

In August 1988, shortly after we returned from Australia, Merlin Tuttle called to welcome me back to the States. After a few minutes of news and gossip, Merlin got to the real purpose of his call. He asked, "How would you like to take a break from your tropical studies and work with me on the lesser long-nosed bat, *Leptonycteris curasoae,* as it pollinates flowers of columnar cacti in the Sonoran Desert?" He had spent some time in May photographing this bat visiting cactus flowers at a place called Bahía Kino in Sonora, Mexico (see map 4). "It's an absolutely gorgeous site located right on the Gulf of California," he enthusiastically continued. "There's a high density of three night-blooming cacti there, and I'm pretty sure we could get some money from National Geographic to study the role of bats in their pollination biology. Does this sound interesting to you?"

Merlin's suggested study came from out of the blue. At the time, I was planning to continue my work with *Carollia* bats and *Piper* plants in Costa Rica. I wanted to move my research to La Selva and study the comparative

Lesser long-nosed bat approaching a cardón cactus flower. Redrawn by Ted Fleming, with permission, from a photo by Merlin D. Tuttle, Bat Conservation International.

ecology of three species of *Carollia* and nearly fifty species of *Piper.* In the summer of 1986, I had spent ten weeks at La Selva, gathering preliminary data on this system while assisted by a crew of Earthwatch volunteers. Just before leaving for Australia in 1987, I submitted a new proposal to the National Science Foundation. I had discussed my research plans with Merlin when we worked together with *Vampyrum* at La Selva in 1987. As we slogged along a muddy trail in pouring rain, he chided me by asking, "Why do you want to work here?" I sometimes wondered about this myself. Although moving up a diversity gradient from Santa Rosa (which has two species of *Carollia* and five species of *Piper*) to La Selva made sense scientifically, I really wasn't sure that I wanted to spend several years working in tropical rain forest again. I was no fonder of getting soaked with rain nearly every day and night, as we did in 1986, than I had been in Panama twenty years earlier.

NSF turned down my La Selva research proposal but gave me a strong message that if I fine-tuned it a bit, they would likely fund my next application. I was just about to do this when Merlin called, asking if I would like to study pollination biology in the desert, the antithesis of the rain forest. I ended our conversation by telling Merlin that I would give his suggestion serious thought and would contact him soon. Then I began weighing the pros and cons of making a major shift in my research direction, at least temporarily. At stake was a twenty-year investment in tropical biology and seed dispersal ecology. I had been working in Costa Rica for the past seventeen years and felt comfortable there. I still wanted to pursue many research questions about fruit bats and their food plants.

But I had to admit that the thought of working in the Sonoran Desert was appealing. Except for a brief visit to Phoenix in 1969, I had never actually set foot in this habitat, but I had a strong mental image of it that dated from my childhood. In the early 1950s I spent hours poring through old issues of *Arizona Highways* at my next-door neighbor's house in Detroit. The spiny stems of teddy bear cholla, the magenta flowers of prickly pear cactus, and the stately silhouettes of saguaro cacti popped into my mind's eye when I thought of this desert. I also had been fascinated by Walt Disney's early nature film *The Living Desert,* with its close-up pictures of tarantulas, scorpions, kangaroo rats, and rattlesnakes. In graduate school I had cared for a Merriam's kangaroo rat, a common Sonoran Desert rodent, while my officemate Jim Brown was in the field. It would be fun, I thought, to actually see K-rats in the wild.

After a few days' deliberation, I called Merlin and said, "The idea of working with *Leptonycteris* in the Sonoran Desert sounds really interesting. How soon can we get together to write a proposal?" Two things had sold me on this idea: first and foremost was the aesthetic appeal of the Sonoran Desert, and second was the appeal of doing experimental studies in pollination biology. It is easy to determine the effect of particular kinds of pollinators on fruit and seed production in their food plants simply by bagging flowers. Pollinator exclusion experiments can quickly answer the question, How important are *Leptonycteris* bats for fruit set in Sonoran Desert columnar cacti? Much more difficult to design are analogous kinds of experiments to answer the question, How important are *Carollia* bats for seedling establishment in their food plants?.

So in late September 1988 I went to Austin, Texas, where Merlin and I quickly wrote a National Geographic proposal to determine experimentally the importance of bats as pollinators of columnar cacti. We felt optimistic about gaining NGS support for at least three reasons. First, National Geographic loved to work with Merlin. He already had written two articles for them, and Mary Smith, the person in charge of scheduling articles at Geographic, was always receptive to Merlin's suggestions about new articles. When we submitted our proposal, we included gorgeous photographs of our proposed study organisms—photos clearly suggesting that a *National Geographic* article based on our research would be visually appealing. The second reason for our confidence stemmed from the conservation slant of our research. Not only were we proposing to work with some of the most picturesque and ecologically important plants in the Sonoran Desert, but their major nocturnal visitor, the lesser long-nosed bat, had just been officially declared "federally endangered" in the United States. Our work would provide timely new information about the functional role of this bat in a desert ecosystem. Finally, in 1988 little work had been done on the pollination biology of *any* columnar cactus (not just Sonoran Desert species), and except for Donna Howell's work on *Leptonycteris* bats as pollinators of flowers of the century plant *Agave palmeri* in southeastern Arizona, relatively little was known about the foraging behavior of this nectar-feeding phyllostomid.

Donna was well known for her work with lesser long-nosed bats and paniculate agaves (century plants). After completing her doctoral studies at the University of Arizona in 1972 and postdoctoral research at Princeton, she worked at the University of Florida, Purdue University, and Southern Methodist University. By 1988 she had left academia and was working for the Nature Conservancy back in Arizona. An articulate and entertaining public speaker, Donna was an effective spokesperson for the endangered status of the lesser long-nosed bat. In the course of her field studies, beginning in the late 1960s, she noted that populations of Lepto (as bat workers affectionately call these bats) seemingly were on the decline in Arizona. In the mid-1970s, she visited all of the locations where *Leptonycteris* had been captured in Arizona and found few bats. Except at Colossal Cave in the Rincon Mountains east of Tucson, where several thousand Leptos had been deliberately displaced during its commercialization in the mid-1960s, reasons for the population decrease were not obvious. Nonetheless, she began to wonder whether a decline in populations of this bat might result in a decrease in fruit and seed set in the *Agave*s they pollinated. She and a colleague, Barbara Roth, surveyed fruit and seed set in century plant populations in southeastern Arizona during 1976–77 and compared their results with seed set in herbarium specimens collected in the 1930s through the 1950s. Their influential (but technically flawed) paper, published in 1981, reported that seed set in bat-visited *Agave*s had declined substantially since the late 1930s.

This evidence, plus a survey of all known Mexican and American roosts used by *Leptonycteris* bats conducted by Don Wilson in the mid-1980s, seemed to indicate that Leptos (both *L. curasoae* and its slightly larger Mexican relative *L. nivalis*) were much less common than they had been a few decades earlier. Wilson and his Mexican field crews, for example, found fewer than twenty thousand lesser long-nosed bats in all of Mexico and Arizona. Howell's and Wilson's observations were influential in convincing the U.S. Fish and Wildlife Service that both species merited "federally endangered" status. This status was officially announced in the *Federal Register* in October 1988.

As Merlin predicted, National Geographic awarded us a grant to study bat-cactus interactions in Mexico in the spring of 1989 with the possibility of extending our study for another year. As always whenever I embark on an extended field trip, I was concerned about finding a good research assistant. A woman named Margaret (Peggy) Horner, who had just received a master's degree at North Carolina State University, was applying to our graduate program for doctoral studies on bats under my guidance. Peggy appeared to have excellent qualifications for working in the desert with Merlin and me. We were looking for someone with radio-tracking experience to help us radio-track Leptos. Peggy had radio-tracked black bears in the mountains of North Carolina for her master's research. She had also worked with bats in Costa Rica during an OTS course and had lots of experience living and working in the field, including a stint in the Peace Corps in Sierra Leone. I was convinced that Peggy would be an excellent research assistant and was pleased when she accepted my offer to join our field party in 1989.

Figure 26. The 1989 field crew; front row: Peg Horner, Mary Schantz, and Tracy Truppman (from left); back row: Janet Debelak, Ted May, Ben Watkins, David Dalton, Ginnie Dalton, and Yar Petryszyn (from left). Photo by T. Fleming.

Merlin used his extensive contacts through Bat Conservation International to round up a crew of volunteer field assistants and a place to live at Bahía Kino. Our first field season at Kino Bay set the pattern for much of my subsequent work there over the next decade. This pattern included using highly motivated volunteers, recruited from all over the United States, as field assistants (figure 26). College students, an artist, an art historian, a financial adviser, a court reporter, a university administrator, a steelworks painter, an engineer, a medical technician, and the owner of a medical equipment company, among others, have spent from two weeks to three months working in the desert with me. Later in our studies, I invited Mexican students and biologists to join our research team. For two years in the mid-1990s, a lively group of Mexican biology students traveled a day and a half by train and bus from Mexico City and Guadalajara to join "el Grupo Fleming" for work and fun at Kino Bay during Semana Santa (Easter week). In addition to fieldwork, we cooked elaborate Mexican meals, gave miniseminars, and taught each other new dance steps.

What did we actually know about the pollination biology of Sonoran Desert columnar cacti and their major nocturnal pollinator, the lesser long-nosed bat, when we began our first field season in 1989? For one thing, we knew that the lesser long-nosed bat is not the exclusive pollinator of these plants.

Studies of two of our three target species, saguaro and organ pipe, conducted near Tucson, Arizona, in the late 1950s indicated that diurnal flower visitors such as white-winged doves and honeybees are also effective pollinators. Pollinator exclusion experiments had shown that diurnal visitors actually are responsible for considerably more fruit set in saguaro, whose flowers stay open nearly twenty-four hours, than nocturnal visitors. Despite these results, it was still widely believed that bats are the major pollinators of these two cacti. This belief probably stemmed from the fact that flowers of both species first open at night and conform well to the "bat pollination flower syndrome." This syndrome, or suite of floral characteristics, includes nocturnal opening, large size, light-colored petals, and lots of nectar and pollen. Many tropical plants that are bat-pollinated also have flowers that conform to this syndrome.

In 1989, it was well-known that the lesser long-nosed bat is only a seasonal visitor (a kind of "reverse snowbird") to the Sonoran Desert, where it feeds on the nectar, pollen, and fruits of columnar cacti. In April, females of this species form maternity colonies in caves and mines in southern Arizona and northwestern Mexico after migrating from somewhere farther south in Mexico. Adult males seldom accompany females on their northern foray. Females give birth to a single pup in mid-May, and the maternity roosts disband after the babies are weaned in midsummer. In late summer and early fall, postlactating females and juveniles show up in upland habitats in southeastern Arizona, where they feed exclusively on nectar and pollen produced by flowers of paniculate agaves. Donna Howell reported that these bats often forage in flocks when feeding at agaves. By the end of September, most Leptos have left Arizona and northwestern Mexico for parts south.

Despite this general understanding of its seasonal cycle, there still were many unanswered questions about the behavior and ecology of this bat. For example, did females undergo two pregnancies a year, as is the case in many tropical phyllostomids, or only one? Mexican researchers had reported finding pregnant and lactating lesser long-nosed bats in southern Mexico in December and January, suggesting that this species is indeed polyestrous. Other major questions were: Where do the Sonoran Desert migrants spend their winters? How far do females migrate to get to their spring maternity roosts? What are their migratory routes, and what kinds of flowers do they use to fuel their migrations? Where and when do matings take place, and why don't males migrate? And finally, what factors are responsible for the population declines in this species? Donna Howell had suggested that overharvesting of agaves to make bootleg tequila in the Sierra Madre of western Mexico had reduced the lesser long-nosed bat's food supply. Was this true, and how dependent is Lepto on the flowers of paniculate agaves and columnar cacti for food? Providing answers to these questions would motivate our research with this bat for the next decade.

My first field season in the Sonoran Desert began in mid-April 1989, when I flew to Tucson with my usual large load of field gear. Compared with lush green Miami, dusty, barren Tucson with its miniscule humidity was something of a shock to me. The Santa Catalina Mountains provided a pretty backdrop for the flat city, but I immediately missed the color green. At the airport, I met Peggy Horner, and together we picked up our field vehicle, an old International Carryall that I was renting from a friend, Mike Rosenzweig, an evolutionary ecologist at the University of Arizona. Having a rugged and reliable vehicle to work with in Mexico had been a major concern. My university has no vehicle fleet, and it is impossible to rent a field vehicle from a commercial firm in Tucson and drive it into Mexico for an extended stay. So when Mike offered to "loan" us his vehicle for the cost of getting it back into operating condition, I breathed a sigh of relief. "Old Betsy" provided adequate, if rather dusty, field transportation for us during our first field season.

Peggy turned out to be a first-rate research assistant. Of medium height with short strawberry blond hair, Peg brought a high level of enthusiasm and professionalism to the project. We hit it off right from the start and had fun beginning to learn the desert flora and fauna together. Before leaving Tucson, we visited the Arizona–Sonora Desert Museum for our first exposure to desert biology. I had first learned about this outdoor living museum from reading *Arizona Highways* and had wanted to visit it ever since I was a kid. I wasn't disappointed by my first visit. I found the combination of botanical garden and zoo to be very appealing. One of the museum's most outstanding features, from my viewpoint, is the presence of as many desert animals *outside* the cages as inside. Gila woodpeckers, cactus wrens, curve-billed thrashers, and white-winged doves regularly perch on the tops of saguaro cactus branches and ocotillo stems. *Sceloporus* lizards and antelope ground squirrels dash between cholla cactus stems. Rattlesnakes and even an occasional mountain lion prowl the museum grounds after dark. Unlike at so many zoos, at the ASDM you really feel that you are in a natural setting. The museum, which has undergone considerable expansion and renovation in recent years, is one of my favorite places in Tucson, and I always begin each field season with a half day visit to it with my new field assistants.

After some last-minute shopping, Peg and I were ready to head to Kino Bay, where we would rendezvous with Merlin. On our way out of town, we picked up one of our volunteer field assistants, Tracy Truppman, who was visiting relatives in downtown Tucson, and loaded her gear into our packed truck. But when we tried to start Betsy, we discovered that we were dead in the water—Betsy's "new" starter had just died. Six hours later, we were finally on the road for the 370-kilometer drive to Kino. We easily passed through Mexican customs at Nogales and then drove straight south on four-lane highway to Hermosillo, the capital of Sonora, before turning west on Highway 16 for the final hundred kilometers to the Sonoran coast. In the dark, it was hard to see the countryside, but our noses told us that we were passing through farmland. The air reeked of fertilizer, and we could also smell each of the large chicken coops that we passed. Where was the desert? I began to think. Finally, just before reaching Kino, we saw a magical sight: a rising moon illuminated a dense forest of giant cardón cacti along the south side of the road. Despite the late hour, my spirits began to soar. Perhaps Merlin was right. Perhaps Kino Bay was something special after all.

Once we reached Bahía Kino, we continued along the coast road to Kino Nuevo, located just north of the small fishing village of Kino Viejo. Founded in the early 1970s, Old Kino has a population of about three thousand people, most of whom live in modest houses along dusty streets. Compared with the rural Costa Rican towns that I knew well (Cañas and Liberia), Old Kino is a large economic step down. It has one Pemex gas station, a couple of *llanterias* (tire repair shops), a few *abarrotes* (small grocery stores), a couple of *tortillerias* and *pescaderias,* and a few *antonjitos* (taco shops) and *restaurantes.* Its only bank closed in 1991. Calle Doce, a farming center midway between Hermosillo and Kino Viejo, is the closest town of any size with a bank and a somewhat larger variety of commercial shops. In sharp contrast with Kino Viejo, Kino Nuevo consists of a string of beach houses owned by wealthy Mexicans (mostly from Hermosillo) or gringos. A small group of retired Americans lives permanently in New Kino, which has a couple of motels and trailer parks for its winter population of northern snowbirds and sports fishermen.

It took us a while to find Casa Muralla, the house owned by a Tucson lawyer that we were renting for the field season. I finally discovered the house by spotting Merlin's photography tent inside its large two-car garage. We knocked loudly on the garage door and eventually woke Merlin up. He was accompanied by David Yetman, a prominent Tucson environmentalist who had agreed to introduce us to the Kino area, and Pat Morton, who had completed her graduate studies in Wisconsin and was now the director of education at Bat Conservation International. Over cold beers we all talked into the early morning about our recent adventures.

After a couple hours' sleep, we began our tour of the desert surrounding Kino Bay. Our first stop was a patch of desert just outside town that contained the three species of columnar cacti we would be studying. I had seen saguaro growing profusely in and around Tucson and had met organ pipe at the Arizona–Sonora Desert Museum. Thus, the only new species for me was cardón, which was now in full flower. Superficially resembling saguaro, cardón plants bear more branches and are much more massive. Some individuals reach a height of twenty meters and weigh several tons. Unlike saguaro, cardón branches are not heavily armed with clusters of long, sharp spines. Instead, their spines are small and innocuous. Another difference is that most of cardón's creamy white flowers are borne on the sides of branches rather than on branch tips. Saguaro and organ pipe, a much smaller plant featuring many relatively thin branches radiating from a central root system, were just beginning to flower.

After checking out the three cactus species, we continued along a rough gravel road that goes up the coast to Punta Chueca, a Seri Indian village some twenty-eight kilometers north of Kino. The Seris have occupied coastal Sonora from Guaymas to Puerto Libertad for centuries. Until recently, these desert hunter-gatherers and fishermen spent part of each year living on Isla Tiburón, the largest island in the Gulf of California, about twenty kilometers off the Mexican mainland just west of Bahía Kino. Now they are confined to a few poor enclaves on the mainland, where they struggle to maintain a cultural identity in the face of a hostile Mexican populace. Seri women are renowned for their basket-making artistry. Using the fibers of the *haat* shrub (*Jatropha cuneata*), they weave tight baskets in a wide variety of sizes (from fist-sized to human-sized) that are highly prized by collectors. In the 1960s, Seri men began to carve animal figures from the dense, deep-red wood of the ironwood tree, one of the Sonoran Desert's most important shade trees. When the Mexicans discovered there was a market for ironwood carvings, they took over this cottage industry and began to decimate ironwood populations around Kino with their chainsaws. Sadly, piles of branches and sawed-off stumps are now all that remain of ironwood populations in many parts of coastal Sonora. Another victim of this kind of excessive exploitation is mesquite, an important desert shade tree that is being converted into charcoal for grilling purposes. The recent demise of large populations of ironwood and mesquite in Sonora emphasizes the sad reality that deforestation is as serious a problem in the desert as it is in tropical forests.

About halfway to Chueca, we turned off on an even rougher narrow road that led across beautiful desert flatlands between two of the many massive brick-red basaltic hills that dot the coastal plains of Sonora. We were now on a ten-thousand-hectare ranch owned by the Cabrera brothers from Hermosillo. David Yetman had already introduced Merlin to the Cabreras, who were establishing a trailer park in New Kino and had given us permission to conduct our research on their extensive lands. Cattle had wandered widely in much of the desert around Kino through the 1970s, but they were now absent from most ranches owing to the lack of flowing wells. A ways along this road we stopped and climbed a steep trail up a hillside that led to a shallow abandoned mine. Merlin had watched *Leptonycteris* bats using this mine as a night roost a year ago. He suggested that we might want to conduct our cactus studies near here, since we knew that nectar bats were in the area at night. So far, this was the only place around Kino where he had seen numbers of these bats.

We headed back to Kino as the desert began to heat up. After breakfast, we unloaded our vehicle and began to unpack our field gear and personal belongings. This was Peg's and my first opportunity to see our temporary "field station." The fully furnished house we had rented was a substantial stone fortress consisting of three steplike stories from street level down to the yellow sand beach. The bottom story included a huge garage, in which we stored all of our field gear, plus a kitchen and dining room/living room. The upper two stories, reached by stone outdoor stairways, contained bedrooms and bathrooms. Behind the top level inside a high stone wall was a trailer for the house's caretaker and his family, which included two young children and a small dog. Lupita, the caretaker's wife, would be our cook; we hired her to prepare two meals a day—a midmorning breakfast of fruit salad, cereal, and tortillas and a substantial midafternoon supper. We were each supposed to scrounge for food if we got hungry between meals. On this meal schedule, I knew I would not get fat this field season.

Just outside the house's lower level was the wide beach and the sparkling blue waters of Mar de Cortés, or the Gulf of California. A diverse array of seabirds—a couple species of gulls and terns, cormorants, brown pelicans, brown boobies, and magnificent frigate birds—constantly cruised up and down the coast in front of us. Schools of common dolphins, their arched gray backs glistening in the sun, sometimes passed within a hundred meters of shore. Old Kino was a couple kilometers' walk south along the beach, and New Kino stretched for about eight kilometers north of us. Our part of the beach had relatively few houses, but this has changed over the years, as private development continues to fill in vacant beachfront property. The southern tip of Isla Tiburón was to our west, and we were often treated to spectacular sunsets as the sun, a giant orange globe, slipped behind Tiburón's southern mountains. A small island, Isla Alcatraz, its rugged peak and sides whitewashed with bird guano, lay offshore to the south.

After we were settled, Merlin, Peg, and I began to plan the field season's research activities. Our list of things to do included a number of plant-related activities—weekly counts of flowers and fruit on a series of twenty tagged individuals of the three cactus species, determination of the rate of nectar secretion and the timing of pollen release of their flowers, and pollinator exclusion experiments. On the animal side, we planned to determine the visitation rates of nocturnal and diurnal pollinators to cactus flowers and to radio-track *Leptonycteris* bats. These activities would have us working at the coolest times of the day—late afternoon, often all night, and early morning. Except under unusual circumstances, we expected to be out of the sun from about 10 A.M. to 4 P.M. each day. With its reduced emphasis on daytime work, this field schedule differed considerably from that of my Santa Rosa years.

In addition to helping with data gathering, Merlin planned to photograph bats and assist two film crews that would be filming desert bats during the month he was at Kino. Our large garage became a film studio with the arrival of George and Cathy Dodge, cinematographers for the television program *Wild America.* While the Dodges were filming captive Leptos visiting cactus flowers, another team of wildlife filmmakers working for the company Survival Anglia—Dieter and Mary Plage—pulled up in their heavily loaded van. The Plages and Merlin were about to embark on an ambitious film project entitled *The Secret World of Bats,* which would take them around the world over the next two years. Work on that award-winning film began in our garage at Kino Bay in April 1989.

I was rather amazed to see the interest of wildlife cinematographers in our bats and plants, an interest that initially arose at the urging of Merlin Tuttle, I'm sure. I had worked in Costa Rica for many years without once being contacted by a wildlife film unit. A couple of National Geographic photographers had visited Santa Rosa while I worked there, but not to photograph *Carollia* eating *Piper* fruits. As soon as I changed systems (and joined up with Merlin), however, filmmakers began to beat a regular path to our field station door. After the two teams in 1989, a two-person crew from the British Broadcasting Company showed up to film Leptos visiting cactus flowers in 1990, and a three-man crew from the Japanese Broadcasting Company did a forty-minute television video on our research in 1995. By 1996, when two German colleagues, Uli Schnitzler and Elisabeth Kalko, joined me to study Lepto's vocal behavior using infrared videotape and high-frequency recording equipment, I was an expert at coaxing these cooperative flower-visiting bats to perform for the camera. I could get freshly captured bats to visit cactus flowers at the snap of my fingers on their first full night in captivity.

As fascinating as it was to watch and help Dieter, Mary, and Merlin film three kinds of desert bats—the California leaf-nosed bat and pallid bat as well as *Leptonycteris*—as they fed at cactus flowers (Lepto) or captured crickets *(Macrotus, Antrozous)* in Merlin's screen tent, our real job was collecting data. Since cardón was well along in its blooming season, our first priority was to work with that species. On 21 April, we drove up the coast in the late afternoon to a site Pat Morton named Madrugada (early morning) near the Lepto night roost to begin our first pollinator exclusion experiment. The density of cardóns was high in this area, but working with them was a bit precarious. The ground here was quite rocky, and the density of flowering cardóns was highest on hillsides, not on level ground. This terrain made working on a twenty-foot aluminum extension ladder somewhat nerve-wracking (figure 27), especially in the dark. None of us was an expert yet at climbing ladders perched against swaying cactus branches, so our first few days of this kind of work definitely had their thrills. In the ensuing years, I have become a confident cactus climber (except at saguaro #6 at Seri Flats. We have to place our ladder nearly vertical to reach its flowers, and I often feel like I'm going to topple backward when I'm working at that plant). I have never fallen off a ladder or had a ladder dangerously shift when I work with tall cactuses. Ironically, the only time I've fallen off a ladder was in my own front yard while I was pruning a gumbo limbo tree. Unfortunately, Marcia was holding the ladder when I slipped, and I nearly broke her nose on my way down!

Our exclusion experiments with cardón and the other two species involved several treatments. Our first treatment was a control, flowers that we tagged but left unbagged and available for visits by both nocturnal and diurnal pollinators. Our second treatment was "nocturnal pollinators only," in which we tagged about-to-open flowers before sunset, left them uncovered at night, and covered them with bridal veil netting before sunrise. Only bats and moths had access to these flowers. Our third treatment was "diurnal pollinators only," in which we tagged and bagged flowers before sunset and removed the netting before sunrise, thus allowing only birds and insects to visit them. Our fourth treatment was "diurnal insects only," in which we tagged and bagged flowers before sunset and removed the bags but replaced them with chicken wire "cages" before sunrise, thus excluding birds but not bees from the flowers.

Figure 27. Ted climbing a cardón cactus, assisted by Peggy Horner, near Kino Bay, 1989. Photo by P. A. Morton.

It may seem like a trivial matter, but the manner in which we tagged our experimental flowers caused a bit of a controversy within our field crew. Because we were dealing with large, robust tubular flowers eight to ten centimeters in length, we wanted to use substantial tags that would remain attached to cactus branches for a couple of months before being removed. We quickly learned that ravens and packrats sometimes played with our tags, so we needed to anchor them firmly into plants. Thus, we made sturdy tagging apparatus by attaching an aluminum tag, on which we wrote a flower treatment and plant number, to an- eighteen-gauge wire anchor, which we stuck into the cactus branch near the flower. Tracy Truppman, who was a staunch animal rights activist, thought we were hurting the plants with our tags. To assuage her fears, we eventually did an experiment with organ pipe flowers to demonstrate that our method of tagging did not "hurt" plants and cause them to abort their flowers.

With the start of the exclusion experiments, our lives began to settle into a daily routine that included tagging and bagging (some) flowers just before they opened in the late afternoon and then revisiting those flowers around 4 A.M. the next morning to remove or add netting or wire cages as appropriate. This involved waking up before 3:30 A.M., grabbing a quick bite to eat, and then driving along bouncy desert roads to our plants. After a couple of days, the forty-five-kilometer round-trip to Madrugada twice a day began to get very tedious. That drive plus Madrugada's difficult terrain inspired us to look for another field site closer to Kino Bay. In a day or so, we settled on a pretty patch of desert much closer to home that I dubbed Tortilla Flats. This site was level and much less rocky and had good densities of all three cactus species. At its northern edge was a short hill, which we nicknamed Cerro Catedral (Cathedral Hill) because of its high density of organ pipe cacti. When we netted at the series of abandoned mine shafts in this hill, we discovered that a small colony of California leaf-nosed bats, an insectivorous phyllostomid, was living in one shaft and that Leptos used these shafts as night roosts. The round-trip drive to Tortilla was only twenty-nine kilometers over dusty but much smoother roads. On 24 April, Tortilla Flats became our main study area for all of my work at Kino Bay.

Although we now knew the locations of two *Leptonycteris* night roosts, we didn't know where Leptos were roosting by day. We got our first lead on this when Polly Boyle, a long-time resident of Kino, told us about a bat-filled cave in Sierra Kino, just north of Kino Nuevo on the Chueca road. Reaching an elevation of 450 meters above sea level, Sierra Kino was the tallest volcanic massif in the Kino Bay region. Merlin, Peg, and another assistant first visited the cave on 28 April and found several thousand Leptos roosting there. They also spotted a herd of seven desert bighorn sheep farther up the mountain.

Located about seven kilometers west of Tortilla Flats, the Sierra Kino cave became a site we visited regularly each field season to census its bats. To get to it, we followed a narrow trail, originally made by bighorns, up a rocky barranca, jumping from rock to rock in places during the ascent. Twenty minutes of uphill exertion brought us to the cave, which consisted of a pair of relatively small chambers angling into the rocky hillside. Several small fig trees grew around the cave entrances, a sure sign that fruit-eating bats occasionally roosted here. Canyon wrens, Gila woodpeckers, two species of vultures, and a great horned owl were the avian residents on this mountainside. As we caught our breath, we could gaze downhill to the extensive flatlands below. Except for the Chueca road, we saw no signs of humans for hundreds of square kilometers from this bat lair. I quickly came to view this cave site as a noble place inhabited by a noble bat.

By the end of April our study was in full swing. In addition to the cardón exclusion experiment, we were beginning to measure nectar secretion rates and to stake out cardón plants to document the visitation rates of nocturnal and diurnal pollinators. The nectar work involved bagging a series of flowers on different plants just before they opened and then removing the accumulated nectar every two hours with a syringe until the flowers closed the next day. We worked in pairs during this task, which had us trudging around the desert carrying an extension ladder. One person steadied the ladder and recorded data while the other one climbed up with syringe in hand (or, in my case, in my mouth) to collect a nectar sample. We used a hand refractometer—an optical device that can be used to directly read the percentage of sucrose equivalents of nectar—to measure temporal changes in nectar sugar concentration. My assistants and I quickly became adept at managing all of the nectar tasks on a ladder perched solidly against a cactus branch. Between rounds of nectar collection, we slept in the desert, either on Old Betsy's dusty seats or on the ground on foam pads. Before beginning my desert research, I never imagined that I could fall soundly asleep nearly instantly and wake up just as rapidly when the alarm on my wristwatch told me it was time for another round of nectar sucking. The cool, clear desert nights, I discovered, were perfect for short bouts of sound but dream-filled sleep.

We also worked in two-person teams to watch animals visit cactus flowers at night and during the day. Night observations involved watching a series of flowers on one plant through a night-vision scope. Sitting in a beach chair, one member of the team would stare into the phosphorescent green biocular screen of the night scope for two hours, recording the time and identity of each flower visitor on a data sheet. At the end of two hours, we would switch places, and the previous observer would take a nap. After sunrise, both members of the team continuously used binoculars to watch groups of flowers until they closed, noting the visits of daytime pollinators.

Sitting quietly in the dark at night or in the shade of a mesquite or ironwood tree during the day proved to be an excellent way to learn the rhythms of the desert and its animal inhabitants. During all of April and most of May, the desert was very cool (in the low fifties) at night. It didn't take long, sitting motionless under a clear, star-drenched sky, for an intense chill to set in. We ended up wrapping ourselves in blankets to keep warm. The cool temperatures helped keep us awake while we waited for bats or moths to visit our target flowers. It took only one cardón stakeout to discover that *Leptonycteris* bats were late-night feeders. From 8 P.M. until almost midnight (the first two shifts), bats would occasionally fly past cardón plants, but they

Figure 28. Lesser long-nosed bat *(Leptonycteris curasoae)* approaching a cardón flower. Photo © Merlin D. Tuttle, Bat Conservation International.

seldom stopped to feed. Whoever drew the midnight to 2 A.M. shift (and we began to fight for this time slot!) got to see most of the bat feeding action each night. Before that time period, we had to amuse ourselves by counting stars and listening to the yips and howls of distant coyotes, the ever-present, sharp *kyew!* calls of elf owls, and the occasional, low-pitched hoots of great horned owls. My Sony Walkman loaded with tapes of John Coltrane, Pat Metheny, Steely Dan, Paul Simon, and Sting helped me while away the lonely hours. But once the Lepto feeding frenzy began, all attention was riveted on the cactus flowers.

I have now watched hundreds of visits to cactus flowers by *Leptonycteris* bats and never tire of this rapid, magical interaction (figure 28), which is over in the blink of an eye. But to describe what it's like to see Leptos visiting cardón flowers for the first time, I will defer to Mari Murphy, a former editor of Bat Conservation International's quarterly magazine *Bats* and one of our early volunteers on the project. On the night of 8–9 May she wrote on her data sheet:

> At 2410 a moth entered [cardón] flower #1. Five minutes later I heard my first bat overhead—noisy wingbeats, not too high, headed for a cardón behind me. Five minutes later a single bat entered the night vision scope field from below, flew up to the observation cactus, ap-

> peared to pause slightly to inspect flower #2, continued past the cactus, and flew down out of view. For the next hour there was little activity except for a few moth visitations; heard several bats overhead in flight toward another cardón.
>
> Intense bat activity occurred from 0120 to 0140. A group of three bats approached from the same flight pattern as the first "scouting" bat exactly one hour earlier. During this period, flower #2 was pollinated twice. Some bats appeared quite light-colored (covered with pollen?), from head to chest. Bats were heard flying overhead—perhaps the same group circling around my plant and then heading towards the cardón behind me. From the first definite flower visit on, things happened rapidly. Sometimes bats visited my plant in groups of 3 to 5, sometimes singly. Flowers were pollinated every 3–7 minutes, and bats often hit one flower twice in one activity period. The most intense activity occurred towards the latter part of the 20 minute period. Bats pollinated flowers seven times from 0134 to 0140, and then the activity stopped. I saw nothing else visit cactus flowers during the rest of my observation period [until 0200].

Mari's observations are pretty typical of what we have observed many times at cardón and the other cactus flowers. *Leptonycteris* bats appear to spend the early part of the evening "scouting out" flowers that they will revisit later to feed. Lepto is a loud, fast flier, and its wingbeats often announce its presence before it is seen. We have used these loud wingbeats as a means of censusing bat activity around cacti in recent years. Once they begin to feed, Leptos circle plants, often flying rapidly among branches for several minutes, before approaching a particular flower from below, stalling in midair as they jam their faces up to their shoulders into the corolla tube, and then dropping backward away from the flower. Unlike smaller tropical glossophagine bats, Leptos do not hover delicately at flowers. Their approach is much more physical—"Forget the finesse, just let me get the nectar I'm after!" they seem to be saying. The entire flower visitation process takes about four hundred milliseconds, according to our videotapes of captive bats.

After the bats returned to their day roosts, flower visitations dropped off until the day shift took over, just before sunrise. Sunrise in the desert is a thing of great beauty. A pink or orange glow in the east usually precedes the actual appearance of the giant red globe above the horizon. Then a gentle pastel light spreads over the sparse desert vegetation, turning creosote bushes and flowering mesquite and paloverde trees into shimmering masses of color. The gnarled trunks of elephant trees, a relative of the gumbo limbo trees in my front yard, become copper-colored torches, and in years of good ironwood flowering, Tortilla Flats is awash in splashes of lavender in the soft early-morning light. Claude Monet would have loved to paint this scene.

Accompanying the visual treats of sunrise are the sounds of desert birds. First sounds are the low-pitched cooing choruses of dozens of white-winged doves roosting in a mesquite bosque far to the southeast of Tortilla Flats. Then the scolding *churrrr, churrrr* calls of Gila woodpeckers and the sharp *kyeea! kyeea!* of flickers nesting in the branches of giant cardón cacti join the chorus. Next are the crowing "songs" of male Gambel's quails standing tall on elevated perches. Finally, the bubbly, warbling songs issued by strawberry-hued house finches join the palette of sound.

Early-morning bird visitations to our cactus flowers were rather sporadic. At one time or another, we saw nearly every common species of desert bird, including ravens and quails, stick its bill into one of our cactus flowers. A conspicuous exception to this was the phainopepla, a beautiful black bird with light wing patches and a cardinal-like crest that feeds exclusively on the orange fruits of parasitic mistletoe plants growing in the feathery crowns of mesquite trees. Woodpeckers of three species and house finches were regular visitors to cactus flowers. White-winged doves often perched on the tops of cactus branches but were common visitors only to saguaro flowers. They appeared to dislike perching on cardón flowers, located on the sides of branches, and were much more confident dealing with the "landing platforms" formed by masses of white flowers on the tops of saguaro branches. Three species of hummingbirds (Costa's, black-chinned, and broad-billed) were frequent flower visitors at some times but not at others. Their migratory movements made them spotty pollinators. Finally, nearly ubiquitous in their attendance at cactus flowers were feral honeybees. The desert literally hummed when bees were feeding at cacti and other flowering bushes and trees.

Although our attention was focused on pollinators, our cactus observation periods also introduced us to many other desert animals. As I had hoped, at night we occasionally saw kangaroo rats hop into the open to scratch rapidly with their front paws in the soil for buried seeds. The desert floor was riddled with the entrances to the burrow systems of K-rats and their smaller relatives, spiny pocket mice. Once I watched a white-throated woodrat climb up a cardón branch and remove one of the flowers I was observing. Their massive, messy stick nests were common at the bases of mesquite trees and organ pipe cacti. Diurnal mammals included desert cottontails and their long-eared relatives, antelope and black-tailed jackrabbits. As they loped along, stiff-legged, jackrabbits reminded me of arthritic old men. Only when they shifted into high gear did they become desert speedsters. More secretive than the rabbits were antelope and round-tailed ground squirrels. These rodents sometimes climbed to the tips of very spiny branches of organ pipe cacti to eat flowers or maturing fruit. They clearly were experts at avoiding this plant's armory of sharp spines.

Reptiles were also our constant field companions. Zebra-tailed lizards, pale tan in color with a black and white striped tail curled over their backs whenever they froze in their tracks, were the most common diurnal lizards. Constantly on the move as they searched for food, dark brown western whiptail lizards shuffled noisily in the litter at the bases of trees and shrubs. Larger than most lizards was the desert iguana, whose pale pink color seemingly allowed it to disappear into thin air when it dashed into the safety of a bush. If we were lucky, once or twice a field season we would encounter a beautiful pink and black, thick-tailed gila monster, waddling along and oblivious to our presence.

Unlike our constant contact with lizards, we had relatively infrequent contact with snakes. Western coachwhips were the most common diurnal snakes that we encountered. This species is polymorphic in color. Some individuals are solid black, others salmon pink, and still others a blend of black and salmon. These long slender snakes invariably made a mad dash to escape when they were first spotted. Harking back to my childhood habits, I couldn't resist trying to catch a few of these beautiful snakes. I was surprised to find that they didn't attempt to bite after I had them firmly in hand for a few minutes.

I came to the Sonoran Desert expecting to see rattlesnakes, and I wasn't disappointed. The first rattlesnake buzz that I heard was that of a tiny sidewinder, tightly coiled under a *Jatropha* bush. Most of the sidewinders we encountered seemed to be rather high-strung and, despite their small size, buzzed loudly when we were still some distance away from them. I saw their S-shaped, looping tracks in sandy roads and trails much more often than I saw the snakes themselves. It also was easy to find the straight, wide tracks of larger rattlesnakes in the sand. Just as frequently, however, we found the owner of those tracks. Western diamondbacks in the one-meter size range were by far the most common snakes that we encountered, day or night. One of my assistants, Sierra Hayden, saw one nearly every day during our 1995 field season. Unlike sidewinders, diamondbacks were much more laid back. They seldom rattled, and we simply let them go their solemn way whenever our paths crossed.

My most notable encounter with a diamondback occurred in 1992 in a patch of desert about fifty kilometers south of Kino Bay. It was late afternoon, and I was busily collecting cardón flowers and wasn't paying particular attention to the ground. I stopped abruptly, however, when a large diamondback began to buzz loudly. I first looked at my feet before trying to spot the snake and found that I had nearly stepped on a twitching round-tailed ground squirrel. The diamondback, coiled into a defensive position with its head and neck elevated in a classic S-loop, was about a meter away from the squirrel, which it had struck just before I arrived. When I didn't move, the snake slowly crawled backward into the shade of a creosote bush, where it continued its loud buzzing. When I revisited the snake an hour later, it was still buzzing. The ground squirrel was now stiff and covered with harvester ants. I felt guilty about possibly causing the snake to lose its meal to the ants.

By early May our data gathering was proceeding smoothly, but some of our cardón results puzzled us. The first puzzle came from our nectar secretion studies. As part of her doctoral work, Donna Howell had demonstrated experimentally that *Leptonycteris* regularly ingests pollen and digests its cellular contents, mostly amino acids, as precursors for protein production. She argued that by ingesting pollen, Lepto does not need to feed on insects to obtain nitrogen-based compounds for their proteins. It instead relies entirely on plant resources for all of its food. Knowing that pollen is important for Lepto's nutrition, we wanted to determine when anthers release their pollen and how much pollen they produce. Merlin and Tracy were the first team to sample cardón nectar and pollen, and they reported that flowers on one of their sample plants appeared to produce no pollen, even though they contained hundreds of anthers. From then on, we paid special attention to pollen production and quickly discovered that quite a few plants in the Tortilla Flats cardón population bore pollenless flowers. We called these plants duds in terms of their pollen production.

Our second puzzle came from the fruiting behavior of many of our cardón plants. A few days after we treated them, many of our experimental flowers began to fall off—they were aborting before setting fruit. We had expected to see flower abortions from some experimental flowers, especially those that had not been visited by bats. But in addition to experimental flowers, many of our control flowers, which had been readily available to all pollinators, began to abort. We watched in dismay as most of our experimental flowers eventually aborted without setting fruit. What was going on here? Why was our cardón experiment failing? Tracy thought she knew the answer. Her hypothesis was that the cactus plants were rejecting their flowers because they were being injured by our wire tags. Change to a more humane method of tagging and then the cactuses won't reject their flowers, she suggested.

I was not convinced that Tracy's hypothesis had much merit and began to delve into the flowering biology of cardón in more detail in an attempt to come up with an alternate explanation for our results. One of the first things Peg and I did was to determine the pollen status of our twenty phenology census plants, which included some of our experimental plants. We classified each plant as a "dud" or a "heavy" (i.e., heavy pollen producer). Then an assistant and I visited each of these plants and counted the number of developing fruits that it bore and the number of aborted flowers that lay on the desert floor beneath it. When I summarized the data, an interesting pattern emerged. The plants fell into two groups that corresponded to their pollen status. Fruit set in the duds was much higher than fruit set in the heavies. Relatively few aborted flowers lay at the base of the duds, whereas the ground was littered with aborted flowers beneath the heavies" On 8 May I wrote in my field notes, "It looks like some plants [the heavies] are acting like males whereas others [the duds] are acting like females. Is cardón evolving towards dioecy [separate sexes]?"

With the discovery of this pattern, I became a serious student of the evolution of plant breeding systems, something that I never dreamed would happen when I first signed on to the Sonoran Desert project. The literature that I had read about cactus reproductive biology indicated that nearly all species, including saguaro and organ pipe, whose breeding systems had been briefly examined in the late 1950s, are hermaphrodites. That is, all of their flowers are "perfect" and contain pollen and ovules; all individuals have both male (pollen production) and female (fruit production) reproductive abilities. As is true in most groups of flowering plants, nonhermaphroditic breeding systems, such as dioecy or gynodioecy (separate female and hermaphroditic individuals), are rare in the Cactaceae. Dioecy was known to occur in three species of *Opuntia* and in one species of *Echinocereus.* Gynodioecy was known to occur in a couple of species of *Selenicereus* and *Mammilaria.* None of these plants were large-statured columnars. Finding an "odd" breeding system in what is perhaps the world's largest cactus was totally unexpected. By the end of the 1989 field season, I thought cardón had a gynodioecious breeding system, with duds being females and heavies being hermaphrodites. Fruit set appeared to be substantially higher in females than in hermaphrodites, which accounted for the high abortion rate of our experimental flowers. We had used more hermaphrodites than females in our experiment.

In contrast with our "disastrous" results in cardón, results of our pollinator exclusion experiment with saguaro, a known hermaphrodite, were more straightforward, though Merlin was not pleased with them. Abortion rates of control flowers were quite low, and the treatment that had the highest fruit set was "diurnal pollination." Contrary to popular opinion, birds and bees appeared to account for much more pollination than bats in saguaro, just as researchers had reported in Arizona over two decades earlier. Like saguaro, organ pipe also behaved as a conventional hermaphrodite. But results of our organ pipe pollinator exclusion experiment also displeased Merlin because diurnal pollinators, mostly hummingbirds, again accounted for more fruit set than bats. Merlin, it turned out, had gone into this project expecting to demonstrate that *Leptonycteris* bats were the most important pollinators of cardón, saguaro, and organ pipe cacti. When our experimental results didn't support his expectations, he became frustrated and began to wonder if we shouldn't be working in another area where bats were more common. However, since the lesser long-nosed bat was supposed to be an endangered species, it wasn't obvious where that other area might be. Our results certainly supported the idea that the scarcity of Leptos in the Kino Bay area reduced their importance as pollinators of night-blooming columnar cacti.

It actually took me two field seasons to work out the exact form of cardón's breeding system. And the answer to our original question, Why are plants aborting their flowers? turned out to be far more interesting than I had suspected in early May 1989. Instead of being gynodioecious, cardón is *trioecious* (separate male, female, and hermaphroditic plants) at Bahía Kino. Its population actually contains four sex types, only three of which produce sex cells. In addition to males, females, and hermaphrodites, its population contains "neuters," plants whose flowers contain neither pollen nor ovules. I discovered the existence of males and neuters by following a tip I received from Deborah Charlesworth, with whom I began corresponding in the fall of 1989. An evolutionary biologist then at the University of Chicago, Deborah is one of the world's leading experts on the evolution of plant breeding systems. In one of her letters, she suggested that I check out the condition of the ovaries of hermaphrodites to see if some individuals might lack ovules and hence be males. When I did this in 1990, I found that, sure enough, although all "hermaphrodites" had well-developed ovaries containing lots of funicular (placental) tissue for attaching their tiny pearl-like ovules to the ovary wall, some individuals lacked ovules. These individuals were males, which by definition produce only pollen and cannot reproduce by seed. When we watched the fate of flowers on male plants, we found that they all aborted. While dissecting and measuring the flowers of many cardón plants at Tortilla Flats, I also discovered the fourth sex class, the neuters. These rare individuals were pollenless, like females, and their ovaries lacked ovules, like males. Though they produce lots of flowers each year and undoubtedly live a long time, these individuals never reproduce. They are the eunuchs of the cactus world!

So the answer to the question, Why did our first cardón pollinator exclusion experiment fail? turned out to be: because many of our experimental plants were males and, in one case, even a neuter. By dumb bad luck (but not really, as I explain below), we had chosen the wrong plants for our 1989 experiment. When we repeated the exclusion experiment in 1990, this time carefully choosing our subjects from among females, which represented about 37 percent of the population, and hermaphrodites, another 37 percent of the population, we found that bats and birds accounted for nearly equal proportions of fruit set in cardón. Because bats have first crack at open flowers, however, they probably account for most fruit set in this species at Kino Bay. At last we had results that would please Merlin.

By the end of our 1990 field season, we had a detailed picture of cardón's odd breeding season. Most of this work involved documenting differences in the flowering and fruiting biology of the four sex classes. These classes turned out to differ in a host of reproductive features, including flower size, sugar concentration of their nectar, and seasonal flower and fruit production. Compared with hermaphrodites, which is undoubtedly the ancestral sex class in cardón, females bear smaller flowers that produce richer nectar, and they produce 60 percent more fruits and seeds per season. Males produce 60 percent more flowers, per night and per season, than hermaphrodites. Differences among the sex classes in nightly flower production was the major reason we inadvertently chose the wrong individuals as our experimental subjects in 1989. We were naturally attracted to plants bearing lots of about-to-open flowers when we set up our experimental treatments each afternoon. Unfortunately, many of these plants turned out to be males. Finally, unlike saguaro and organ pipe, flowers of hermaphrodites of cardón are self-compatible, and many of their seeds are the products of self-fertilization.

This dispassionate listing of the differences in the floral biology of cardón's sex classes does not convey the excitement and fun I had posing questions and then answering them with fieldwork. Perhaps it was only because I was moving into a new area of biology, but the cardón work, especially during the 1990 field season, really *was* exciting for me. For one of the first times in my scientific career, I was working with a novel system whose theoretical predictions were clear-cut and could be easily tested. The big theoretical questions were: What are the relative fitnesses of males and females compared with hermaphrodites? What allows males and females to coexist with hermaphrodites? Is trioecy, which is an extremely rare breeding system in plants, an evolutionarily stable breeding system (that is, will it persist indefinitely through time), or is cardón in the process of evolving a dioecious breeding system from an ancestral hermaphroditic system?

To begin to answer these questions, all I had to do was go out to Tortilla Flats and closely observe and experiment with my plants. Each time I gathered data, I immediately analyzed it to see how the unisexual individuals (males and females) were doing compared with the bisexuals (hermaphrodites). One night, Cathy Sahley, one of my doctoral students who had just joined our field team, and I hauled an extension ladder around the desert to sample nectar production rates and nectar sugar concentration in flowers of thirty-three plants. This night was unseasonably warm, and a strong dry wind sucked a substantial fraction of our body water from us. But in the excitement of unveiling a new pattern, we hardly noticed the weather. By midnight, we knew that females produce richer nectar than the other sex classes, a study that I have since replicated under less stressful conditions. It took me over a day to rehydrate fully after our exertions on that hot, dry night.

By the end of the 1990 field season, we knew that cardón has a trioecious breeding system and that males and females have higher relative fitnesses, in terms of pollen and seed production, respectively, than hermaphrodites. We also knew that hermaphrodites are self-compatible, and that they can set fruit by self-fertilization, even in the absence of any pollinator visits. Though we had learned a lot about the odd breeding system of this giant cactus, we still didn't know whether cardón is in evolutionary transition toward dioecy. Furthermore, we didn't know whether cardón is trioecious throughout its geographic range, which includes the coastal region of Sonora from Guaymas, about 130 kilometers south of Bahía Kino, to Puerto Libertad, 100 kilometers north of Kino, and most of Baja California. My cardón research in later years would focus on these questions.

Although we spent a lot of time studying cactus reproductive biology in 1989–90, we didn't neglect the chiropteran side of our research project. As I've indicated, our earliest work with *Leptonycteris* involved photographing captive bats visiting cactus flowers. During the course of this and other work, I became totally enamored by this bat and its amazing lifestyle, which differs strongly from virtually all of the phyllostomid bats I had studied in Panama and Costa Rica. Two things about its biology immediately set Lepto apart from most tropical phyllostomids—its gregarious roosting habits and its migratory behavior. Unlike most phyllostomids, which tend to roost in small colonies, Lepto often roosts in colonies containing tens of thousands of individuals, as recent research has indicated (to be discussed in chapter 10). This aspect of its behavior was not totally clear in 1989, and large colonies of this species were virtually unknown then. When Merlin discovered a couple of thousand Leptos in the Sierra Kino cave in April of that year, we thought this might be the largest known colony left of this endangered species. Fortunately, we were way off the mark on this conclusion.

In addition to being highly gregarious, *Leptonycteris* is much more strongly migratory than other phyllostomids. Most tropical phyllostomids tend to be very sedentary. Only a few species, including females of the short-tailed fruit bat in western Costa Rica, are known to undergo seasonal habitat shifts that sometimes involve altitudinal migration. But long-distance latitudinal migrations appear to be restricted to few species, including both species of *Leptonycteris* and the Mexican long-tongued bat *(Choeronycteris mexicana)*—species whose geographic distributions are among the most northern in this family of bats.

An additional difference between *Leptonycteris* and tropical phyllostomids became obvious when we first netted them for photographic purposes. Lepto is a calm, gentle bat, unlike its much smaller relative, the common long-tongued bat, which is a high-strung, snarly bat that is quick to bite when you take it out of a mist net. When caught in a net, most Lepto individuals lie quietly waiting to be removed. Whereas I never handle newly captured tropical phyllostomids without gloves, we always handle freshly caught Leptos (as well as *Macrotus californicus,* another gentle desert-dwelling phyllostomid) bare-handed. Leptos seldom attempt to bite, and if they do bite, their small teeth rarely break the skin.

Lepto's appearance matches its lovely disposition. Tawny brown in color, it has large, intelligent-looking eyes, foxlike ears, a slightly elongated snout, and a small nose leaf. It is virtually tailless and has a much-reduced tail membrane. The second largest member of the glossophagine clade of phyllostomid bats, nonpregnant females weigh about twenty-three grams and males weigh about twenty-six grams (compared with eleven grams for the common long-tongued bat). More impressive than its size, however, is Lepto's obvious strength. For its size, it is quite a muscular bat, for reasons that will soon become obvious. Lepto in the hand feels like a well-conditioned Olympic athlete. On the wing, it flies extremely fast, and its long, narrow wings slice through the air with a kite-rippling sound. A final notable characteristic is its slightly sweetish body odor, undoubtedly the product of its sugary-sweet diet. On numerous occasions I have been able to detect its distinctive smell as a Lepto barreled past me overhead. Downwind, *Leptonycteris* roosts emit an exaggerated version of this odor; their roosts are distinctive because of their fruity bouquet.

Our major hands-on work with *Leptonycteris* in 1989–90 was a study of its foraging behavior using radiotelemetry. We didn't begin this project until early June 1989, by which time few Leptos were living in the Sierra Kino cave. We had been censusing the number of Leptos leaving this cave at sunset every two weeks and watched their numbers decline until there were only about fifty bats in the cave in late May. We had no idea where the two thousand or so bats that had been there a month earlier had gone but assumed they had migrated to some other location just before giving birth to their babies. Instead of netting at the Kino cave to capture our first radio-tracking subjects, we captured and tagged three of them as they flew into the Cerro Catedral night roost. Since two of the three individuals were still lactating, we figured that wherever they were roosting by day, they couldn't be too far away. We outfitted each female with a 0.9-gram transmitter and released her into the dark night to resume foraging. By 2:30 A.M. all three bats had disappeared from "sight," heading west toward the coast. We assumed they were heading back to the Kino cave.

The following evening, I climbed up to the Sierra Kino cave to do an exit count and to monitor the exit of our radio-tagged bats with the help of Janet Debelak, a recent graduate from Trinity University in San Antonio, who would end up working for Bat Conservation International later in 1989. To our surprise, none of the three bats was in the cave. They had apparently roosted elsewhere that day. Peg and her fiancé, Ted May, who had joined our field team in mid-May, were able to detect each of the bats foraging north and west of Cerro Catedral later in the evening, and they again flew west when they were done foraging. Over the next couple of days, the four of us began to explore on foot and by truck the countryside north and east of Sierra Kino, looking for another cave that housed our tagged bats by day. The basaltic hills along the coast contained dozens of small caves, but we received no radio signals from any of them. We feared that we were looking for three very small needles in a very large haystack.

On 7 June, we finally solved the mystery of the alternate day roost. That evening Ted was stationed on a hill near the coast, Peg was on the Chueca road farther up the coast, and Janet and I were on a hill west of Cerro Catedral. Ted first detected the bats and relayed the news to the rest of us by walkie-talkie. *"Bats 5 and 8 are coming from Isla Tiburón!"* he shouted excitedly. These foxy ladies were living by day in a maternity roost on Tiburón and were commuting over twenty-four kilometers one way to feed on the mainland. We were astounded by this discovery. Surely these twenty-three-gram bats didn't have to fly this far just to find a few cactus flowers to feed at, did they? We eventually radio-tracked nine bats in 1989 and fourteen in 1990. Nearly all of these individuals commuted to the mainland from one or two roosts on Tiburón, and they routinely flew a total of about a hundred kilometers each night. Our first two tagged bats were no flukes.

By radio-tracking on hills along the coast and a bit farther inland, we were able to construct in detail the flight paths of our Tiburón-roosting bats. From a distance of about nineteen kilometers away, which Fred Anderka, the manufacturer of our transmitters, tells me is a detection-distance record, we could follow the bats as they flew north along the eastern coast of Tiburón and then across to the mainland near Punta Santa Rosa, the point closest to Tiburón. The bats then flew south through a series of valleys that led them to their foraging areas around Cerro Catedral and Tortilla Flats. They arrived in their feeding areas around 10 P.M. and left for the coast, this time flying in a more direct beeline toward Tiburón over open ocean, between 2 and 3 A.M.

In 1990, we supplemented our tracking data with a light-tagging study. Small glass balls filled with cyalume, the same chemical that makes fireflies light up, were first glued to the backs of bats in 1975 by Ed Buchler, who was studying the foraging behavior of the little brown bat in upstate New York. In 1976 Ray Heithaus and I had tried to study the foraging behavior of light-tagged *Glossophaga soricina* as they visited flowers of calabash trees in a savanna at Santa Rosa. We had absolutely no success with this technique, however, because our tagged bats quickly disappeared into the nearby forest. Our light-tagging success rate in the desert, by contrast, was nearly 100 percent. Not only could we visually follow bats for up to nearly two hours at a time (the cyalume loses its brightness in about four hours), but our maximum detection distance turned out to be about five kilometers. The desert obviously was an ideal habitat in which to light-tag bats.

On two nights in mid-April, I sat on the highest point of Cerro Catedral until 2 A.M., anxious to watch illuminated bats foraging for the first time. Before settling down to work each night, however, I was able to enjoy the sunset. Like sunrise, sunset is another wonderful time to be in the desert. By 5 P.M. in April, the low afternoon sun bathes everything in a soft golden glow. This is when the columnar cacti are most beautiful, their branches and spines backlighted with a golden haze. Shortly before sunset, the desert's first bats—western pipistrelles—take flight. These tiny, butterfly-like bats are North America's smallest species. Early in the evening they fly erratically a few meters above the desert floor, chasing tiny beetles. A few minutes after sunset, North America's largest bat, the western mastiff, emerges from its day roosts, probably narrow cracks in rocky hills, to feed. These large, fast-flying molossids weigh about sixty grams and fly a hundred meters or higher while hunting for large beetles above the desert floor. They issue harsh, widely spaced audible calls as they rocket along. A few high-pitched *ping-ping-pings* and they are out of earshot. By the time the sun has fully set and the night sky twinkles with a billion stars, other bats, including the big brown bat and the Mexican free-tailed bat, are out hunting insects around and above columnar cacti.

We light-tagged six *Leptonycteris* each night and were able to watch their movements in considerable detail. By luck, four of the six bats foraged around the base of Cerro Catedral on the first night, and I was able to see and record much of their behavior. In my field notes, I wrote that "Lepto flight appears to be 'controlled' rather than 'erratic' with lots of 'systematic' and thorough coverage of small areas [a few hectares]. They appear to fly among several flowering plants, swooping up and tightly circling them, rarely hanging up to rest." After thoroughly covering one group of cacti, they would move on to another nearby group, eventually covering an area of about a square kilometer in a night. On a couple occasions our light-tagged bats flew high above the desert for up to two kilometers between two feeding areas. Once they began feeding, however, they flew within about ten meters of the desert floor around and between plants.

I thus spent two chilly nights closely watching our mammalian "fireflies" through my binoculars as they flew back and forth in tight loops through their feeding areas. From our radio- tracking studies, we already had a pretty good idea how large an area they used in their nightly feeding. This area ranged from about one-half to two and one-half square kilometers a night. But until we attached glowing glass balls to their backs, our only images of foraging bats consisted of fleeting glimpses of their visits to cactus flowers during our stakeouts and the *beep-beep-beeps* of radio signals, whose direction and intensity we recorded on our data sheets. Having their ephemeral yellow paths temporarily etched onto our retinas connected us much more strongly to their world. Once we had seen their paths illuminated by cyalume, it was much easier to picture them flying from one plant to another, something that we could easily do with diurnal pollinators.

By the end of the 1990 field season, we had a detailed picture of the foraging behavior of the lesser long-nosed bat. In addition to its long commute flights, two of its other behavioral characteristics struck us as being quite unusual. The first was the large amount of flying that it did early in the evening before it actually began to feed. Like many other bats, individuals of *Leptonycteris* leave their day roosts just after sunset, around 8 P.M. in May. But unlike virtually all other early-flying bats, they do not begin to feed regularly for another four hours. Then, after about two hours of feeding, they wing their way back to Tiburón. We wondered why Lepto didn't wait until closer to midnight to leave its roosts, thereby saving energy and reducing its exposure to nocturnal predators. We eventually rationalized this behavior by postulating that these bats need time to look for good feeding sites within their relatively large feeding areas each night. They do their flower searching early in the evening. We also noted that, unlike tropical phyllostomids such as the short-tailed and Jamaican fruit-eating bats (as well as the common vampire), which often retreat to their day roosts whenever a bright moon is above the horizon, *Leptonycteris* is not lunar-phobic. That is, its foraging activity is not suppressed by bright moonlight. Our radio-tagged bats' arrival and departure times to and from their feeding areas did not differ from one part of the lunar month to another. This suggested to us that Lepto's risk of predation must be quite low in the desert.

Even if it needs to spend some time searching for promising feeding sites each night, why does Lepto wait until nearly midnight to begin feeding? Flowers of cardón and organ pipe secrete nectar as they open, which is right at sunset (those of saguaro open considerably later and stay open much longer the next day). Why doesn't Lepto start feeding earlier in the evening if both nectar and pollen are available well before midnight? Our explanation for this behavior relies on optimal foraging theory, the theory that originally served as the conceptual framework for our studies of *Carollia perspicillata* at Santa Rosa in the mid-1970s. This theory states that animals should choose to eat only those food items that give them a net energetic profit, after accounting for the costs of searching for and handling the food. By waiting until midnight to begin feeding, Lepto was behaving as though cactus flowers had low profitability the first four hours they were open. The bats seemed to be waiting, albeit by burning fuel on the wing rather than by hanging up quietly in a roost, until flowers had accumulated enough nectar to make repeated visits profitable. In the case of cardón, flowers apparently need to accumulate about 0.8 milliliters of nectar—about eight flower visits worth according to a study we did with captive bats—before their flowers are profitable enough to visit. This "threshold" amount of nectar was reached in most cardón plants between 11 P.M. and midnight each night, the time when bats began to feed.

The existence of a threshold volume of nectar per flower suggested to me that Lepto's feeding schedule is controlled by the rates at which cactus flowers secrete nectar. The earlier cactus flowers fill up with nectar, the earlier *Leptonycteris* bats will begin to feed. Alternatively (and less plausibly), Leptos may be late-night feeders because their guts simply cannot process carbohydrate-rich food before midnight. Perhaps they need time to "warm up" before they begin to feed, in the same way that some people cannot eat breakfast until they have been active for a while.

An easy way to discriminate between these two hypotheses is to inject a full night's quota of nectar into freshly opened flowers early in the evening and see whether bats change their feeding schedule at these experimental flowers compared with their schedule at control flowers. If the hypothesis involving nectar secretion rate is correct, then bats should begin feeding earlier at experimental flowers than at control flowers. If it is wrong, then their feeding schedule will not change. So far, I have been able to conduct this experiment only once, so I do not yet have conclusive support for the secretion rate hypothesis. But the results are suggestive. In 1994, a young Mexican biologist, Jorge Vargas, and I staked out a cardón plant in the flatlands below the Sierra Kino cave for two nights. On the first night we documented bat visitation patterns to seven control flowers on one branch. As in the past, these flowers were not visited until 11 P.M. Between then and 2 A.M., they were visited an average of 4.4 times apiece. At 7:30 the next evening, we added 1.5 milliliters of artificial nectar to four newly opened flowers on the same branch. One or more bats began feeding at those flowers at 8 P.M., and our experimental flowers received an average of 14.4 visits apiece during the night. These preliminary results provide strong support for the nectar secretion hypothesis, and I will be surprised if additional replicates do not continue to support this hypothesis. I believe that nectar secretion rates control the flower visitation pattern of *Leptonycteris.*

The other unusual feature of Lepto's foraging behavior is the large size of its foraging areas in relation to its energetic needs, which we have estimated as being about forty-two kilojoules (or about ten kilocalories) per day. To obtain this much energy, Lepto needs to make from 50 to 133 flower visits, depending on which cactus species it is feeding at on a particular night. Forty-two kilojoules is not much energy when you place it in the context of total nectar production per hectare at Tortilla Flats. Early in the cactus blooming season, when cardón is in full flower, nightly nectar production is about 1,100 kilojoules per hectare. Later in the season, when organ pipe is the major flower source, nightly nectar production drops to about 350 kilojoules per hectare. These values indicate that a (nonreproductive) *Leptonycteris* bat can obtain all the energy she needs in an area no greater than a fraction of a hectare throughout the cactus blooming season. Even with several bats foraging in the same general area, she could still stick to less than half a hectare and find enough nectar (and pollen) to meet her needs.

Our radio-tracking and light-tagging data clearly indicated that lesser long-nosed bats didn't pay much attention to our calculations. Instead of confining their foraging activity to a small area, they foraged over areas ranging from 25 to 250 hectares. When Peg Horner took a crew of assistants out to estimate the number of flowers in the foraging areas of several of our radio-tagged bats in 1990, she found that these areas contained from 4,600 to 15,900 open flowers—enough energy to support potentially up to two thousand bats for one night! Our overall conclusion from these calculations was that *Leptonycteris* bats certainly did not appear to be energy-limited in the Kino Bay area and that individuals forage over much larger areas than necessary simply to meet their daily energetic needs.

We do not yet know why Leptos forage over such large areas each night, but we suspect it has to do with this bat's flight style. Unlike most phyllostomids (and more like pteropodid bats), *Leptonycteris* bats are built for fast, long-distance flight. They have relatively long, narrow wings, and their wings bear more weight per unit area than the wings of most other phyllostomids. These two design features result in fast, low-cost flight. With such a design, it costs a Lepto only about seven or eight flower visits to recoup the energy it spends flying over twenty-four kilometers from Isla Tiburón to Tortilla Flats. Given such a cheap flight cost, it hardly matters whether Lepto flies fifty or five hundred meters to the next plant to visit a flower. In fact, by flying fast between plants it probably reduces its vulnerability to predators such as great horned owls. Whatever the real reason behind the disparity between the sizes of its actual and theoretical foraging areas, Lepto is an exquisite flying machine, one that seems to be ideally suited to foraging in the wide open spaces of the Sonoran Desert.

After working with *Leptonycteris* bats and columnar cacti for two field seasons in the Sonoran Desert, I was hooked on this habitat and its flora and fauna. By the end of our 1990 field season, I had no desire to abandon the desert work and return to studying tropical frugivorous bats. This new study system turned out to be much more scientifically exciting than I (or Merlin Tuttle) ever envisioned. We had no idea that we would discover a rare breeding system, trioecy, in a giant cactus. In the years to come, our cacti had even more surprises in store for us, including the discovery of a highly specialized moth-cactus pollination system. Nor had we anticipated that *Leptonycteris* would be such a long-distance forager. Even the pollinator exclusion experiments produced unexpected (and for Merlin, initially unwanted) results. Good scientist that he is, Merlin has long since become reconciled to whatever results our observations and experiments produce. His involvement in our collaborative research ended in 1990, when he refocused his enormous energies and talents entirely on bat conservation. But at least once a year, he reminds me about how proud he is that he introduced me to the study system involving *Leptonycteris* bats and columnar cacti. And just as regularly, I thank him for calling me that afternoon in August 1988.

10 Along the Nectar Trail

Marcia came to visit me in the middle of our first field season at Kino Bay. Like me, she had never been to the Sonoran Desert before and wanted to see the new plants and bats I was studying. She froze with me in the desert at night while we slept between rounds of nectar sucking. One sunny morning we climbed past the Sierra Kino cave to the top of this 450-meter-tall basaltic massif. From its top, swaying in a stiff wind, we could see nearly forever—to the sparkling blue waters of the Gulf of California to the south, to the hazy mountain peaks of Isla Tiburón in the west, and to Tortilla Flats in the cactus-filled lowlands in the east. From this aerie we could see almost the entire daily home range of a typical lesser long-nosed bat—a vast area spanning kilometers of open ocean as well as many square kilometers of desert scrubland. Back at sea level, we swam in the still-cool waters of the gulf and walked barefoot along the water's edge of the beautiful New Kino beach.

On one of these walks I described to her my vision of what it was going to take to understand the population biology of this bat. I began this

Baja California scene with cacti and boojum trees. Drawing by Ted Fleming.

soliloquy by reminiscing about the "good ole days" when I was studying the short-tailed fruit bat in western Costa Rica. "I could literally walk to three *Carollia* roosts in about two hours as well as traverse a typical bat's feeding area in less than an hour. Many males lived within a few kilometers of their birthplace their entire lives," I exclaimed in amazement. Not so with the lesser long-nosed bat. Our radio-tracking data would reveal that these bats typically fly over a hundred kilometers each night and have huge feeding areas. We didn't yet know how far female bats migrate to get to their Sonoran Desert maternity roosts each spring, but this distance had to be orders of magnitude greater than the distances female *Carollia* migrate when they leave Santa Rosa for the uplands in the dry season.

"To truly understand Lepto's ecology and conservation needs," I continued, "we'll have to travel extensively in Mexico to find its winter roosts. Then we'll have to figure out its migration routes and the plants that it uses to fuel its migration. Is it an obligate cactus and agave feeder throughout its geographic range, or does it have a broader diet away from the Sonoran Desert? " Knowing the potential distances involved and the rugged terrain that lay between Kino Bay and south-central Mexico, I knew that answering these questions would be a daunting task. But then I had an even more audacious thought. I recalled that the lesser long-nosed bat had a southern subspecies that lived in arid parts of northern Venezuela and eastern Colombia as well as on the Netherlands Antilles islands of Aruba, Bonaire, and Curaçao. "Wouldn't it be neat," I asked, "if we could do parallel studies in Venezuela to determine Lepto's role as a pollinator of tropical cacti? I wonder if it's also a migratory bat down there?"

As I rambled on, I'm sure Marcia must have thought I was crazy—that I had probably been out in the desert sun too long. However, she kept her thoughts to herself and just listened to my ideas that day. But as this grand scheme began to unfold over the next few years, she cheerfully joined me for long forays into southern Mexico and Baja California. As our desert research became more widely known, Mexican colleagues began to participate in these studies. My graduate students also became infected with my enthusiasm for the desert and joined our growing research team. By the tenth anniversary of our first field season in Kino, our studies had covered much of Mexico and the drylands of Venezuela and Curaçao. Much as I had hoped in 1989, we were well on our way to understanding what makes this nectar-feeding bat tick and its conservation needs. We had also gathered enough information to convince most bat biologists that this species was not truly endangered in North America. Its addition to the U.S. Endangered Species list in 1988 had been premature.

One of the first large-scale questions that we tackled was determining how dependent Lepto is on cactus and agave food resources. Vertebrate ecologists have traditionally used a variety of techniques to study the food habits of animals. Under the right circumstances, direct feeding observations will give you this information. This method works best with large, diurnal creatures such as lions and zebras in open habitats such as the African Serengeti. Unfortunately, those of us studying small, secretive critters have to be a bit more invasive into their lives to learn their food habits. The ultimate degree of invasiveness, of course, is to capture and kill animals and examine their stomach contents. I resorted to this method way back in Panama but haven't killed bats for diet information for decades. Instead, we routinely take pollen swabs from a bat's fur or take fecal samples from live bats for this information. But studying a species' diet this way—especially over broad geographic areas and at different times of the year—is labor intensive and expensive. To get a general overview of the diet of the lesser long-nosed bat throughout Mexico, I proposed that we use a new technique that is elegant, efficient, and powerful—carbon stable isotope analysis.

Anthropologists and ecologists have been using carbon stable isotope analysis to study the food habits of ancient human populations as well as those of contemporary animals in a wide variety of habitats since the late 1970s. This technique takes advantage of the fact that tissues of different kinds of plants contain different ratios of two stable (nonradioactive) isotopes of carbon, ^{13}C and ^{12}C. The former (heavier) isotope is much rarer in the atmosphere than the lighter one. Most flowering plants use the Calvin or C3 photosynthetic pathway, which strongly discriminates against atmospheric ^{13}C during photosynthesis. Their tissues thus contain a low ratio of ^{13}C:^{12}C. Any animal eating C3 plant tissue, or eating animals that have eaten C3 plants, will also have a low ^{13}C:^{12}C ratio. A minority of flowering plants, including many tropical grasses such as sugar cane and corn as well as succulent plants such as cacti and agaves, have different photosynthetic pathways (C4, or Hatch-Slack, in the case of grasses; CAM, or crassulacean acid metabolism, in the case of cacti and agaves). These plants do not discriminate as strongly against ^{13}C during photosynthesis, and hence their tissues contain much higher ratios of ^{13}C:^{12}C than C3 plants. Animals (including Paleo-Indians eating corn) whose diets include C4 or CAM plants will also have higher ratios of ^{13}C:^{12}C in their tissues than if they had eaten

C3 plants. Therefore, the carbon isotope ratio in animal tissue can be used to determine the relative importance of C3 and CAM or C4 plant material in its diet.

As soon as I began my desert research, I realized that I could use carbon stable isotopes to determine the importance of cactus and agave food resources in Lepto's diet. I owe this insight to my University of Miami colleague Leo Sternberg, a plant physiologist who specializes in stable isotope research. Leo's lab is right around the corner from mine, and we had been discussing possible uses of stable isotopes in bat research for several years before I began my desert work.

The beauty of working with carbon stable isotopes is that you don't need much tissue to conduct your analyses. A single bat toe, for instance, contains enough muscle tissue (and carbon atoms) to obtain a ^{13}C:^{12}C ratio when you run purified carbon dioxide, obtained from your sample by high-temperature combustion and vacuum distillation, through a mass spectrometer. This whole laboratory process looks terribly "scientific," because it involves bubbling cauldrons of liquid nitrogen, a fancy vacuum line with lots of intricate glass tubing, and, for the final analysis, a large complicated machine (the mass spectrometer) that counts atoms that differ slightly in their mass. But the technique is pretty easy to learn. I must admit that I have never actually done stable isotope analyses myself. I leave them to my lab assistants and grad students, who are usually keen to learn new techniques.

So when I first went to the Sonoran desert, one of my goals was to collect small snippets of muscle tissue—a *Leptonycteris* bat toe here, a tiny piece of breast muscle from cactus-visiting birds or whole honeybees there—for carbon stable isotope analysis. It may sound cruel to snip off a bat toe, but the alternative—killing the animal for tissue samples—is far worse. Bats can easily operate with nine toes, and the wound heals quickly, as I learned when working with short-tailed fruit bats in Costa Rica. We removed one toe from many individuals of *Carollia* for genetic studies and found no difference in the survival rates of clipped vs. unclipped individuals.

Our first stable isotope study addressed the question, When and where in Mexico does *Leptonycteris* specialize on CAM plants? For this study, we used only twenty-five toes taken from Kino Bay bats. The bulk of our 135 tissue samples came from long-dead museum specimens collected at various times and places throughout Mexico. We were able to use museum specimens, because carbon isotope ratios do not change as the tissue in these specimens dries out (you are what you eat—in death as in life). Several different museums in the United States and Mexico donated bat toes for this study. I even received a series of toes, neatly packed in cotton in individual glass vials, from a Polish colleague, Voytek Woloszyn, who had collected *Leptonycteris* bats in Baja California in the early 1980s. I wonder what U.S. Customs officials must have thought when they received a package labeled "bat toes" from Poland. My oldest toes by far were from bats collected by Edward Goldman and Edward Nelson in Michoacán and Jalisco in 1892 and 1897. I'm sure it would have astounded those two field biologists to learn that someone was extracting carbon atoms from toes of bats a hundred years after they had been collected. This makes me wonder what biologists will be extracting from the Nelson-Goldman specimens a hundred years from now.

It was interesting to hear the reactions of different museum curators when I called them to request permission to snip a single toe off some of their specimens. My request at first stymied Don Wilson at the U.S. National Museum. In 1990 this museum did not have a policy regarding the (partial) mutilation of their valuable specimens. I gather the museum curators must have had a quick powwow, because it wasn't long before Don gave me the green light on my request. Dave Schmidly at Texas A & M apparently didn't blink an eye in giving me permission on my first call. Bob Timm at the University of Kansas burst into laughter when I made my pitch. "You want to do *what* with bat toes?" he giggled. By the time he had heard my whole story, though, his tune had changed. About stable isotope analysis he exclaimed, "Geez, that's a really neat technique. I've never heard of such a study before." This turned out to be a common reaction from bat biologists that I talked to in 1990.

Results of our study indicated that the presence of CAM carbon in *Leptonycteris* does indeed vary in space and time. During the fall and winter, Leptos from south-central Mexico contain a mixture of C3 and CAM carbon. At this time, they visit flowers of tropical trees and shrubs—species of *Ceiba, Pseudobombax,* and *Ipomoea,* among others, according to published pollen records—as well as cacti and/or agaves. As the bats migrate north in the spring, however, they leave tropical C3 plants behind and concentrate entirely on arid zone CAM plants. This means that they probably are migrating north along a "nectar corridor" composed of spring-blooming columnar cacti, most of which are concentrated along the coastal lowlands of western Mexico. By the time they arrive in the Sonoran Desert, they feed exclusively on CAM plants and continue this dietary specialization throughout the spring, summer, and early fall. When they migrate south, beginning in late August, no cactus flowers or fruit are available in most of western Mexico. But *Agave* plants are in full bloom in a wide latitudinal swath along the western flank of the Sierra Madre from Arizona south to at least Jalisco and Michoacán states. This fall nectar corridor of agaves, we hypothesized, is the fuel source for Lepto's southward migration.

Lesser long-nosed bats living on Baja California, in contrast, presented us with a different pattern. In most months for which we had samples, bats were CAM in composition, indicating that Lepto is a year-round cactus and agave specialist across the Gulf of California from the Mexican mainland. A few years later, one of my grad students, Jafet Nassar, collected Lepto tissue samples throughout the arid regions of northern Venezuela and found the same pattern of year-round CAM specialization. Finally, when we analyzed samples from an island roost off the coast of Jalisco in west-central Mexico, we found a third pattern. These bats, which live in tropical dry forest where cactus and agave densities are low, were feeding mostly on tropical C_3 plants year-round.

So the answer to the question, How dependent is *Leptonycteris* on cactus and agave resources? turned out to be a bit complicated. These bats are CAM specialists in areas where cactus and agave flowers are reliably available, notably in very arid areas where succulent plants are common. In less arid forested regions, their diet is broader and includes a mix of C_3 and CAM plants. From an ecological and conservation perspective, our most interesting and important discovery was the tentative identification of the plants that fuel Lepto's long-distance migrations in Mexico. These plants included lowland coastal columnar cacti in the spring and upland paniculate agaves in the fall.

Our carbon stable isotope research thus provided us with a "big picture" of the diet and seasonal movements of the lesser long-nosed bat in western Mexico. But it did more than this, because shortly after we began studying *Leptonycteris* at Kino Bay, Merlin Tuttle and I discovered that a second species of bat was visiting cactus flowers. Whenever we netted for Leptos at the Sierra Kino cave or the Madrugada night roost, we also caught a few pallid bats *(Antrozous pallidus)*. This bright-eyed eighteen-gram vespertilionid normally preys on large arthropods, including scorpions, which it captures on the desert floor. But some of our individuals—on some nights most individuals—came to night roosts with their faces completely covered with pollen. When I examined this pollen under a microscope, it turned out to be from cardón cacti as well as from *Agave subsimplex*, the only agave in the Kino Bay area. This was exactly the same kind of pollen that we found on the head and shoulders of *Leptonycteris* in April and May. What was going on here? I wondered. Was *Antrozous* a wannabe nectar-feeding bat?

To determine the extent to which *Antrozous* visits cactus and agave flowers, we again turned to carbon stable isotope analysis. In collaboration with my graduate student Gerardo Herrera and an old friend, Jim Findley, then curator of mammals at the University of New Mexico, I conducted another geographic survey of the carbon composition of toe muscle based on a few Kino Bay pallid bats and about a hundred museum specimens. We predicted that if *Antrozous* is a regular visitor to CAM plants, then its carbon composition should be skewed toward CAM carbon during the blooming seasons of bat-adapted cacti and agaves in the Sonoran and Chihuahuan deserts. Outside the geographic ranges of these plants, the pallid bat should be composed of C_3 carbon.

Our data provided only partial support for our predictions. At certain locations—Kino Bay and central Baja California, for example—pallid bats indeed contained substantial amounts of CAM carbon, much more than either the California leaf-nosed bat or the big brown bat, two insectivorous bats from these sites, which we used as "controls" in our study. But at a number of other localities within the Sonoran Desert, pallid bats contained only C_3 carbon. These results suggested that visiting cactus or agave flowers may be a learned behavior in this species. It appears that individuals in some populations have learned that food, probably small beetles rather than nectar, can be found in cactus and agave flowers and is easier to capture than scarce, ground-dwelling arthropods. Given *Antrozous*'s large ears, short snout, and short tongue, it is highly unlikely that it visits cactus flowers to eat nectar. It doesn't yet have the nectar-gathering equipment of a *Leptonycteris.* Nonetheless, it certainly has the potential for depositing pollen on cactus stigmas during flower visits, and hence, like *Leptonycteris,* it is a legitimate pollinator. This intelligent bat appears to be giving us a glimpse of how flower visiting likely evolved in phyllostomid bats, whose primitive species are still highly insectivorous. And in *Antrozous,* we apparently are seeing the early evolution of flower visiting in another bat family, the Vespertilionidae.

Now that we had tentatively identified Lepto's Mexican migration routes, we needed to determine where the Sonoran Desert bats were coming from. Were they spending their winters just south of Sonora in Sinaloa or Nayarit, as some Mexican biologists believed? Or were they traveling farther south to places such as Jalisco or Michoacán, over a thousand kilometers away from their spring maternity roosts? If they were truly long-distance migrators, what hazards did they face during their annual north-south forays?

Prior to our *Leptonycteris* research, the "traditional" method that biologists used to study bat migratory behavior, beginning in the 1930s, involved placing numbered aluminum bands on a large number of individuals in one roost and looking for them in other roosts. In the 1950s and 1960s, for example, Lennie Cockrum and his students at the University of Arizona along with Bernardo Villa Ramirez, the dean of modern Mexican mammalogists, banded thousands of Mexican free-tailed bats in the United States and Mexico in an attempt to trace their migratory pathways between the two countries. Only a handful of these bats were ever recovered. The longest distance between a banding site and a recovery site was about thirteen hundred kilometers, but most recoveries occurred close to banding sites. In the mid-1950s, another group of bat workers banded about fourteen thousand free-tails at Bracken Cave near San Antonio, Texas. Only seven (fewer than 0.1 percent) were recovered away from that site. These studies indicated that banding operations can yield very low results for the time and effort involved.

Over beers at the annual meeting of North American bat researchers in 1989, Jerry Wilkinson and I began discussing the use of mitochondrial DNA (mtDNA), rather than traditional banding techniques, to study the migratory behavior of the lesser long-nosed bat. Jerry and I first met in 1980 in Costa Rica when he was conducting his doctoral research on the social organization of the common vampire bat. Although he did most of his fieldwork at La Pacifica, he also studied the vampire colony living in the Sendero cave at Santa Rosa. Now at the University of Maryland, Jerry had just completed a study of genetic relatedness within colonies of evening bats in Missouri using mtDNA and was very familiar with new techniques that were being developed for DNA analysis. He seemed like the ideal person to collaborate with in a study of the migratory behavior of *Leptonycteris.*

Our proposed study was conceptually straightforward. We would use genetic markers, specifically gene sequences from mtDNA, to look for genetic connections between the maternity roosts in Sonora and southwestern Arizona and the winter roosts located south of Sonora. We chose to use mtDNA as our genetic marker for several reasons. First, unlike nuclear DNA, which undergoes genetic recombination during sexual reproduction, mtDNA does not undergo recombination and is transmitted from generation to generation in most animals through maternal inheritance. That is, we inherit half of our nuclear genes from each of our parents, but all of our mitochondrial genes from our mothers. Second, in the absence of recombination, any mutations that occur within mtDNA will be faithfully passed on from one generation to the next as a new *haplotype.* Owing to the absence of recombination, each haplotype can be viewed as an allele of one giant gene (the entire mtDNA molecule). As a result of maternal inheritance, each haplotype also represents a different maternal lineage (a matriline) within a species. Third, the mutation rates of mtDNA are much higher than those of nuclear DNA. The so-called control region of this circular molecule, where replication and transcription of mtDNA are initiated, is especially prone to mutate and hence helps create new haplotypes that can be used for fine-scale analysis of population genetic structure. Mitochondrial DNA analysis has played a prominent role in determining the migrational history of our own species as we spread "out of Africa" over one hundred thousand years ago (Sykes 2001).

The number of different mtDNA haplotypes (or different matrilines) and their geographic distributions thus provide us with considerable insight into genetic and behavioral processes that are occurring within a species. A low number of haplotypes, for example, suggests either that a species has recently undergone a severe reduction in population size (a "genetic bottleneck"), as might be expected in a truly endangered species, or that its breeding population size has been chronically small, perhaps because of a highly polygynous mating system, in which a few males account for most matings. Conversely, a large number of haplotypes suggests just the opposite—that population size historically has been large and/or that the mating system is not strongly polygynous (i.e., most males are successful fathers). And finally, nonrandom geographic distributions of haplotypes provide strong evidence for direct genetic connections between spatially separated populations.

During 1992 and 1993 my research plan was to travel through as much of the geographic range of *Leptonycteris curasoae* as possible. Funded by a new grant from the National Geographic Society, this travel would eventually take me through much of western Mexico as far south as Acapulco as well as to Baja California and northern Venezuela, the home of the southern subspecies of the lesser long-nosed bat. During my Mexican travels, I planned to check on the population status of *Leptonycteris* to update Don Wilson's surveys of the 1980s. I would also collect tissue samples from Lepto to determine the genetic structure of its Mexican populations and to estimate how much genetic differentiation has occurred between the Mexican and Venezuelan subspecies.

A sabbatical leave during the 1992–93 academic year gave me time for my proposed travels. To help support my sabbatical, I was awarded a Mid-Career Fellowship by the National Science Foundation to spend a year in Tucson at the University of Arizona. This generous fellowship was designed to support senior scientists who wanted to learn new techniques and to pursue new lines of research. In addition to conducting my geographic surveys, I proposed to learn molecular techniques—specifically the extraction, amplification, and sequencing of mtDNA—that I could apply to my bat and cactus studies. The University of Arizona's Department of Ecology and Evolutionary Biology was just opening the new Laboratory of Molecular and Evolutionary Systematics (LMSE) and was eager to teach new molecular techniques to faculty and grad students. I was one of the first students in the department's Molecular Techniques workshop. When I wasn't in the field chasing bats and cacti around Mexico and Venezuela, I was working with DNA molecules in the LMSE.

My first field trip in 1992 lasted from mid-April to mid-May. This trip started with a cross-country drive, which I have now made many times, from Miami to Tucson and then down into Sonora in our brand-new Ford Explorer. Covering a distance of about thirty-eight hundred kilometers, the trip takes four days, not counting time out for a much-needed break in Tucson. This is a far cry from the two and a half hours it used to take me to fly to San José, Costa Rica, and the five-hour drive from San José to Santa Rosa National Park. We originally moved to Miami, in part, to be close to Costa Rica. Back in 1978, I had no idea that my research focus would eventually take a strong westward shift from Costa Rica to southern Arizona and northwestern Mexico.

Despite the distance, I always look forward to this road trip each spring. For one thing, it gives my travel partners and me a chance to see the great vegetational changes that occur across the southern tier of states along Interstate 10. Starting from the solid mass of humanity and concrete in Miami-Dade and Broward counties, the drive up the Florida peninsula takes us through marshy grasslands and sandy pine flatwoods that eventually give way to rolling hills covered with mixed pines and hardwoods as well as cypress swamps in northern Florida. Bayous and swamps filled with trees dripping with Spanish moss dominate the landscape across Mississippi and Louisiana. The highlight along this stretch of the road is the extensive Atchafalaya swamp, just west of Baton Rouge, over which we skim on a long concrete bridge. We usually reach the Texas border around midday of our second day on the road. Texas mile marker 880 tells us that New Mexico is seemingly an eternity away. By lunchtime the next day, however, we have passed through Texas, having negotiated the interstate spiderwebs of Houston and San Antonio, where it is invariably raining, and then the lovely Texas hill country with its miniforests of stunted oaks and junipers west of San Antonio. In moist years, this part of the interstate is bordered by a profusion of pink, yellow, and blue wildflowers. West Texas gradually becomes increasingly dry and desolate, and towns—Junction, Sonora, Ozona, Fort

Stockton, Van Horn, and finally, El Paso—are few and far between. This is Larry McMurtry country, and I recall scenes from *The Last Picture Show* as we speed past west Texas's dusty towns. This is also the land of cattle-denuded rangelands set among widely scattered mountains and sweeping vistas. In the early spring, traffic is sparse and consists mostly of older couples driving large U.S. cars or elephantine RVs toward home in Arizona or California. Finally, we are into the mountainous countryside of southern New Mexico and Arizona. Despite the aridity, the landscape is greener than in west Texas owing to the presence of irrigated agricultural fields. The rugged Chiricahua Mountains—former homelands of Cochise and Geronimo and their Apache clans—slip past south of us as we approach Tucson. It is now late afternoon on our third day on the road, and in the soft afternoon sunlight we pass through an amazing pile of orange granite rocks called Texas Canyon, elevation fifteen hundred meters. In her novel *The Bean Trees,* Barbara Kingsolver describes this place: "It was a kind of forest, except that in place of trees there were all these puffy-looking rocks shaped like roundish animals and roundish people. Rocks stacked on top of one another like piles of copulating potato bugs. Wherever the sun hit them, they turned pink. The whole scene looked too goofy to be real" (1988, 35). A mere hundred kilometers later, we are in Tucson.

In addition to passing through many different vegetation zones, my road trip passes through a variety of human cultures, which I also love to introduce to my traveling companions. In my trips in 1992 and 1994, for instance, I was accompanied by Mexican students who had never traveled cross-country in the States before. Gerardo Herrera traveled with me in 1992 and Jorge Vargas, whom we met in the previous chapter, in 1994. Gerardo had lived in Miami for nearly two years and was used to that Hispanic (Cuban)-dominated kind of American culture. But when we stopped our first night in Pensacola, with its large population of navy and air force personnel, he saw another group of Americans for the first time. We ate in a restaurant called the Country Buffet—all you can eat for a ridiculously low price—full of blue-eyed, blond-haired (and often overweight) Anglos. Much to Gerardo's delight, we stopped for lunch the next day at a Cajun restaurant near Lake Charles, Louisiana, where we ate crawfish gumbo, hush puppies, and blackened catfish. That night, Gerardo and I stopped in a Mexican part of San Antonio, although on most trips I travel well past San Antonio and stop either in Junction or Sonora on the second night. Here in San Antonio, Jorge Vargas was first exposed to Texas cuisine—chicken-fried steak complete with biscuits and insipid white gravy—and among some locals, a thick Texas accent he could not fathom. As we traveled through Texas, we listened to country music on the radio to give him a musical taste of the West. By the time we arrived in Tucson, we had left a variety of strong regional dialects behind and were into a Mexican-influenced culture—a much more familiar environment to most of my travel companions.

Once we left Tucson in 1992, Gerardo and I drove to Guaymas about 150 kilometers south of Hermosillo and Bahía Kino to begin a survey of geographic variation in cardón's breeding system and to collect tissue samples from lesser long-nosed bats. Most of the land between Hermosillo and Guaymas no longer consists of the widely dispersed desert vegetation encountered a hundred years ago by Edward Goldman during his Mexican faunal surveys. Instead, it is now covered with agricultural fields planted in wheat, wine grapes, citrus, sunflowers, and pecans. In 1899 Goldman had seen extensive stands of cardón around Guaymas, but now most of those stands have been replaced by urban development, RV parks, and large glitzy resorts along the coast. This made me begin to wonder how much of Lepto's "nectar corridor" farther south was being similarly destroyed.

In 1992 we didn't know the locations of any *Leptonycteris* roosts in the Guaymas area, but we certainly knew that Lepto was still around. Standing in remnant cardón stands at night, Gerardo and I watched nectar bats inspecting cardón flowers about an hour after sunset. Using radiotelemetry, a Mexican colleague, Francisco "Pancho" Molina, and I discovered where these bats were coming from in April 1999. We tagged our bats in a cardón stand just east of Guaymas. Three days later, a couple of local ranch hands took us to a cave known as Aguajita de Venado (little water hole of the deer) located on the south side of a rugged series of hills about sixteen kilometers north of Guaymas. Booming radio signals coming from the cave told us that two of our three tagged bats were inside. Our third bat was likely roosting in another cave a couple of hills away, said the ranch hands. In late May we revisited the Aguajita cave and counted fifteen thousand bats exiting at sunset. When I briefly looked inside, I saw thousands of pink Lepto pups covering the ceiling of one of the cave's two chambers.

Back at Kino Bay, Gerardo and I obtained our first Lepto samples at the Sierra Kino cave. As in the case of our carbon stable isotope studies, I didn't need much tissue from each bat for our genetic analysis. All I needed was about ten milligrams of wing tissue—a small hemispherical piece that I cut from the trailing edge of one wing using surgical scissors—which I quickly popped into a small vial containing supersaturated salt solution and an antibacterial compound. With this preservative, my samples needed no refrigeration and could safely travel in my backpack. After swabbing the small cut with alcohol, we released each bat virtually unharmed. Ten milligrams of wing membrane contained thousands of cellular mitochondria with the DNA that we would eventually extract, amplify, and sequence for our analysis.

Before returning to Miami, we obtained tissue samples from five roosts in northwestern Sonora and southwestern Arizona. Visiting each of these roosts began to give me a feel for the range of conditions under which *Leptonycteris* lived during the day. After Kino Bay, we next visited Cueva del Tigre, near the town of Carbó, about sixty kilometers north of Hermosillo. For decades, two of my mammalogist friends at the University of Arizona, Yar Petryszyn and Lennie Cockrum, had been visiting this cave, which serves as a maternity roost for a small colony of *Leptonycteris* and a huge colony of Mexican free-tail bats. Yar supplied me with a rough map that took us right to the cave's entrance, located in a relatively small hill rising from flatlands dominated by mesquite and paloverde trees. The cave entrance was nearly thirty meters wide and was guarded by three rattlesnakes (a typical situation, in my experience), which quickly retreated under rocks at our approach. Armed with a butterfly net and cloth collecting bag, I entered the first of three substantial chambers that sloped downhill for about eighty meters. The *Tadarida* colony contained tens of thousands of bats and covered the ceiling of the first two chambers like a bumpy velvet carpet. Flying bats, deep piles of guano, and ammonia fumes were thick in these chambers. The fumes increased in concentration as I moved deeper into the cave. Yar had warned me not to spend more than about fifteen minutes in the cave to avoid harming my respiratory tract. With an ammonia concentration of about sixteen hundred parts per million, this air can quickly be fatal to most mammals, including humans. But many kinds of bats, especially Mexican free-tails, which live in colonies of hundreds of thousands to millions of bats, tolerate this kind of atmosphere their entire lives. Because of this, these bats certainly deserve to be considered "extremophiles"—the term biologists apply to (micro)organisms living under extreme physical and chemical conditions on earth.

By the time I reached the third chamber, where the Leptos lived, I was sweating profusely and was trying hard not to breathe too much of the ammonia-rich air. I saw Leptos flying around me and roosting high up on the ceiling. Despite the stinging sweat in my eyes, I managed to snag seven individuals with my net before quickly retreating to the delicious fresh air at the cave's entrance. After I caught my breath, Gerardo and I processed and released the bats.

The other three roosts we visited that spring were physically much less challenging. Our next two maternity roosts were located about twenty-five kilometers apart in abandoned mines in southwestern Arizona's Cabeza Prieta National Wildlife Refuge and Organ Pipe Cactus National Monument. We were taken to these roosts by Tucson colleagues, including Ginnie and Dave Dalton as well as Yar Petryszyn and Lennie Cockrum. It was now mid-May, and Leptos were giving birth to their pups. During the birthing period, females do not leave their day roosts until several hours after sunset. In fact, at the Cabeza Prieta mine, most of the approximately two thousand bats didn't leave their roost until long after we had processed a few early-departing individuals that we caught in a mist net set at its entrance. About twelve thousand adult *Leptonycteris* were living in the Organ Pipe mine, which turned out to be an eight-hundred-meter-long tunnel running completely through the base of a small hill. In visiting this roost, we camped overnight at the monument's Alamo Canyon primitive campground at the base of the rugged, brick-red Ajo Mountains. Yar, who is a true "desert rat" and expert on desert biology, fixed a tasty dinner of rice, steamed veggies, and barbecued chicken. After supper we hiked two kilometers across gravelly desert, full of flowering saguaros, to the hill. Yar and I independently conducted a partial exit count with the aid of a bright moon before I entered the mine and hand-netted a few bats. In my brief time inside the mine, which was well-ventilated and ammonia free, I could see mother bats with their pink, hairless babies all over the ceiling.

The high point of our roost visits that spring was a trip to a large maternity roost located in the Pinacate Biosphere Reserve just south of the U.S.-Mexico border. In his book *Desert Heart,* William Hartmann introduces us to this region: "A forgotten country straddles the international border where Arizona, Mexico, and the Colorado River meet. It includes the most desolate desert landscapes of North America but also contains some of the continent's most striking geologic features. It is known to only a handful of people. It is the heartland of the Sonoran Desert" (Hartmann 1989, 4). Pinacate is a moonscape of giant sand dunes, cookie-cutter volcanic craters stamped into the bone-dry earth, and massive black lava flows frozen in time. And in the middle of this inhospitable land, over a hundred thousand *Leptonycteris* females have chosen to roost in a large lava tube.

This roost was discovered in 1983 by a Tucson geologist named William Peachey. Bill tucked this find into a corner of his mind, however, until the "*Leptonycteris* controversy" erupted in the late 1980s. This controversy revolved around the population status of *L. curasoae,* which, along with its sister species *L. nivalis,* the U.S. Fish and Wildlife Service declared federally endangered in 1988.

Although many southwestern bat biologists were pleased with the federal listing of both *Leptonycteris* species, Lennie and Yar were not convinced that the lesser long-nosed bat, at least, was truly endangered. Their experiences with this bat did not jibe with those of Don Wilson and Donna Howell, the two people most influential in obtaining the federal listing. For example, they had not seen a steady decrease over the years in the size of the *Leptonycteris* colony at Cueva del Tigre. Furthermore, Yar's observations in Arizona in the late 1980s indicated that Lepto was far more common than Don had reported a couple of years earlier. Three maternity roosts, collectively containing at least fifteen thousand adults in the spring, were known in southwestern Arizona. In August a cave near Patagonia in the uplands of south-central Arizona housed thousands of Leptos rather than the few hundred that Don had reported seeing there in July.

When Bill Peachey told Yar about the Pinacate roost in 1989, the bat was out of the bag. The lesser long-nosed bat was apparently in much healthier shape than its endangered status implied. To publicly air their views, Cockrum and Petryszyn published a peer-reviewed paper entitled "The Long-nosed Bat, *Leptonycteris:* An Endangered Species in the Southwest?" in 1991. In their paper, they reported all known records of this bat by place and time. They concluded that "it appears probable that current populations in the northwestern part of the range of the species are little, if any, decreased from those a quarter century ago. It even has been suggested that populations have increased in the past century because of more suitable roosts being available as the result of mining activity in the area" (22–23).

This conclusion outraged Donna Howell, whose short-lived academic career had been built on ecological and behavioral studies of *Leptonycteris* and its interaction with *Agave* flowers. Having studied it for many years, Donna understandably had a deep emotional attachment to this bat. To support her belief that *Leptonycteris* populations were declining, Donna wrote a paper entitled "The Long-nosed Bat, *Leptonycteris:* An Endangered Species in the Southwest!" and sent it, unpublished, to state and federal wildlife agencies whose territories housed this bat. In it, she disputed recent counts of thousands of *Leptonycteris* living in Arizona and adjacent parts of Mexico, saying that researchers were either grossly miscounting bats or misidentifying bat species (or both). Her not-so-subtle message to the agencies was, "Don't believe Cockrum and Petryszyn, because I don't." Unfortunately, by the early 1990s, Donna's objectivity regarding the population status of *Leptonycteris* was highly suspect. For reasons known only to her, she had backed herself into an indefensible position and refused to consider the possibility that competent observers were legitimately challenging her views. When I talked with her in the summer of 1993, she steadfastly refused to believe that Lepto's numbers were in the hundreds of thousands rather than 15,500, as Don Wilson had reported in 1985.

All of this controversy flashed through my mind as we headed toward the Pinacate roost in mid-May. We drove south of Organ Pipe into Mexico on paved highway for about half an hour, then turned west onto a gravel road and drove another hour, crossing a series of tire-slashing lava fields and rocky washes before reaching a campsite used by Bill Peachey and his friend Tom Beathard whenever they visited the Pinacate roost. We found Bill, Tom, and Tom's wife, Cindy, comfortably seated out of the hot afternoon sun in an *hornita*—a vertical, hollow lava vent—near camp. After talking awhile with them, I left the *hornita* to explore the magnificent landscape before supper. I stood on a vast lava flow that resembled frozen molasses and looked around. To the south I could see the rim of a giant sand dune that emerges from the north end of the Gulf of California. To the north was Pinacate's complex of volcanoes and extensive fields of volcanic cinders. Vegetated with only scraggly brittlebush and occasional ocotillos, this and other lava flows were laced with large and small grottoes and caves, formed by the puckerings and wheezes of cooling lava. These structures undoubtedly served as shelters for Pinacate's wildlife during the blistering summer heat.

After a burrito supper, our crew, which included the Daltons and Yar and Lennie, set out on foot to visit "Seething Beetle Cave," Bill Peachey's name for the lava tube housing the *Leptonycteris* colony. Along much of the way we walked on wide "paths" of jagged lava tiles, another common land form in this region. Our pace was slow because of the precarious footing and in deference to Lennie Cockrum. Now almost seventy-two years old, he hadn't been in the field for several years and walked with the aid of a cane. Despite the rough walking conditions, Lennie was in high spirits and was as anxious to see this roost as the rest of us were.

Seething Beetle is an extensive lava tube, at least four hundred meters long and over six meters wide and two meters tall in most places. Oriented in a north-south line, it lies in a large field of platelike chunks of lava haphazardly strewn in all directions. Bats emerge from the tube through a large hole in its collapsed ceiling. Just above this exit is a flat, altarlike stone on which we found a pile of fifty-seven *Leptonycteris* wings. Bill told us that a female bobcat periodically crouches on this stone and swipes bats out of the air as they leave the roost. She and her cubs eat everything but the wings. When feeding on Leptos, these bobcats appear to be specializing on fast food.

Bats began to leave the roost at 8 P.M. A few *Macrotus* left first, then a steady stream of Leptos emerged, flying along the open tunnel before peeling off in a variety of directions. Most turned west and then north (toward Organ Pipe Cactus National Monument, about fifty kilometers away?) before disappearing from sight in the silver light of a full moon. Yar began an exit count, tallying bats by the hundred—"hunnert . . . hunnert . . . hunnert"—into a small cassette recorder. By 10 P.M. he had counted eighty thousand bats, and they were still steadily emerging. I hand-netted a series of bats—all pregnant females—and Gerardo and I processed them sitting on slabs of lava. Then a couple of us slipped quietly into the tube to look around. In a chamber about thirty meters inside the tube, a few "pinkie" babies were widely scattered on the ceiling. Most females in this roost had not yet given birth. After the babies are born, the ceilings of some chambers are packed solid with pinkies, Bill told me.

The fact that a large number of migrant *Leptonycteris* females choose to roost in the middle of nowhere to have their babies indicates that safe, thermally favorable roosts are a critically important resource for this bat. Distance to the nearest food source, which has to be even farther for the Pinacate bats than it is for the Tiburón bats, appears to be of secondary importance to this strong-flying species. If, as appears likely, some of these bats travel as far north as Organ Pipe to feed, this poses an interesting physical challenge for newly weaned Leptos. Imagine what it must be like having to make a round-trip commute flight of over a hundred kilometers on your very first night of foraging. To be in physical shape to make this trip successfully, young Leptos must exercise their flight muscles intensely in and around the roost before making their first feeding flight. During their early summer visits, Bill and Tom have observed that young Leptos, curious as puppies or kittens, sometimes follow them from the cave back to camp, a distance of about two kilometers. Perhaps this is an example of the preforaging warm-up flights of young Leptos.

At 11 P.M. we began the slow walk back to camp under a full moon. Now the region really did resemble a moonscape washed in shades of white, gray, and black. We didn't need our headlamps to navigate in this open, rugged country. The nighttime air was still a bit nippy in mid-May and a steady breeze further cooled us. Back at camp we relaxed with a round of chips, salsa, and beer before throwing our bedrolls on the desert floor and crashing for the night. We broke camp shortly after sunrise. By early afternoon, Gerardo and I were back in Tucson, in a world entirely different from the Pinacate. By sundown we were in Las Cruces, New Mexico, on the long road back to Miami. But images of the Pinacate and its improbable bat roost persisted in my mind's eye long after we had left the West.

In late August 1992, my family and I drove west to Tucson to begin our sabbatical year. Although Mike helped us relocate, he was now in college, so he didn't accompany us on our Mexican travels. Cara was a junior in high school and conducted her studies by correspondence so that she could travel with us. The timing of our move was lucky. We arrived in Tucson the day before the devastating Hurricane Andrew struck Miami. The storm hit early the next morning, and, fortunately, our neighborhood was spared major damage. Our house ended up with two oak trees on its roof, and most of our extensive landscaping was destroyed. But our friends told us not to bother returning to Miami during the chaos after Andrew. They would take care of house repairs and tree removal for us. We wouldn't see our house and Miami for nearly a year after the hurricane.

My fall fieldwork began with a family trip to the Patagonia bat cave to collect tissue samples in late August. Located about eighty kilometers southeast of Tucson, the Patagonia region consists of rolling hills covered with oaks and junipers. Manzanita, sotol, and paniculate agaves are also conspicuous members of this cooler, moister upland vegetation. It took us about an hour and a half to hike uphill to the cave from our parking spot near Patagonia. The cave is located in a huge, dome-shaped rock whose two chambers are spacious enough to house thousands of bats. In early summer it contains a large maternity colony of the cave myotis. When *Myotis* abandons the cave, beginning in late July, *Leptonycteris* bats move in. At the time of our visit, Yar Petryszyn and I estimated that the cave contained about twenty thousand Leptos; this number included postlactating adult females and male and female young-of-the-year. At the time, Yar and I figured that many of these bats were the same ones that occupy the maternity roosts in southwestern Arizona and northwestern Sonora earlier in the year. We thought that after cactus resources dry up in the desert, many bats move to higher elevations in southeastern Arizona to feed at the flowers of paniculate agaves before migrating south into Mexico. The results of our genetic analysis, however, would reveal a different scenario.

Our major field trip that fall began in October when Marcia, Cara, and I embarked on a five-week tour of western Mexico. Most of our route would be along the nectar corridor that brings female Leptos north to the Sonoran Desert in the spring. We expected to drive on paved roads for most of the trip and planned to stay in hotels or motels on most nights. When we reached Mexico City, Hector Arita, a conservation biologist and bat expert at the Centro de Ecología at the Universidad Nacional Autonoma de México (UNAM), would supply me with maps and directions to most of the known *Leptonycteris* caves in central and western Mexico.

Our first day on the road covered now-familiar territory. On our way to Kino Bay, we stopped at Cueva del Tigre just to confirm that *Leptonycteris* and *Tadarida* had migrated from the cave. Its only inhabitants now were a few California leaf-nosed bats and Mexican funnel-eared bats, and consequently its air was much more tolerable. Early the next morning, we visited the Sierra Kino cave. The desert around Sierra Kino was lush with dew-covered plants bearing red, yellow, and purple flowers, all a legacy of a Pacific hurricane that had struck the Kino region in mid-August. In less than one week, my study site and house had been pummeled by two separate hurricanes. Small birds, including a couple of species of hummingbirds, darted among the plants, and the cool desert air was fragrant with the tangy, resinous smells of *Bursera* trees and the wood-smoke aroma of creosote bushes. A couple hundred *Macrotus* now occupied the chamber that is full of Leptos in the spring.

By midmorning we were heading south along coastal Route 15, past Guaymas and toward Alamos in southern Sonora. Alamos is a colonial city set in the foothills of the Sierra Madre Occidental. In the late seventeenth century, when the mountains around it were being mined for silver, it was an important economic center. In recent years, wealthy Americans and Mexicans have been restoring to their former elegance many colonial houses built around beautiful courtyards full of colorful tropical flowers. We checked into the Casa de Tesoros, a renovated eighteenth-century convent, whose sturdy stone rooms feature high ceilings and old wood and leather furniture. After lunch we drove a few kilometers along backroads into the hills to the former mining town of Aduana, the site of the area's largest silver mines. Columnar cacti, including hecho, a cousin of cardón, were common on the hills between Alamos and Aduana. One of Aduana's abandoned mines houses large populations of several species of bats, including *Leptonycteris.* We used a map drawn by a Tucson herpetologist to find this mine. As was the usual drill during the entire trip, I went into the wet and muddy mine with my trusty butterfly net and collecting bag and captured a few Leptos while Marcia and Cara waited outside. Few Leptos were present in the mine in October compared with the twenty thousand or so that I would see when I returned in February 1993. We processed the bats as the sun was setting. Cara recorded data while Marcia and I weighed, measured, and snipped a bit of wing tissue from our bats before returning them to the mine.

Once we left Alamos, we wouldn't see *Leptonycteris* bats again for many days and hundreds of kilometers farther south in Mexico. As we drove through Sinaloa toward Mazatlán the next day, I began to think about Lepto's nectar corridor, the columnar cactus route that it follows north in the spring. Until we were south of Culiacán, Sinaloa's capital of nearly half a million people, most of the land we passed through was flat agricultural country in which cacti were few and far between. Pretty slim pickings for hungry Leptos, I thought. Below Culiacán, the road entered hilly country covered with dry tropical forest. Scattered in the forest were tall hechos, whose large white flowers undoubtedly serve as feeding stations for hungry Leptos in the early spring. Hecho is broadly distributed along Mexico's west coast and begins blooming in a south-to-north progression in late November or December. With a flower season that lasts until April, it undoubtedly is a key resource for migrant nectar-feeding bats as they begin their northward journey each spring.

As we drove south, a large series of questions about migration passed through my mind. Where are the migrating Leptos coming from? How fragmented is their nectar corridor? Could aerial photographs or Landsat images help us locate the most likely path they take through this part of Mexico? Where do Leptos roost along the way? Do they follow "traditional" routes and know the locations of a series of caves or mines that serve as stopover places? How flexible are they in their choice of stopover roosts? How do young females learn the migration route—from their mothers or other relatives? How far does each bat travel in a night, and, like migrant hummingbirds, do they stop in good feeding areas to fatten up for several days before continuing their northward journey? Is the corridor in danger of becoming so fragmented that bats (and migrant birds) will no longer be able to complete their journeys?

Unlike the many physical and anthropogenic hazards faced by migrating nectar bats, our main hazards along the way were pot-holed stretches of road and patches of heavy bus and truck traffic. In many places in the mountainous regions of central Mexico we crept along behind noisy trucks belching clouds of smelly diesel fumes. Traveling uphill behind these behemoths became our least favorite part of the trip. As we turned southeast from Tepic, the capital of Nayarit, toward Guadalajara, we left the lowland tropics behind and quickly climbed into cool oak and pine forest. The upland hills and valleys around Tequila were covered with endless fields of *Agave tequilana,* the commercial source of tequila. Tequila's undulating fields of blue-gray agaves bordered by orange- and yellow-flowered composites were strikingly beautiful in the warm glow of a late afternoon sun.

From Guadalajara we pushed on to Mexico City in the midst of increasingly heavy traffic. There we stayed in a hotel in the Zona Rosa for several days, doing some sightseeing and, for me, visiting the Centro de Ecología to give a seminar and visit old friends, including Bernardo Villa Ramirez. Now in his early eighties, don Bernardo was born of Aztec ancestry in the town of Teloloapan in the mountainous state of Guerrero, in southwestern Mexico. As a young man, he had traveled in remote areas of Mexico on foot, horse, and mule to collect mammals, just as the U.S. collectors Edward Goldman and Edward Nelson had done a half century earlier. His substantial mammal collection was eventually housed in UNAM's Instituto de Biología. Prior to the late 1940s, major collections of Mexican mammals could be found only outside Mexico, most notably in the U.S. National Museum as a result of the Nelson-Goldman expeditions. Don Bernardo single-handedly initiated the modern study of his country's mammals by Mexicans. In 1966, he published his doctoral thesis, "Los Murciélagos de Mexico," which summarized his vast knowledge of his favorite mammals. To gather information for this book and many other publications, don Bernardo constantly traveled throughout Mexico, always accompanied by a cadre of bright young biologists. Several generations of Mexican mammalogists learned their field techniques under don Bernardo's expert tutelage.

We left Mexico City armed with a new set of maps and instructions provided by Hector Arita and retraced our path through congested Guadalajara before heading south and west past Colima to Jalisco's Pacific coast. This route took us through an extensive *tolvenera* (dust storm) flood plain that was full of wading birds, including great egrets, black-necked stilts, little blue herons, ducks, and a few brown pelicans. Traveling on one of Mexico's newly built toll roads, we passed over deep mountain gorges on a series of bridges before reaching the lowlands with their extensive coconut plantations. Our destination was the Cuixmala Biological Reserve, about eighty kilometers north of Manzanillo, the port where Nelson and Goldman first landed in January 1892 to begin their epic fourteen-year journey around Mexico. Cuixmala is a large tract of dry tropical forest set aside for conservation purposes by James Goldsmith, one of the world's richest men. Until his death in 1997, Goldsmith lived from December to March in a Turkish-style "palace" overlooking the Pacific Ocean on a 120-hectare farm near the biological reserve. The farm has its own airstrip and private collection of exotic wildlife. It is heavily guarded by a small army, which gives the place a feeling right out of a James Bond movie.

Gerardo Ceballos, one of the Centro de Ecología's three mammalian ecologists, convinced Goldsmith to set up a conservation foundation in the late

1980s. As a symbol of their friendship, Goldsmith allowed Gerardo to live in the spacious guest quarters above a ten-stall horse stable when he was conducting his research in the dry forest. Gerardo, in turn, allowed us to stay in these elegant quarters, which turned out to be our best lodging on the trip. Cara, whose passion since the age of eight has been horses, even received permission to ride one of the farm's saddle horses.

We had a fascinating time touring the Cuixmala reserve and adjacent Chamela Biological Station, but our real object, of course, was to survey *Leptonycteris* bats. Hence, on the morning of 29 October, one of Gerardo's graduate students, Cuauhtemoc Chávez, our guide at this site, took us to the rustic seaside town of Perulas, where we hired a *launcha* to transport us to Isla San Andres, nearly two kilometers out in Chamela Bay. A cactus-covered island that serves as a rookery for seabirds, San Andres contains a sea cave that houses thousands of bats. When Don Wilson visited this cave in May 1984, he estimated that it contained fifteen thousand Leptos. At that time, this was thought to be the largest Lepto roost in all of Mexico.

The boat ride out to the island was smooth, but we slightly miscalculated the timing of our visit. At low tide it is possible to land a boat right at the cave's entrance, but today we arrived at high tide. Thus, Cuauhtemoc and I had to jump overboard into chest-deep water and wade ashore, trying to keep our headlights and collecting gear as dry as possible. Once inside the cave, my immediate sensation was the warmth produced by tens of thousands of bats that were roosting on its walls and twelve-meter-high ceiling. Most of these were Leptos, and I estimated that this roost contained at least fifty thousand lesser long-nosed bats. As Cuauh and I explored the cave's long, narrow chamber, waves kept crashing against its outer walls, sending in surges of water that constantly stirred the thousand-peso-sized stones covering the cave floor. After surveying the cave, I netted about twenty-five bats from various places on the walls and placed fifteen of them in a wet collecting bag.

With our sample in hand, we waded back to the boat and immediately returned to shore, where we processed our bats under a palm-thatched *palapa*. Ten of our fifteen bats were adult males; each of them had enlarged testes and appeared to be in breeding condition. These were the first breeding males of this species that I had ever seen. Two of our bats unfortunately died in the bottom of the bag. When Cara, Cuauh, and I dissected them back at the stable, we got a nice surprise. Under its skin, the chest and back of each bat was covered with a thick layer of white fat. I had skinned a fair number of phyllostomid bats in my life but could not remember ever seeing an obviously "fat" bat before. These two Leptos were simply loaded with fat. How can a bat that flies long distances to feed on nectar and pollen (or fruit in season) acquire enough energy to store some of it as fat? I wondered. Is heavy fat deposition a seasonal phenomenon, perhaps in preparation for migration?

The next day, I called Gerardo Ceballos in Mexico City to thank him for letting us stay in such wonderful quarters. I also told him of my observations at the San Andres roost and suggested that we conduct a year-long study of its *Leptonycteris* population. In addition to estimating colony size and sex ratio each month, we could collect a few individuals for detailed anatomical examination and for stable isotope analysis. He enthusiastically accepted my suggestion, and between March 1993 and February 1994 he and Cuauhtemoc visited the cave nine times to gather data. They sent me their bat carcasses for isotope analysis in Leo Sternberg's lab.

Results of this study were fascinating. For one thing, we discovered that the size and sexual composition of this *Leptonycteris* colony changed seasonally. From January through June, the cave contained five thousand to fifteen thousand males and few females. In the latter half of the year, large numbers of males and females arrived at the roost, swelling its size to about seventy-five thousand individuals in November. Male testis size was small early in the year and then increased to maximum size in October through December. As I suspected, bats were lean during the first half of the year (the dry season) and carried substantial amounts of fat in October through December. Beginning in December, roost size decreased when most females and many males left for parts unknown. Finally, unlike the annual "carbon cycle" that I had previously documented for *Leptonycteris* in mainland Mexico, our carbon stable isotope data told us that these bats were feeding mostly on C3 plants year-round.

When we put all of these data together, it became obvious what was going on in this particular roost. The San Andres cave is a mating site for *L. curasoae*. Females come to this roost for one reason—to get pregnant. Though we haven't yet studied their behavior in detail, Leptos must be mating up a storm in this Jaliscan cave in November. In my imagination, I can picture the scene inside the cave. Fueled by margaritas (tequila) and mariachi music, these bats are having a *fiesta de amor*. Females then leave in December and eventually have their babies someplace else, beginning in mid-May. Our genetic analyses would give us a strong clue about where many of these females end up having their babies.

From Cuixmala, our *Leptonycteris* odyssey took us back toward Guadalajara. We were joined on this leg of the trip by Jorge Vargas, the twenty-seven-year-old master's student at UNAM, mentioned briefly earlier. Originally from Campeche, in Yucatán, Jorge is a short, stocky fellow of Mayan ancestry. For his master's thesis, he was studying the use of caves by bats in the mountains of Tamaulipas, in northeastern Mexico. In 1992 he didn't speak much English, so we communicated mostly in Spanish. He and Cara rode in the back seat of our packed Explorer as we toured the uplands of central Mexico looking for Lepto roosts.

Our first stop was the village of Ajiijic, on the north side of Lago Chapala, about thirty kilometers south of Guadalajara. Jorge made a few inquiries, and we were quickly directed to a cave called La Mina, on the northeast side of town. A short walk up a hill brought us to its entrance, a triangular hole about three meters wide. The entrance dropped some six meters into a large chamber, in which we could see bats flying around. Rocks at the entrance were covered with yellow splats of pollen-rich feces, so we knew that the cave must sometimes house *Leptonycteris.* Without a ladder or climbing gear, though, we couldn't easily enter the cave, so we draped a short mist net across the entrance to capture bats for our tissue samples.

Bats began to leave the cave shortly after sunset. The first two species out were the ghost-faced bat and Davy's naked-backed bat, two insectivorous members of the Mormoopidae (mustached or leaf-chinned bats), which commonly share roosts with the lesser long-nosed bat in parts of its range. All three species are highly gregarious and form colonies of tens of thousands of individuals. Their numbers are so large that their collective metabolism produces tremendous amounts of heat, making the microclimate of their roosts very hot. Hot caves make wonderful incubation chambers for baby bats, and they also reduce the diurnal metabolic costs of adults.

Within a few minutes the leaf-chinned bats completely filled our net, forcing Jorge and me to remove it from the entrance. While we cleared the net, I gave Marcia my butterfly net and asked her to snag bats as they left the cave, beginning at 7 P.M. In short order she had captured seventeen Leptos (eleven females and six males), the third species to leave the cave. We processed our bats downhill near our car while under attack by a swarm of pesky mosquitoes. From the rate at which they left La Mina cave, we estimated that tens of thousands of bats were present at the time of our visit. *Leptonycteris* was at least as common as the other two species.

After Ajiijic, we ran out of luck finding Leptos for another long stretch of time. We hunted in Michoacán in dry forest south of Uruapan but found only small colonies of Waterhouse's bat and gray sac-winged bats in caves that local people told us about or took us to. Marcia's lasting memory of this part of the trip was our driving into small towns and Jorge's jumping out of the Explorer to ask people on the street about the locations of bat caves. Such locations were common knowledge wherever we went, but these leads didn't take us to *Leptonycteris.*

My most vivid memory of this part of the trip was don Salvador Montiel, a delightful old farmer who led Jorge and me to a large cave at the bottom of an extensive barranca one sunny afternoon. Don Salvador was seventy years old. Short and pudgy, he sported a grizzled stubble of beard, a mouth that was missing its lower front teeth, and wispy white hair peeking out from under a beat-up straw hat. On his tough, leathery feet he wore old sandals. Despite his age, he was sharp as a tack and had a wonderful sense of humor. To get to the cave, we hopped across a series of stones in a wide river. In telling us how deep the water gets in the rainy season, don Salvador kept jumping up and down in the shallow water with his arms upraised. With a twinkle in his eye, he was obviously having a merry time educating us about his neighborhood. We, in turn, told him about the lives of Mexican bats and their ecological importance. Unfortunately, though the cave was as large as he said it would be, it contained only a small colony of gray sac-winged bats. When we arrived back at don Salvador's modest house, Marcia and Cara introduced us to a young boy who claimed he was Jesse James. While Jorge and I were cave hunting with don Salvador, young Señor James had entertained the women with wild tales of the Old West.

During our stay in Michoacán, we spent part of a day in Patzcuaro, on the south shore of Lago Patzcuaro, one of the highest and most picturesque lakes in Mexico. In his posthumously published monograph of his biological travels in Mexico, Edward Goldman wrote: "Lake Patzcuaro, perhaps the most beautiful in Mexico, is crescent-shaped, about 15 miles long and 2 to 3 miles wide. A very irregular outline is due to its setting in an amphitheater of high hills with spurs and buttresses that break the shore line and form deep bays" (1951, 195). This region of Mexico is inhabited by Tarascan Indians who have built a terraced town on Isla Janitzia in the middle of the lake. Hector's notes indicated that Leptos have been found in a cave near the top of this island. We arrived at Patzcuaro late in the afternoon and took a commercial launch to the island. At the dock we found a group of four young barefoot boys who knew the way to the cave. Since we had only an hour before the last launch returned to the mainland, we told the boys that we needed to hurry to the cave. Thus started a mad dash up a cobblestone street past colorful Indian markets, past young girls tending goats, to the top of the island. Jorge had no trouble keeping up with the kids, but the Flemings definitely were not used to running uphill at an elevation of twenty-four hundred meters. I puffed along behind Jorge, and Marcia and Cara fell far behind me. At the island's top, we had to crash through masses of tall daisies and bee balms growing over a muddy trail before tiptoeing along a narrow ledge (don't look down, for God's sake!) that led to the cave.

The cave had a wide entrance but turned out to be rather small, its floor measuring about ten meters in diameter. Its single cone-shaped chamber was absolutely packed with Mexican free-tailed bats. I tried to walk partway into the cave to see if any Leptos were hiding among the free-tails, but a solid wall of ammonia fumes blocked my way. The air here was far worse than that of Cueva del Tigre, back in Sonora. From the cave's entrance, I scanned the bat population and estimated that the cave contained at least two hundred thousand free-tails but no Leptos. Marcia and Cara soon arrived, out of breath as I had been, and peeked into the cave before we all turned around and trudged back into town to do some gift shopping prior to our launch shuttle back to Patzcuaro.

From Patzcuaro, we headed east through Morelia toward Toluca. The first part of this route, from Morelia to Mil Cumbres, was spectacular. We twisted and turned along a narrow, two-lane mountain road through majestic pine forest. Unfortunately, the day was heavily overcast so our views of the deep mountain valleys and distant peaks were restricted. Nonetheless, this was by far the most beautiful stretch of road that we encountered in Mexico. At the road's summit, a marker indicated that it was built between 1929 and 1932. As we descended from Mil Cumbres toward Ciudad Hidalgo and Toluca, we entered deforested land dotted with clean montane towns that reminded me of the rural highlands of Panama and Costa Rica. We searched in vain for *Leptonycteris* caves around Tuxpan and in the overcrowded Valle de Bravo before traveling south from Toluca toward Ixtapan de la Sal and Tonatico, west and southwest of Mexico City.

Then my good health temporarily deserted me. Early in the morning of 8 November, I woke up in our Ixtapan motel room with intense abdominal pains. I initially thought I was coming down with a case of *turista,* something I rarely experience in Latin America, but then forlornly realized that I was suffering from anti-*turista.* Days of long car rides, irregular meal schedules, and dirty Mexican public restrooms had resulted in a partial intestinal blockage, the kind that I had been experiencing about once a year as an aftermath of my intestinal surgeries of 1984. Those operations had left my abdominal cavity full of spidery adhesions that tended to snag my small intestine during times of stress or other intestinal upsets.

Except in 1986, when my gut was really clamped shut by adhesions, which required another operation, I have managed to rid myself of these blockages by undergoing "decompression" in the hospital. This procedure involves having a nasogastric tube snaked through your nose into your stomach—a most unpleasant experience—where it slowly drains the stomach of liquids and gases and allows the intestine to deflate enough to work itself free from the adhesions' stranglehold. A couple of days of decompression, complete with morning and afternoon X-rays to monitor the state of my gut, are usually enough to eliminate my gastric distress.

Given my history of intestinal blockages, my Miami internist has never been pleased with my foreign travels. "What if you should get a blockage overseas?" he would ask. Well, now I was faced with this reality. I spent a day in bed while Marcia, Cara, and Jorge checked out another cave near Ixtapan. Cara's memory of that day is sitting in the pitch-black cave with vampire bats flying around her. Claustrophobic Marcia waited patiently outside the cave for Cara and Jorge to emerge. My condition hadn't improved the next morning, so we quickly packed up, and Marcia drove us through thick fog back to Toluca. Luckily, we spotted a modern hospital, Centro Medico de Toluca, as we reached the southern edge of that large city. I was admitted into the hospital, and after looking at my first series of X-rays, Dr. Tonatuih Moreno told me what I already knew: I had a partial blockage. If it didn't begin to clear today, he said, he would be forced to recommend surgery.

Fortunately, I made steady progress under the same decompression routine that I receive in Miami and managed once again to dodge the scalpel. Contrary to my Miami internist's fears, I received excellent medical treatment and enjoyed interacting (mostly in Spanish) with the nurses and doctors. While I lay in the hospital listening to American pop music coming over the PA system, Marcia had her own little drama in Toluca on two counts. First, she had to navigate in congested traffic to find a motel in an unfamiliar city. And second, she had to arrange for the delivery of a new credit card from the States. In a span of about five seconds of inattention, I had lost my wallet to a pickpocket on the Metro subway in Mexico City, forcing us to cancel our credit cards. We were currently traveling without any plastic money. My admission to the hospital had been a bit dicey because of our inability to guarantee payment until our new credit card arrived. Nearly two days and numerous international phone calls after our arrival in Toluca, Marcia picked up the new card at the airport. Actually, our financial situation wasn't completely desperate, because Daniel Piñero, the director of UNAM's Centro de Ecología, kindly offered temporary payment of our bills when he learned of my predicament. Fortunately, we didn't need to take Daniel up on his generous offer, and I was released from the hospital with all financial obligations cleared up on the morning of 12 November.

Though my gut was working fine (despite the rumor that quickly spread among Mexican biologists that I was dying of a stomach disorder), I was weak from the blockage and lack of food. We had planned to drive as far south as Chiapas to visit a cave near San Cristóbal de las Casas, the site of the 1994 Zapatista uprisings, but several more days on the road now seemed like a bad idea to all of us. So we decided to shorten our trip and return to Arizona after visiting one more *Leptonycteris* site about a day's drive from Toluca. Another Mexican colleague, Rodrigo Medellin, from the Centro de Ecología, eventually collected the Chiapas samples for me. As soon as Jorge Vargas rejoined us, we left Toluca with Marcia at the wheel and headed south toward Chilpancingo, the capital of Guerrero.

After driving awhile through hilly country full of winding roads, we entered the new dual-lane highway that zips people from Mexico City to Acapulco. A long stretch of this road passed through nectar bat heaven. Just south of Iguala, near don Bernardo's birthplace, the hillsides were covered with morning glory trees full of white flowers. Next, an arborescent cactus similar to hecho appeared on the steep hills. This cactus was soon joined by a second, smaller species of treelike columnar cactus. Finally, telephone-pole-like *Cephalocereus* cacti appeared, sometimes covering steep hillsides in dense monocultures.

Our goal near Chilpancingo was Gruta Juxtlahuaca, a tourist cave that houses substantial populations of bats. To reach Colotipa, the town nearest the cave, we drove on a rough road that passes through two small towns in a broad valley planted in corn. Columnar cacti covered some of the hills, but closer to Colotipa, they were replaced by moister forest containing tropical legumes and *Bursera* trees. An old friend from Santa Rosa in Costa Rica, the weedy tree *Muntingia calabura,* whose strawberry red fruits are eaten avidly by bats, was common along the roadside. At Colotipa, Jorge engaged a young farmer and part-time tourist guide named Sergio to take us to the *gruta,* located a few kilometers from town over an even rougher road.

Once we reached the cave, whose entrance is covered by an artistic iron gate, Marcia and Cara elected to stay outside in the shade while Jorge, Sergio, and I went bat hunting. Carrying a Coleman lantern, Sergio led us about 250 meters along a muddy path past thick columns of dripping stalactites. Then we made a right turn and started up a muddy, slippery ramp to where most of the bats roost. We first encountered a warm chamber containing thousands of tiny orange funnel-eared bats. Then we came to an area that housed thousands of western long-tongued bats. Still farther up the ramp, we found ghost-faced bats and, roosting high in pockets in the ceiling, several dense clusters of *Leptonycteris.* By now, I was sweating heavily, and my glasses were fogged with humidity. I didn't have enough strength to swing my butterfly net, so Jorge was elected to try to catch bats. We taped my net to a mist-net pole. Then, by climbing up a mud-slick side ridge, he was able to maneuver into position to net a few Leptos.

We emerged from the cave into the warm sun after about an hour and a half of exertion. Although all I had done was struggle to stay upright in the mud, I was exhausted and dehydrated. Sergio and Jorge were also soaked with sweat. After drinking copious amounts of Gatorade and eating candy bars, we processed our bats. Once we released the bats, we packed up and started on the first leg of our four-day trip back to Tucson. We dropped Jorge off in Mexico City and then drove straight north through broad expanses of Chihuahuan Desert in central Mexico toward Saltillo and Monterrey. On the outskirts of a couple of towns, we passed a number of odd roadside stands manned by Indians. Draped over long horizontal sticks were the skinned carcasses of many rattlesnakes. We must have seen hundreds of dead snakes. I wondered over how wide an area and with what techniques did these people harvest their prey? After spending a night near Saltillo, we continued north before turning west toward New Mexico and Arizona, just outside of San Antonio, Texas. When we left Colotipa, we were in the warm tropics. Three nights later we drove along Interstate 10 into a pretty west Texas sunset drenched in reds and grays. The Texas hill country gave us a glimpse of fall colors—oak trees covered with leaves of red and yellow—and reminded us that we were again back in the temperate zone.

If standing on the top of Sierra Kino gave me a good overview of the daily home range of a lesser long-nosed bat, standing just below Pico de Aguila, a five-thousand-meter peak in the Venezuelan Andes, gave me a good sense of the expansive geographic range of this species' southern subspecies. At this altitude I was in *paramo* vegetation, high above the columnar cactus and agave zones that cloak the lower slopes and dry valleys of the northern Andes. Vegetation in the *paramo* is dominated by different kinds of fleshy plants, including silvery-leafed Espeletias, senecios, lupines, and orange-flowered Indian paintbrushes. At this elevation, I had trouble breathing, and it seemingly took an eternity to walk a couple hundred meters back to the jeep I was traveling in. My travel companions on this sabbatical trip included Jafet Nassar, a young Venezuelan biologist who intended to pursue graduate studies at the University of Miami, and Pascual Soriano, a bat biologist at the University of the Andes in Mérida, who was studying bat-cactus interactions in an arid valley just south of that city.

Jafet and Pascual introduced me to the haunts of *Leptonycteris* in the arid tropics of northern South America in December 1992. We had just spent several days in the coastal lowlands around Coro and on the Paraguana Peninsula, a bulbous, islandlike chunk of land that remains connected to the Venezuelan mainland by a thin ridge of sand dunes. From the Paraguana Peninsula, you can see the desert island of Curaçao, about seventy kilometers offshore in the southern Caribbean Sea. Paraguana, Coro, and land to the south in Falcón and Lara states is prime cactus real estate. Several species of spindly columnar cacti, only distantly related to Mexican columnar cacti, form dense *cardonals* covering hundreds of square kilometers. The exclusive pollinators of these plants are two glossophagine bats, the lesser long-nosed bat and its smaller cousin the southern long-tongued bat. On the Paraguana Peninsula, we visited two large "hot" caves filled with thousands of funnel-eared bats, common mustached bats, and ghost-faced bats, as well as *Leptonycteris.* In December, one of these caves contained mostly male Leptos, whereas the other contained mostly females. Alexis Arends, a physiological ecologist who lives in Coro, told me that he had seen Leptos mating in the "male" cave in November, at about the same time as mating is taking place in the Isla San Andres cave, off the coast of Jalisco. As in northern Mexico, female Leptos give birth in mid-May in Venezuela and Curaçao.

From Pico de Aguila, we wound our way south through the mountains to Mérida, a pleasant city nestled in the Río Chama valley at an elevation of about 1,360 meters. Here we visited several dry valleys full of columnar cacti. Pascual's studies indicated that Lepto is only a seasonal visitor to these valleys, which suggested that the lesser long-nosed bat is also migratory in Venezuela. Our later genetic studies would confirm that, like its northern subspecies but unlike the southern long-tongued bat, the lesser long-nosed bat likely undergoes extensive seasonal movements in northern Venezuela. These movements are undoubtedly driven by seasonal changes in the availability of cactus flowers and fruit, as well as agave flowers. Instead of migrating, the long-tongued bat stays put year-round and switches to eating the fruit of tropical trees when cactus and agave resources disappear.

Our last major field trip during my sabbatical year was a three-week trip to Baja California—a far different part of Mexico from what I had ever seen before. To me, the land called Baja California has always seemed somewhat mystical. Images that come to mind when I hear that name include dusty desert road races, fabulous deep-sea fishing, and desolate country that was still relatively untouched by the hand of man in the late twentieth century.

To Edward Goldman and Edward Nelson, who traveled the length of the peninsula on horseback between May 1905 and February 1906, Baja must have evoked all kinds of feelings—from elation at the panoramic views from the Sierra San Pedro Mártir in the north, to desperation when their supplies of food and water ran dangerously low in Baja's central desert, and, finally, to relief when they reached Cabo San Lucas just after Christmas. In Baja they could finally travel without fear of harm from other people, and the landscape that they traversed was unlike any other they had encountered in their extensive Mexican journeys—granitic mountain massifs in the north, dense cardón forests amid giant boulder fields near the middle, and lava-strewn pathways in the south.

By the spring of 1993, when four of us traveled for three weeks in Baja, the trip to the Cabo region could be made in relative comfort. A two-lane road now runs the thirteen-hundred-kilometer length of this peninsula, whereas for about half of the twentieth century, a road from the north extended along the Pacific coast only as far south as Ensenada, about seventy-five kilometers from Tijuana. Beginning in 1973, this rugged land has been accessible not only to intrepid explorers driving sturdy trucks and jeeps (or riding horses or burros) but also to increasing numbers of ordinary people driving ordinary cars or, worse yet, lumbering recreational vehicles. The accessibility inevitably will take a toll on the natural beauty and unique biological resources of this enchanted land, as it has everywhere else in the world. As the naturalist Joseph Wood Krutch once noted, "Baja California is a wonderful example of how much [good] bad roads can do for a country" (1961, 14).

Our two-car caravan—our maroon Ford Explorer and a bright yellow Datsun hatchback driven by Sandrine Maurice, a French postdoctoral student who was working with me on the evolution of cardón's breeding system—left Tucson for Baja on 8 April. We weren't really equipped to do any off-road traveling but did plan to camp during part of the trip. Hence, in addition to an extension ladder, a set of aluminum poles that could be bolted together to reach high cactus flowers, and mist-net poles, our field gear included camping equipment and a fair amount of food. From talking with people in Tucson who had traveled extensively in Baja, we knew where we could reliably find unleaded gas and safe drinking water, so the logistics of the trip were easy. Our plan was to drive straight to the southern tip of the peninsula and then sample the sex ratios in cardón populations on a slow trip north. I knew the location of only one *Leptonycteris* roost, but roost hunting in this unpopulated countryside would not be a high priority on this rather brief trip.

We crossed into Baja from the oak woodlands of southern California at Tecate and then drove through lush valleys covered with vineyards amidst granite hills as far south as Ensenada our first day. The next day we continued south along Mexican Highway 1 through chaparral before encountering desert vegetation, including stocky, yellow-flowered coastal agaves, candelabra cacti, and pithaya agria cacti, just north of El Rosario. Below El Rosario we entered boojum and cardón country for the first time. In 1951 Edward Goldman described this strange new landscape: "At this point the wonderfully picturesque combination of fan palms, cirios, giant cactuses, yuccas, and agaves, made a landscape suggesting the strange plant life of some remote geological period" (Goldman 1951: 57).

Boojums (*cirio* in Spanish) are members of the ocotillo family. Their tall, candlelike trunks are covered with a profusion of short, thin branches that bear small leaves only after it has rained. In contrast with the more familiar ocotillo, whose whiplike branches are graced with red flowers in the early spring, boojum's upper branches bear tubular white or yellow flowers in July and August. Boojums and cardóns, the two tallest plants on Baja, covered the hills and valleys in this part of our trip. Just before arriving at Cataviña, our second day's destination, we encountered boulder country. Massive granite boulders were strewn singly and in piles all over the countryside, as though giants had been playing a game of geological billiards. Tucked among these rocks was a lush array of desert plants whose sparse canopy was composed of boojums and cardóns.

We camped at a run-down trailer park at Cataviña for the night. Operating from years of experience as a Girl Scout leader, Marcia quickly organized the campsite into an efficient unit, and we all pitched in to cook and clean up after a supper of spaghetti, canned corn, and instant chocolate pudding.

From Loreto, Baja's first capital on the gulf coast, we recrossed the peninsula to Ciudad Constitución, passing lots of skinny, nonflowering cardóns, before turning southeast across the vast alkaline Plains of Magdalena toward La Paz. We continued another hundred kilometers south of La Paz through subtropical thorn forest, in which tall hechos jutted up through the canopy, to the coastal town of Los Barriles, near the Tropic of Cancer. There we set up camp in a modern RV park full of American sports fishermen and their families. Barriles' beautiful beach was lined with their vans, trucks, and boats. I wanted to stop at Barriles because it was close to a *Leptonycteris* roost that I'd heard about from Voytek Woloszyn, a Polish bat biologist who had worked in this part of Baja for a year in the early 1980s.

After breakfast the next morning, we set out to find this roost, which was supposed to be located near an old chapel between Santiago and Buenavista. After following a false lead to Santiago, we backtracked to Buenavista, where a policeman told me how to find *"la capilla vieja y panteón."* A few kilometers down a sandy road through a substantial population of cardóns and we were at our destination—a small red-roofed chapel next to an old graveyard surrounded by a dilapidated picket fence. Just east of the chapel was a gully with a low cave entrance at its north end. Donning my headlamp and carrying my usual bat-capture gear, I slithered several meters on my belly through a narrow tunnel, hoping not to meet a rattlesnake along the way, before entering the cave's single chamber. The cave, which was about thirty meters long by a meter and one-half wide and two meters tall, was densely packed with female Leptos with their recently weaned young. The birthing season here was about two months earlier than the one on the Mexican mainland. I conservatively estimated that this colony contained at least thirty thousand bats. After hand-netting a small sample, I quickly exited the cave to avoid further disturbing the bats. Outside, Marcia, Cara, and I processed the bats in our usual efficient fashion in the shade of our Explorer. That evening we returned to the cave at sunset to watch the exit flight. The air was filled with the sweet smell of Lepto as thousands of bats rocketed past us en route to their feeding grounds.

The next day we drove to Cabo San Lucas for lunch. Not the thriving tourist mecca that it is today, Cabo in 1905, according to Edward Goldman,

> contained about a dozen families who occupied a group of adobe houses on a flat just back of the broad sandy beach and only a few hundred yards easterly from the base of the cape. At the time of our visit the inhabitants were making a livelihood mainly by gathering tan bark from small paloblanco trees . . . and shipping it to San Francisco. A few date palms among the houses, the tropical jungle behind, and the blue sea in front, flanked by the bold headland forming the cape, unite to make this a most picturesque spot. (1951, 39)

Now, however, that same spot in Cabo San Lucas is full of luxury hotels, condominiums, shopping malls, and discotheques. Its population in 1970 was fifteen hundred but now numbers over fifteen thousand. Golf courses are springing up all over in former stands of cardón. Only the monolithic, wave-eroded rocks at *finisterra* (land's end) have not changed in this region in the nearly one hundred years since Edward Goldman noted its compelling beauty.

From Los Barriles we retraced our route north, stopping at various locations to survey cardón populations and to confirm that lesser long-nosed bats were visiting their flowers. Our nicest campsite of the entire trip was located at Bahía Concepción. Beginning about eighty kilometers north of Loreto, this sheltered bay of azure water is ringed by white sand beaches and steep basaltic hills. It was a spectacular place to spend two days camping on the beach. We also camped near Bahía de los Angeles and again at Cataviña before heading back to the States.

Once I had collected all of my *Leptonycteris* tissues, it was time to begin the DNA analyses. Although I wasn't directly involved in these analyses, I could easily visualize what was going on in Jerry Wilkinson's lab. This work entails the extraction, purification, and amplification of specific pieces of DNA through the polymerase chain reaction (PCR). Invented in the mid-1980s by Kerry Mullis, PCR has revolutionized the field of molecular genetics. It can take a minute amount of DNA and multiply (amplify) it millions of times so that its nucleotide sequence—its pattern of ACGT-base pairs—can be determined. With this process, we were able to use milligram amounts of *Leptoncyteris* tissue for DNA extraction without harming the bats. Without it, we would have had to kill bats and preserve their livers in liquid nitrogen to obtain enough DNA for analysis.

Our PCR products, each containing about five hundred base pairs worth of mtDNA, were sent to the University of Georgia's Molecular Genetics Instrumentation Facility for automated sequencing. Included in our fifty-six samples were forty-nine from thirteen *Leptonycteris* roosts in Mexico and Arizona, four from a roost in Venezuela, two from *Leptonycteris nivalis* from central Mexico, and one from a common long-tongued bat that Jerry had captured in Costa Rica. A primitive member of the subfamily Glossophaginae, *Glossophaga* and its DNA would serve as an "outgroup"—an indication of the ancestral mtDNA condition in nectar-feeding bats—in phylogenetic analyses of our genetic data.

Both electronic and paper copies of the sequence data ended up back in Jerry's lab for analysis. I was there in late June 1994, when we first ran the data through two different computer programs to see which bats and sites are most closely related. Computer analysis of morphological, chromosomal, and biochemical data in evolutionary studies is routine these days. Development of phylogenetic analysis programs was just beginning when I was in grad school in the late 1960s. Prior to this, systematists of earlier generations, including my doctoral adviser, Emmett Hooper, as well as Charles Handley and Edward Goldman, used more subjective means, usually based on morphological data, to assess evolutionary relationships among mammals.

Early in our analysis, it appeared that there might be two major matrilines of the lesser long-nosed bat in Mexico, a coastal matriline and an inland Sierra Madran matriline. If this pattern held up upon further analysis, it would imply that bats probably use two different migration routes to get to southern Arizona. But it would take a considerable amount of computer analysis by Jerry, who is gifted at this sort of thing, to tease out the actual patterns in our data. Prominent among these analyses were comparisons of our phylogenetic trees, which contained geographic information, with randomly constructed trees lacking this information. Using these comparisons, we could rigorously test the hypothesis that observed genetic similarities between sites were not simply the product of chance associations.

Results of our genetic analyses reconfirmed what I had been thinking for some time—that the mild-mannered but athletic *Leptonycteris curasoae* is a fascinating bat. One of our basic questions was, How much haplotype diversity exists in its populations? The answer was, Lots. Thirty-two of our forty-nine Mexican samples were different haplotypes, some of which differed by as much as 3 percent in their DNA sequences. A 3 percent difference between haplotypes within a species as mobile as Lepto is very unusual in vertebrates, in which within-species sequence divergence tends to be less than 1 percent. High haplotype diversity and substantial sequence divergence imply that Lepto's "genetically effective" (breeding) population size is very large—on the order of fifty thousand to one hundred thousand individuals according to theoretical calculations—and that, in the recent past, it has not undergone a strong genetic bottleneck, which results in a loss of genetic variation. Its current numbers may be lower than they were in Nelson and Goldman's time, but *L. curasoae* clearly is not on the brink of extinction the way truly endangered species such as the California condor and black-footed ferret are.

Analysis of our samples further revealed a set of geographically nonrandom genetic connections between northern and southern roosts, which supported our original suspicion that two migration pathways exist in western Mexico. We found close genetic similarities and/or shared haplotypes between two sets of roosts. From south to north, one set includes Juxtlahuaca, Isla San Andres (the mating cave in Jalisco), Baja California, Kino Bay, Cueva del Tigre, Pinacate, and the two maternity roosts in southwestern Arizona. We hypothesized that bats move between these roosts along a coastal migratory pathway, basically the one composed of spring-blooming columnar cacti that I proposed in 1993. Interestingly, our samples from the Isla San Andres cave in Jalisco shared haplotypes only with northern roosts—specifically, Kino Bay, Cueva del Tigre, and Organ Pipe Cactus National Monument—and not with the other geographically closer southern roosts. At least some (many?) of the females that leave the Isla

San Andres roost in December must migrate over fourteen hundred kilometers north to have their babies in the spring. We finally know the source of some of Arizona's migrant Leptos.

The other set of roosts included Chiapas, Ajiijic (Lago Chapala), Aduana (Alamos), the Patagonia cave, and a mine in the Chiricahua Mountains of southeastern Arizona, which I sampled in May 1993. We hypothesized that bats move between these roosts along a route on the western flank of the Sierra Madre, probably following the blooming seasons of *Agave* plants both north and south. Incredibly, the presence of the same haplotypes in Chiapas and southern Arizona suggests that some bats might actually migrate the entire length of Mexico, a one-way distance of about thirty-two hundred kilometers. In a paper reviewing the seasonal movements of lesser long-nosed bats based on capture records, Lennie Cockrum had suggested that most of the bats roosting in southeastern Arizona don't arrive there until July. The results of our study clearly indicated that the bats colonizing southeastern Arizona in the summer and the bats that form maternity roosts in southwestern Arizona in the spring come from at least two different places in Mexico.

Finally, we used our full data set to reach some broad evolutionary conclusions. Our phylogenetic analysis indicated that the current classification of *Leptonycteris* into two species and the existence of two subspecies in *L. curasoae* (i.e., a Mexican and a Venezuelan subspecies) is well-founded. Furthermore, assuming that nucleotide sequences in the control region of mtDNA of different bat taxa diverge at a rate of about 10 percent per million years (a value found in the house mouse), we could estimate the divergence times of the two species of *Leptonycteris* from each other and from *Glossophaga*. Our sequence data, for example, suggested that *Glossophaga* and *Leptonycteris* last shared a common ancestor some 2.4 million years ago; that the two species of *Leptonycteris* separated about one million years ago; and that the two subspecies of *L. curasoae* diverged about half a million years ago.

Modern molecular techniques and phylogenetic computer algorithms are powerful tools that give us increasingly more detailed and more profound views of the evolution of life. Deep views of the evolutionary process provided by the fossil record and described by George Gaylord Simpson, Loren Eisley, and Alfred Sherwood Romer had originally inspired me as an undergraduate biology student to pursue a career as a professional biologist. In my collaboration with Jerry Wilkinson, I have finally come full circle. I am once again thinking about broad evolutionary patterns and processes that produce new species. What geological and evolutionary events, apparently taking place in the late Pliocene and Pleistocene geological epochs, produced bats of the genus *Leptonycteris?* How did *Leptonycteris* end up in northern South America? Did it migrate there along a corridor of arid-zone vegetation running from Mexico along the Pacific coast of Central America to northern Venezuela? Remnants of that corridor are still visible today as far south as western Costa Rica. How long has *L. curasoae* been moving around Mexico on an annual circuit, spending part of each year in lowland dry tropical forests, then arid, cactus-dominated deserts, and finally the cooler mountain slopes where agaves abound?

These evolutionary questions inspire a new but worrisome set of questions based on contemporary events. How long will it be before the botanical links in this annual migratory chain are broken? before the distances between patches of thorn forest and desert in Sinaloa and Sonora are too great for even the mighty Lepto to traverse safely in a night's flight? before the fabric of dry tropical forest has been reduced to tatters of its former self, reduced to remnants too small to supply the bats that converge on the Chamela Bay roost in the fall with enough nectar and pollen to consummate their reproductive imperatives? Last and most important for the long-term survival of *Leptonycteris* and its kin, how can we explain the grandeur of Lepto's annual cycle, including its dependence on geographically disparate populations of columnar cacti and agaves and the importance of protecting its roost sites and food plants, to people such as don Salvador Montiel and his progeny and neighbors in rural Michoacán? Conservation, like charity, ultimately begins at home. Unless local communities value their local biological diversity, that diversity is likely to be lost.

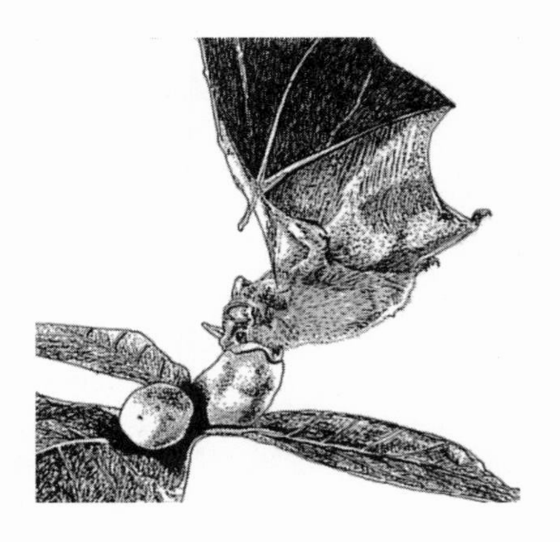

11 In the Blink of an Eye

Over the years, I've spent a fair amount of time staring at the phosphorescent green screen of night vision scopes, waiting to see bats interact with plants. These devices, originally called snooper scopes when they were first used in the Vietnam War, amplify ambient light thousands of times to form a bright image under low-light conditions. With a little supplementary light—for example, from infrared light, which is invisible to mammals—the viewing scene at night can be as bright as day. I first used a night vision scope to watch bats harvest fruit from plants in the mid-1970s at Santa Rosa National Park in Costa Rica. On different nights, I trained our scope on fruiting *Piper* bushes or fig or *Cecropia* trees to watch short-tailed fruit bats and their relatives grab fruits. More recently, I have spent many nights watching lesser long-nosed bats visit cactus flowers in Mexico and Arizona. No matter what the situation, I have always been impressed with how little time most plant-visiting bats spend in contact with their food plants. Except for large flying foxes, which often camp out for hours in the canopies of flowering or fruiting trees, these bats harvest their food literally in the blink of

Common tent-making bat grabbing a fig fruit. Redrawn by Ted Fleming, with permission, from a photo by Merlin D. Tuttle, Bat Conservation International.

an eye. About all you see on the screen during this process is a brief flurry of wings. Then they are off again to visit another flower or to take their fruit to a night roost to eat.

Viewed individually, these brief encounters might easily be dismissed as having little or no ecological or evolutionary importance. But viewed collectively, these fleeting interactions, repeated night after night throughout the world's tropics and subtropics, are far from insignificant. Because they affect the reproductive success of plants, these contacts play an exceedingly important role in the economy of tropical nature. But these brief interactions are highly vulnerable to human interference. In the blink of an eye, they can be disrupted. In many instances, all it takes is the drop of one match to destroy a roost of plant-visiting bats. Whenever this happens, bat-dependent plants suffer reduced seed set or seed dispersal. Whenever they lose their chiropteran allies, these plants are also imperiled.

I personally became aware of the vulnerability of cave-dwelling bats to human destruction during my second field season at Kino Bay. In 1990 we were still getting to know the area and were asking local Mexicans if they knew the locations of bat caves. Other than the Sierra Kino cave and another one they had taken us to in 1989, they really didn't know of any other caves in the area. As we talked, one of the men pulled a color photo of a char-broiled *Leptonycteris* out of his back pocket. It turned out that he and a friend had hauled some old tires up to the entrance of the Sierra Kino cave last summer (after we had left the field), doused them with gasoline, and set them on fire. They ended up asphyxiating dozens of lesser long-nosed bats, whose bodies they tossed behind a rock. When we checked out their story, sure enough we found many Lepto skulls and wing bones right where they said they'd be. "Why did you kill the bats?" we asked. Because they thought they were *vampiros,* they answered. Once they understood the ecological importance of lesser long-nosed bats, the men promised to spread the word about these beneficial bats and to stop burning out bat caves. Since then, I have not found overt evidence of bat destruction at the Sierra Kino cave.

In June 1999, my Mexican colleague Pancho Molina and I visited another reported Lepto roost along the Sonoran coast about halfway between Kino Bay and Guaymas. Located in an abandoned mine just outside the village of El Colorado, the roost currently housed a few California leaf-nosed bats. But the mine floor was covered with dusty guano containing thousands of shiny black cactus seeds—a sure sign that *Leptonycteris* bats roost here, probably in the summer after bats have left their maternity roosts. At the mine's entrance was a series of circular steel cables, the remnants of a pile of steel-belted tires. When asked why people had burned tires at this roost, our guide replied that it was fun to watch bats fly during the day. I mentally noted that a bit of bat education was also needed in this village.

Not all of the people inhabiting coastal Sonora view bats with a destructive bent. The Seri Indians, the group of nomadic hunter-gathers that originally lived along the Sonoran coast, have been able to live peacefully with bats for centuries. In fact, many of the bat caves in this area and on Isla Tiburón are used as sacred ceremonial caves by the Seris. The entrances to these caves, including the Sierra Kino cave, are decorated with ochre-colored hand prints and human stick figures. In the spring of 1998, a Seri elder named Roberto Molino took Pancho and me to one of those sacred caves located 120 kilometers north of Kino Bay, near the Seri village of El Desemboque. Nearly fifty-two years old, Roberto had been born on Isla Tiburón, the seasonal homestead of many Seris before they settled in the mainland villages of Punta Chueca and El Desemboque in the 1950s and 1960s. Roberto had last visited the sacred cave on Cerro Pelón when he was only eight years old. He remembered that the cave had contained a colony of bats, but he didn't know what kind of bats they were. Although they are excellent naturalists, the Seris do not have detailed knowledge about animals that have no direct economic value to them. Their knowledge about bats, therefore, is pretty rudimentary. When we found the cave, we were elated to discover that the entrance to the smaller of its two chambers was overflowing with cactus seed guano. The cave also reeked of the sweet, fruity smell of Lepto. Rather than climb into the narrow chamber and risk disturbing the bats that I could hear flying around inside, I suggested that we wait until sunset and do an exit count. Nearly five thousand Leptos left this cave just after sundown. The Seris time their visits to ceremonial caves to periods when the caves are empty. In this way, there is no conflict between humans and bats, and a kind of time-share arrangement allows the two groups to coexist peacefully. Would that all human cultures could coexist this easily with bats.

Before we examine the conflict between humans and plant-visiting bats in a bit more detail, let's review what is at stake on the biological side of this conflict. How many different kinds of plant-visiting bats are there, and how many plants do they pollinate and disperse? What are the ecological and evolutionary consequences of these interactions? What might happen if plant-visiting bats were to disappear from the face of the earth?

The world's mammal fauna contains about 270 species of plant-visiting bats—about 105 species of New World phyllostomid bats and 165 species of Old World pteropodid bats. A little over one-quarter of all bats thus are fruit and/or nectar eaters. Of these, about 46 species (34 phyllostomids and

12 pteropodids) are anatomically specialized for visiting flowers. Generally small in size, they have elongated snouts, long tongues, and teeth that are reduced in size and number. The other 224 species are mostly frugivorous, although many of them also sometimes visit flowers. These bats tend to be larger and have more robust dentition, including daggerlike canines, than their nectar-feeding kin. Regardless of their diet, many plant-visiting phyllostomids are cave dwellers, whereas most pteropodids are foliage roosters.

Species of tropical or subtropical plants whose flowers are pollinated by these bats undoubtedly number in the thousands. In his extensive review of bat-flower interactions, for example, Klaus Dobat (1985) indicates that over 750 kinds of plants are pollinated by bats worldwide. Otto von Helversen (1993) has estimated that about 1 percent of all neotropical flowering plants (i.e., 800 to 1,000 species) are bat-pollinated and that the New World tropics contain about four times more bat-pollinated species than the Old World tropics. These plants include understory herbs, palms, and shrubs as well as vines and subcanopy and canopy trees and their epiphytes. Some of the ecologically most important plant families in their habitats (e.g., the legume, balsa, catalpa, eucalyptus, and banana families in tropical forests; the cactus and agave families in arid habitats) contain many bat-pollinated species.

We have less precise knowledge about the different kinds of fruit eaten by bats, but again number of species is undoubtedly in the thousands. As in the case of bat-pollinated flowers, bats eat the fruits of a wide variety of trees, shrubs, vines, and epiphytes. Two of the dominant families of plants in the understory of many tropical forests—the *Piper* and nightshade, or tomato, families—contain many bat-dispersed species. Several important families of canopy trees, including figs, palms, eucalyptus, cashews, and sapodillas, also produce bat-dispersed fruits. Seeds of many columnar cacti in arid habitats are also dispersed by bats.

How dependent are these plants on bats for successful pollination or seed dispersal? This question is especially important from a conservation viewpoint, because the more dependent a plant is on one particular kind of pollinator or disperser, the more strongly affected that plant will be if its pollinator or disperser disappears. The degree of specialization or sole dependence on bats varies tremendously among these plants. At one end of this spectrum are plants that are highly specialized for bat pollination or dispersal. Plants that depend on bats for pollination tend to be more specialized than plants that use bats as seed dispersers. Those having features such as nocturnal flower opening (and closing) and specialized floral anatomy that excludes other potential pollinators are ensured that bats will be their only pollinators. Examples of this kind of floral specialization in the New World include calabash trees, *Bauhinia* shrubs, *Markea* vines, and the bromeliads of the genus *Vriesia.* Examples of flowers that are specialized for bat pollination in the Old World include various species of wild bananas, baobab, durian, *Oroxylum indicum,* and mangroves in the family Sonneratiaceae. At the other end of the specialization spectrum are plants such as kapok and *Inga* trees, whose flowers are pollinated by birds and insects as well as bats. As we've already seen, eucalyptus trees in Australia are also visited (and presumably pollinated) by bats and many kinds of birds.

Careful experimental studies indicate that degree of bat dependency in certain groups of plants varies geographically. For example, columnar cacti in the arid zones of tropical Mexico, Venezuela, and Curaçao are exclusively or nearly exclusively pollinated by bats. A similar situation occurs in tropical paniculate agaves. In subtropical North America, however, bat dependency is not exclusive in columnar cacti and agaves. Our studies at Kino Bay, for instance, show that cardón, saguaro, and organ pipe are effectively pollinated by birds and bees as well as by bats. Liz Slauson has obtained similar results working with *Agave palmeri* in southeastern Arizona.

In general, bat dependency tends to be lower on the fruit side of bat-plant interactions than on the flower side. To be sure, some tropical plants appear to be specialized for bat dispersal. New World examples of this specialization include *Piper* and *Solanum* shrubs and *Andira inerma* trees. As Elisabeth Kalko and her colleagues point out, New World figs tend to fall into two groups: bat-dispersed and bird-dispersed. Fruits in these two groups differ in size, color, and degree of fruiting synchrony. Compared with "bird" figs, "bat" figs occur in a wider range of sizes, are green when ripe, and ripen synchronously on trees. Old World examples of specialized bat fruits include mangoes, guavas, jackfruit, and wild bananas—many of the kinds of fruits that are important table fruits in Asian markets. On the other hand, many tropical fruits tend to be eaten by a rather wide spectrum of vertebrates. Our studies in dry tropical forest in Costa Rica, for example, showed that certain species of figs, *Cecropia, Chlorophora,* and *Muntingia* are eaten by many species of mammals and birds. Similarly, in the Sonoran Desert, the seeds of columnar cacti are dispersed by many species of birds and mammals in addition to lesser long-nosed bats.

Even when plants are either pollinated or dispersed by a variety of different animal species, however, it is likely that some species are more important for plant reproductive success than others. Thus, although flowers of the Australian tree *Syzygium cormiflorum* are visited during the day by honeyeaters (nectar-eating birds) and at night by the Queensland blossom bat, bats account for twice as much fruit set as birds. Furthermore, because this bat is a wide-ranging forager, it is more likely than birds are to carry *Syzygium* pollen between trees living in different forest fragments. This result may highlight a general feature of bat pollination. Nectar bats generally tend to be highly mobile, and this mobility can lead to extensive gene flow through the movement of pollen within and between populations of bat-visited plants.

Relatively high mobility and a tendency to fly into and across open areas such as treefall gaps and abandoned pastures also characterize the foraging behavior of many fruit-eating bats. Because of these behaviors, bats often produce seed shadows (spatial patterns of seed "rain," or deposition) significantly different from those produced by birds. Ecological studies in tropical America and Africa, for example, have shown that most of the seeds falling into seed traps placed in treefall gaps, pastures, and natural savannas are deposited at night by bats. Compared with most fruit-eating bats, many forest-dwelling frugivorous birds are much more reluctant to enter or cross open habitats. As a result, it is common in Latin America and elsewhere in the tropics for newly exposed soil—whether it is the result of natural disturbances, such as floods, landslides, and treefalls, or human-caused disturbances, such as fire and land clearing—to be quickly colonized by small-seeded, bat-dispersed plants. Fruit-eating bats thus play a critical role in the regeneration of tropical forests. They truly are the prime movers of the seeds of early successional plants.

I was able to witness this colonization process first-hand at Santa Rosa National Park, beginning in 1979. That year, the park's main road was paved, and during this process, a wide strip of bare soil was exposed on both sides of the road. As soon as the rains began, a variety of seedlings began to sprout beside the road. Closer inspection revealed that many of those seedling species were very familiar to me; they included those that routinely occur in the wet-season diet of the short-tailed fruit bat. Thus, two or three species of *Piper* began poking up out the soil, along with seedlings of *Solanum, Cecropia, Muntingia,* and *Chlorophora.* I could even spot old night roosts of bats because of the high density and diversity of these plants in the soil beneath them. Over the next few years, I was able to watch some of these fast-growing plants reach reproductive maturity and begin to feed new generations of bats.

A much longer-term study of the colonization of disturbed land by bat- or bird-dispersed plants comes from several generations of botanists that have worked on the islands of Krakatau in Southeast Asia. These three small islands, which are located between the much larger islands of Java and Sumatra, were completely sterilized by a volcanic eruption in 1883. Dust from this eruption remained airborne and darkened skies worldwide for at least two years. According to Robert Whittaker and Stephen Jones (1994), about 40 percent of Krakatau's current flowering plant flora arrived there via bird or bat dispersal. Flying foxes and their kin have been important dispersers of about forty species of over-water colonists. They have likely played an especially important role in the colonization and subsequent dispersal of several species of figs. More generally, Paul Cox and his colleagues have argued that these large bats play a key role in the preservation of biological diversity on many isolated islands in the South Pacific through their pollination and seed disperser "services" (Cox et al. 1991).

These examples suggest that it would be a mistake to conclude, solely on the basis of a species list of who pollinates or disperses a particular plant, that all pollinators or dispersers have the same impact on a plant's reproductive success. Indeed, not all pollinators or seed dispersers are created equal, and not all species are interchangeable in their ecological function. In most plant-pollinator or seed-disperser interactions, some species are likely to be much more effective than others. In many cases, it appears that bats are often more effective in these roles than other kinds of animals, the major reason being their greater mobility (and their gentler treatment of seeds compared, for example, with that of monkeys).

Bats, then, are ecologically important pollinators and seed dispersers in many tropical and subtropical habitats. If plant-visiting bats were to disappear from the face of the earth, the lives of many plants and many habitats would be negatively affected. The loss of these bats would have nontrivial ecological and, in the long run, evolutionary consequences. While I, and many other biologists, find these ecological and evolutionary reasons sufficient to argue for the conservation of plant-visiting bats (and all other kinds), much of the rest of mankind seems to feel differently. In the Western world at least, a species' or habitat's value must be expressed in economic terms before people will pay it heed, before they will value its conservation. In *The Arrogance of Humanism,* David Ehrenfeld expresses this "conservation dilemma": "The difficulty is that the humanistic world accepts conservation of Nature only piecemeal and at a price: there must be a logical, practical reason for saving each and every part of the natural world that we wish to preserve. . . . Species and communities that lack an economic value or demonstrated potential value as natural resources are not easily protected in societies that have a strongly exploitative relationship with Nature" (1981, 177).

I noted a relatively minor example of our tendency to place an economic value on wildlife when we were on sabbatical in Tucson, Arizona, in 2000. Kartchner Caverns, a spectacular series of chambers filled with geological wonders, were opened for public visitation in the fall of 1999. The caverns were an immediate hit, and nearly all of the daily tours became booked solid for months in advance. The human-bat conflict in this instance was in the form of a seasonal maternity roost of the cave myotis, one of the Southwest's most common insectivorous bats. Local bat biologists recommended that certain parts of the cavern system be closed to human traffic during this colony's maternity period in June and July. This prompted the state to reply that partly closing the cavern could result in the loss of about a million dollars in tourist revenue. Amused to see that an obscure bat finally had a price on its collective head, I was prompted to write a letter to the editor of the *Arizona Daily Star:*

> The recent estimate of $1 million in lost ticket sales if Kartchner Caverns is closed during the Myotis velifer maternity period is the latest in a long line of examples of our tendency to view native plants and animals solely in economic terms. One of Arizona's most common bats, the cave myotis has undoubtedly been using Kartchner Caverns as a safe maternity roost for thousands of years. Until recently, I'm sure no one ever viewed this . . . species as negatively affecting the economy. Now that Kartchner has become a "cash cow," this species suddenly has considerable economic impact. Simply because it discovered Kartchner long before cavers did, it is now in danger of being vilified for reducing state revenues and for disrupting people's vacation plans.
>
> Like the ferruginous pygmy owl, the cave myotis is blameless in the debate between development and protection of Arizona's biological heritage. These and many other species deserve safe nesting or roosting sites. Several decades ago, a maternity colony of lesser long-nosed bats, one of the major pollinators of saguaro cactus flowers, was excluded from Colossal Cave during its commercial development. Let's not repeat this biological insensitivity by eliminating the cave myotis from Kartchner Caverns during its commercialization. (*Arizona Daily Star,* 8 April 2000)

Bat Conservation International takes the economic impact of bats seriously and frequently uses available economic data to argue for bat conservation. In the case of plant-visiting bats, for example, Marty Fujita and Merlin Tuttle have reported that in Southeast Asia, pteropodid bats are responsible for pollinating or dispersing the seeds of at least 289 species in 59 plant families. They noted that at least 448 products, ranging from drinks and dyes to medicines and timber, come from 186 of these species. Some products, such as durian fruits and petai seeds, have market values of millions of dollars annually. At least as lucrative are the timber and small wood products created annually from 105 plants on their list. They concluded that Asian economies stand to lose substantial amounts of money if they allow pteropodid bats to be extirpated.

Despite the undoubted ecological and economic importance of flying foxes and their kin in natural and human systems of the Old World tropics, they have been steadily persecuted in many areas to the point where their populations currently are only a fraction of their former sizes. One of the main causes of pteropodid declines is hunting pressure. Flying foxes are large bats (some species weigh a kilo or more), and in crowded Southeast Asia, including many South Pacific islands, these animals represent important sources of protein. Human consumption of these animals is a long-standing tradition. On some South Pacific islands, it has probably led to the extinction of several species. Karl Koopman and David Steadman, for instance, have reported that prior to human occupation, the island of 'Eua in the Tonga group supported three species of flying foxes. Now only one species occurs there. Steadman has also documented the extinction of many species of birds, probably through the direct hand of man, on many South Pacific islands (Koopman and Steadman 1995). Hunting pressure plus the occasional effects of cyclones have led to serious population declines in flying foxes living on Guam and American Samoa in recent decades.

In addition to being eaten as food, flying foxes have considerable medicinal value for some populations. The Chinese living on Sarawak, for example, consider bat meat to be good for curing asthma, kidney ailments, and general malaise. Overzealous "pest control" of flying foxes living near fruit orchards in tropical Asia and Australia has also had a strong negative impact on local bat populations. As Francis Ratcliffe (1932) observed decades ago in eastern Australia, the economic impact of flying foxes is often exaggerated to justify persecution of these animals. Nonetheless, fruit farmers throughout the Old World tropics consider bats to be vermin and continue to kill thousands of them annually to "protect" their crops.

Being microbats, New World plant-visiting phyllostomids are not often hunted for food. A person would have to be pretty desperate to consume ten- to fifty-gram bats regularly (although some human societies have routinely eaten small bats and songbirds). Early in my career, when I was studying the ecology of tropical rodents, I ate each of my study species at least once just to see how they tasted. This really put off Francisco, our cook at La Selva, who could barely tolerate pan frying a couple of spiny pocket mice for me. When I became a full-time bat ecologist, however, this habit stopped.

I just didn't think it would be worth it nutritionally to fry up a mess of Jamaican fruit-eating bats.

While they probably are safe from hunting pressure because of their small size, tropical phyllostomids (and other bats) have the misfortune of being associated with or mistaken for vampire bats—a problem not faced by pteropodid bats. Because of this guilt by association, hundreds of thousands of innocent bats have perished during nonselective vampire bat control programs involving the dynamiting or burning out of bat roosts in caves and mines. Bat Conservation International learned early on that the key to bat conservation in the New World tropics is to educate people about proper vampire control. Selective control of vampires now involves the use of anticoagulant chemicals, either spread on the bats themselves as paste or injected into domestic livestock being "bled" by vampires. These highly social bats groom each other, which exposes several individuals to the paste applied to one bat. They also tend to feed repeatedly on the same victim and can acquire fatal doses of anticoagulant in the process. Most important, bats don't have to be killed wholesale to reduce vampire populations in areas where they pose an economic threat.

In addition to being maligned in Latin America because of the *vampiro* problem, phyllostomid (and other) bats are sometimes slaughtered because of the *chupacabra* problem. *Chupacabras* are mythical creatures that kill other animals (and people?) under the cover of darkness by complete desanguination. Descriptions suggest that they may be half-man, half-bat. A drawing of one that I saw in Mexico actually bore a rather striking resemblance to the current Mexican president! Whereas *vampiros,* which are far too small to desanguinate large animals totally in one feeding, are a chronic problem, *chupacabras* only occasionally undergo population outbreaks, typically during times of economic hardship in Latin America, a Mexican friend once told me. The last outbreak of *chupacabras* in Mexico and elsewhere in Latin America, for example, occurred in the mid-1990s, a time of widespread economic distress. At that time, we could hear heated discussions about these creatures on Radio Sonora as we drove around Kino Bay. When Pancho Molina and I began searching for bat caves around Guaymas in 1998, people told us that many caves had been burned out recently at the height of the *chupacabra* scare. By the spring of 2000, this scare had disappeared from Mexico, but people still mentioned *chupacabras* wherever we looked for caves in western Mexico. A restaurant owner in Sinaloa told my traveling companion, Tom Vandevender, and me that although he didn't really believe in *chupacabras,* he had seen their handiwork. On a nearby ranch he had seen a flock of sheep that had been completely gutted in a totally bloodless fashion. What else could have done such a thing? he asked us.

As if hunting pressure and outright persecution were not enough, plant-visiting bats also face the universal threats experienced by all wildlife—habitat destruction and habitat fragmentation. Loss of feeding and roosting grounds (in the case of bats living in hollow trees or tree foliage) through land conversion to agricultural fields and pastures and through timber extraction is rampant in the tropics and subtropics. In some places in southeastern Australia, for example, eucalypt forests and other habitats providing food and roosts for flying foxes have been reduced to one-tenth of their former area, and many flying fox colonies have disappeared or decreased markedly in size in recent decades. Similarly, rain-forest habitat on South Pacific islands has steadily disappeared with growing human populations, and island carrying capacities for flying foxes are now much reduced.

An excellent example of this "island effect" comes from the doctoral research of my student Sophie Petit. Sophie studied interactions between nectar-feeding bats (the lesser long-nosed and southern long-tongued bats) and three species of columnar cacti on the island of Curaçao. By conducting aerial surveys of cactus densities and distributions, she and her collaborator Leon Pors estimated that the island presently can support a total population of only about two thousand nectar-feeding bats during the summer maternity period, when their energy demands are greatest (Petit and Pors 1996). They concluded that further destruction of cactus populations would place these bats dangerously close to the brink of extinction. Clearly, cactus conservation (plus cave protection) is desperately needed on Curaçao to prevent these extinctions from occurring.

Along with forest destruction and land clearing, forest fragmentation—the creation of forest "islands" in a sea of pasture or agricultural land—also threatens the persistence of plant-visiting bats and many other kinds of wildlife. These islands often still contain food resources and potential roost sites for bats, but the question remains, How fragmented can a forested (or any other wild) landscape become and still support healthy bat populations? Because of their mobility, bats are less likely to suffer from habitat fragmentation than other animals (for example, primates and many kinds of forest understory birds), but it is still possible that this form of landscape modification will have a negative effect on them.

Two recent studies indicate the possible consequences of forest fragmentation for plant- visiting bats. Mark Schulze and colleagues have compared the abundance and diversity of phyllostomid bats in continuous forest and forest fragments in Tikal National Park, Guatemala. While the overall abundance and diversity of bats did not differ between the two habitat types, the proportional representation of different kinds of bats did. In continuous forest, their mist-net captures were dominated by two relatively large species of *Artibeus*—species that feed on the fruits of canopy trees. In contrast, captures in forest fragments ranging from about three to two hundred hectares in area were dominated by small species, principally a species of *Carollia* and the little yellow-shouldered bat—species that feed on the fruits of early successional plants. These authors noted that large canopy trees producing fruits eaten by *Artibeus* bats have been selectively removed from forest fragments whose understories are now dominated by light-demanding species such as *Piper* shrubs. A change in the availability of different kinds of fruit resources has thus led to a change in the composition of bat communities in these fragments (Schulze, Seavy, and Whitacre 2000).

Somewhat different results were reported by Jean-François Cosson and colleagues in their study of changes in the structure of bat communities following the creation of forest islands when a river was dammed in French Guiana. In the first three years after the islands were created, both abundance and diversity of bats declined in the forest islands but not in nearby continuous forest. In contrast with the Guatemalan study, understory-feeding bats were affected more strongly by this form of fragmentation than canopy-feeding bats (Cosson, Pons, and Masson 1999). The absence of selective logging in French Guiana probably explains, in part, why these studies report different results.

These and other studies clearly indicate that fragmentation has an effect on the composition of bat communities. This effect tends to be selective: some species become more common (relatively speaking), whereas others become less common after fragmentation. And in the case of plant-visiting bats, who benefits and who loses depends on the nature of vegetation changes that accompany fragmentation. Overall, however, fragmentation studies are in their infancy, and it is too early, in most cases, to know the final outcome of this anthropogenic perturbation. Theoretical studies and a few long-term studies with birds clearly show that fragments will lose species through time. The original diversity of plants and animals contained within an area of continuous forest will inevitably decline through time as a result of forest fragmentation. Bats will be no exception to this general trend. Bat diversity in a fragmented world will inevitably be a shadow of its former richness. In such a world, plant-visiting bats that feed on the fruits of light-demanding, early successional plants will probably prosper, whereas canopy-feeding species will suffer.

Outright persecution and habitat destruction thus appear to be the greatest threats faced by plant-visiting and other bats today. Persecution of bats often stems from ignorance, and the key to reducing this threat has to be education of the general public worldwide about the beneficial aspects of bats. Instead of viewing bats as malevolent creatures and enemies, people need to view them in a positive light—as our allies in the control of insect pests or in the propagation of plant populations. Once bats are viewed positively, then it might be possible, though I fear that I'm being terribly idealistic here, that bats can be appreciated simply because of their biological uniqueness. Bats are, of course, the world's only flying mammals. Nearly a thousand species strong, they have been around for over fifty million years, have evolved some pretty amazing lifestyles, and are exquisitely adapted for flying in cluttered environments in total darkness. No other vertebrate group has come close to achieving this level of evolutionary success in the nocturnal aerial realm. To reduce the abundance and diversity of these animals substantially would be yet another atrocity that our short-sighted and dangerously exploitative species has inflicted on the natural world.

All is not lost for these fascinating animals, of course. The number of bat species known to have gone extinct in historic times is small—fewer than two dozen, according to the International Union for the Conservation of Nature and Natural Resources (IUCN). On the basis of current information, which in the case of bats is highly incomplete, only about sixty-two species (6.2 percent of all bats) are considered by the IUCN to be endangered. But these numbers are no cause for complacency. An ever-increasing human population, especially in tropical parts of the world, where the diversity of bats in general and plant-visiting bats in particular is highest, coupled with increasing material expectations and resource consumption, does not bode well for most of wild nature.

Concern for bats and their conservation has increased markedly in recent decades. As recently as 1973, the venerable *National Geographic* magazine was able to publish an article insensitively titled "Bats Aren't All Bad" without raising a storm of protests. In less than a decade, the tone of its bat articles, now featuring the photos and research of Merlin Tuttle, had changed completely. Gone were the snarly faced images of bats that frighten the general public. These have been replaced by more benign countenances and images of bats behaving in interesting ways. Over this time period, many professional biologists, as well as small armies of enthusiastic amateurs, have "gone to bat" for the protection of bats around the world. National and international legislation protecting bats has been enacted. International traffic in flying foxes as luxury food items in Asia, for instance, is now prohibited under the Convention on International Trade in Endangered Species of Wild Fauna and Flora (CITES). The amount and quality of educational materials promoting greater understanding and appreciation of bats also increase yearly. When our children were growing up in the 1970s and the early 1980s there was only one children's book—*Hattie the Backstage Bat*—that we could read to them. Now, the Bat Conservation International sales catalog routinely sells several bat books for small children, including the classic *Stellaluna,* along with a variety of educational materials based on bats and their conservation.

The BCI catalog also includes four children's book written in Spanish: *Marcelo el murciélago* (about the Mexican free-tailed bat), *Valentin, un murciélago especíal* (about the common vampire bat), *Don Sabino, el murciélago de la ciudad* (about urban bats), and *Flores para Lucía, la murciélaga* (about the lesser long-nosed bat). Written and illustrated by two talented Mexicans (Laura Navarro and Juan Sebastian Barberá Durón, respectively), these books are part of the binational Programa para la Conservación de Murciélagos Migratorios de México y los Estados Unidos de America (PCMM). Founded in 1994, PCMM is a coalition of U.S. and Mexican scientists and educators that is coordinated by BCI and the Instituto de Ecología of the Universidad Nacional Autonoma de México. Aims of the program, which until recently was directed by BCI's Steve Walker and UNAM's Rodrigo Medellin, include determining and safeguarding the population status of three important migratory bats (the Mexican free-tailed and lesser and greater long-nosed bats); identifying and protecting important roost caves of these species; and educating the Mexican people about the importance of these and other species of bats.

The educational portion of this program has been especially impressive. Not only has Laura Navarro produced four engaging books highlighting different aspects of the behavior and ecology of important bats, she and her assistants have also created a set of teaching materials, including puzzles, coloring books, toys, and games, that have been effective in changing the attitudes of Mexican schoolchildren toward bats. These materials are strategically introduced to school kids living in towns and villages near major bat caves, so that children and their parents will develop a sense of community stewardship toward their local bats. The pilot run of this program, which took place near Cueva de la Boca in the northeastern state of Nuevo León, has been spectacularly successful. In the period 1995–2000, the Mexican free-tailed bat colony living in this cave increased in size from about 100,000 to over 1.5 million bats, largely as a result of the cave's protection from human disturbance. Finally, Navarro and Rodrigo Medellin have produced a series of twenty fifteen-minute-long radio programs on bats and their conservation that have been aired on Radio Educación throughout Mexico. Entertaining and scientifically accurate, these wonderful programs have been heard by thousands of Mexican schoolchildren.

On the scientific side, PCMM has made steady progress in documenting the migratory behavior of its three target species. For *Leptonycteris* bats, for example, we now know the locations of several new maternity roosts, and Rodrigo and his coworkers have documented the seasonal population dynamics of several roosts in central and southern Mexico. He and I are collaborating on fine-tuning our understanding of the genetic structure and migratory behavior of the lesser long-nosed bat using mitochondrial genetic markers. Our overall goal here is to develop a map of the seasonal genetic connections between different roosts within Mexico and between Mexico and the southwestern United States. Such a map will be of fundamental importance for identifying multiple migratory pathways and for developing a binational conservation plan for protecting this highly mobile species.

The combination of sound basic science, effective education, and realistic conservation planning espoused by PCMM makes this program an excellent model that can be exported to other parts of the world to further the protection of bats. Both science and humanity benefit from such a multifaceted approach. To be sure, interesting new information about the natural history and behavior and ecology of bats will emerge from this approach. But more important, in the long run, will be the education of a much larger portion of mankind about the fascinating lives of our earth's flying mammals. With increased education and understanding, there is hope that the number of extinct or endangered species of bats on the IUCN list will remain small.

Let's let Lucía, *la murciélaga,* a young lesser long-nosed bat who has just safely completed her first season's migration from the Sonoran Desert, speak for all bats:

> Estaba en casa, pero se sentia muy distinta. Ya no era una pequeña. Ya sabía sobrevivir por si misma y sabía también que la vida está llena de experiencias maravillosas. Ahora sí, Lucía estaba lista para enseñarle a su futura cría todo lo que su sabia madre le había enseñado.
>
> *Lucía was home, but she felt quite different. She was not a little bat anymore. She knew how to survive on her own, and she also knew that life was full of wonderful experiences. One day, she would teach her own baby the things her mother had taught her. (Navarro 2000, 38)*

More and more, the fate of Lucía and all her kin now rests in our hands. To avoid extinction, Lucía needs an Ecological Bill of Rights that includes access to safe roost sites, extensive feeding grounds, intact migration corridors, and a pesticide-free environment. These "inalienable rights" certainly are not unique to bats. In one form or another, all creatures great and small on earth have a similar set of requirements. Our species simply has to do a better job of sharing the earth's resources with all of its other inhabitants.

Epilogue

Late January 1996 marked the thirtieth anniversary of my work with tropical bats. I celebrated this milestone in a cattle pasture near the village of Montepío, located about two hundred kilometers southeast of Veracruz, on Mexico's Gulf coast. Receiving nearly six meters of rain a year, Montepío lies in a region of hills, valleys, ancient volcanoes, and crater lakes, a region that, until recently, was covered with the neotropic's northernmost extension of lowland rain forest. Not as rich in species as lowland rain forests in Costa Rica and Panama, this forest contains (or at least did before deforestation) about 150 species of trees, 250 species of birds, and 90 species of mammals, including 38 bats. Sadly, most of the forest around Montepío is now gone. Remaining forest occurs as scattered fragments surrounded by cattle pastures and fields planted in corn, beans, yucca, and bananas. Subsistence farming and cattle raising dominate the landscape and local economy in this part of Mexico.

I was in coastal Veracruz in the company of two Mexican biologists, Vinicio Sosa and Jorge Galindo. One of my former doctoral students, Vinnie is now a research biologist at the Instituto de Ecología in Jalapa, the capital of

Mexican free-tailed bats exiting Bracken Cave, Texas. Drawing by Ted Fleming.

Veracruz State. He conducted his doctoral research on seed dispersal and seedling establishment in three species of columnar cacti at my study site at Bahía Kino. Vinnie was now codirecting Jorge Galindo's doctoral research on the role that bats play in the regeneration of rain forest in abandoned pastures. In the preceding few years, two of Vinnie's colleagues had studied the role played by birds in bringing seeds from forest fragments into pastures, but nothing was known about the importance of bats in this kind of reforestation.

That night in the cattle pasture we were netting bats around two isolated standing trees at its center. With their flat-topped canopies and straight, epiphyte-laden boles, these trees, which are now a common feature of the landscape in this region, are a gaunt reminder of the former glory of the tropical rain forest. Dan Janzen has called such trees the living dead. However, this designation may not necessarily be appropriate around Montepío. The bird studies, for example, have shown that isolated standing trees are common stopping places for many individuals of canopy-feeding species on their way from one patch of forest to another. Frugivorous birds feed heavily on their fruits and presumably carry some of their seeds away when they leave them. They also deposit many kinds of seeds under them during their stops. When fencing excludes cows from around the bases of such trees, a rather diverse miniforest springs up under their canopies.

We set two mist nets in the shape of an L about ten meters from the base of each tree. One was a fig tree that should attract hoards of bats when it is fruiting; that day it was barren. The other was a "control" tree that does not produce fleshy fruits attractive to birds or bats. When we set our nets before sunset, I heard many familiar bird sounds from my three decades in the tropics: the harsh *ee-ahhhhhhh* of a roadside hawk, the gurgling clacks and bubbles of an oropendola (a sound I first heard near Jaqué, Panama), the piercing squawks of a flock of *Amazona* parrots, and the rattling cry of a lineated woodpecker. In the distance, a lone howler monkey roars in the dying sunlight. Vinnie wonders, "Is that the last *sarahuato* in this part of Mexico?" The other species of native monkey, the spider monkey or *mono araña,* was shot out for food decades ago. Howlers owe their continued existence here and throughout Latin America to their rather poor-tasting flesh.

We opened the four nets at sunset. A first quarter moon depressed bat activity in the open and made our nets quite visible to most bats. In five hours we caught only ten bats of five species. By luck, I happened to remove the only two individuals of short-tailed fruit bats that we captured. One of these defecated a sweet-smelling mass of *Piper* pulp chock full of tiny seeds onto my gloved hand. How many times had this happened to me in the past thirty years? Jorge systematically took data on each bat, marked it with a color-coded series of small plastic rings on a plastic necklace, and released it. The moon had set, and the pastoral landscape was partly obscured by swirling mist as we walked back to our truck. The isolated standing trees truly appeared to be ghosts in this setting. One or more common mustached bats, hunting insects a meter or so above the ground, silently sliced through our headlamp beams as we walked along single file. The insects attracted to our lights were so tempting that the bats seemed unaware of, or are at least unconcerned by, our presence.

The next night we set two nets diagonally across a narrow stream in a forest corridor near last night's isolated standing trees. In marked contrast with our low catch of last night, we caught nearly forty bats of ten species. Like I-95 and its rush hour traffic every morning in Miami, the dark and leafy streambed was full of bats commuting to their feeding areas at sunset. The three of us worked steadily for an hour, removing bats and placing them in individual cloth bags. Jorge would process the bats when the capture rate slowed down. Two of the species we caught were new to me, at least as captures away from their day roosts. The ghost-faced bat has a grotesque series of skin folds encircling its face, which give it a countenance that Chester Gould, the creator of Dick Tracy and his archenemy Pruneface, would surely envy. Its cup-shaped ears and tiny eyes complete a face so ugly that it is actually cute. What is the function of all those skin folds? I wondered. How and in what habitats does this bat capture insects? The other bat, a large species of *Myotis,* is much more conventional, and hence less interesting, in appearance. To Jorge's delight, four of our bats were recaptures that he had banded at individual standing trees up to three months before.

Back at our motel in Montepío, we sat on a porch drinking a Corona (a wimpier beer than Coors!) while watching an insectivorous bat fly back and forth in front of us and listening to the Gulf's surf pound against the beach. We talked about the modifications Jorge would have to make to his research plan to develop it into a full-blown doctoral dissertation. At least ten years older than I was when I began my doctoral research, Jorge was a seasoned field biologist. Vinnie and I were confident that with patience and perseverance, he would learn a lot about the movement patterns of fruit-eating bats and their role as seed dispersers in this highly disturbed landscape.

The next day we packed up, and after a delicious breakfast of Mexican scrambled eggs, spicy empanadas filled with cheese, fresh, thick tortillas, and rich *café de olla* at doña Reyna's modest open-air restaurant, we headed back to Jalapa, some 280 kilometers to the northwest. For the first few kilometers we drove along a dusty gravel road that becomes deeply rutted with mud during the long rainy season. Except for a short stretch that passes through the Los Tuxtlas Biological Station, which is owned by the Universidad Nacional Autonoma de México in Mexico City, the road was surrounded by open pastures and fields. The Los Tuxtlas section passed through a green tunnel of thick vegetation—the jungle that originally covered this part of Mexico. A diversity of *Piper*s and wild gingers choked the forest understory. Spindly *Cecropia*s and a host of other trees formed the canopy overhead. Too soon we passed back into bright sunlight and left the forest and its dwindling wildlife behind.

I first visited the Los Tuxtlas field station in 1984 at the invitation of Alex Estrada and his wife, Rosy Coates-Estrada, two of the station's research ecologists. The Estradas and I shared a common interest in seed dispersal ecology, and we eventually co-organized two international symposia on this topic. In 1984 the thirty kilometers of road from Catemaco north to the station were unpaved, and a fair portion of it still passed through forest. When my family and I visited the Estradas in December 1988, it took us nearly three hours to negotiate those thirty kilometers of mud and ruts in a rented sedan. In 1996, all but the last ten kilometers from Catemaco to Los Tuxtlas were paved, and most of the forest had been cut. This is the one part of the world where I have personally seen the rapid loss of rain forest. Los Tuxtlas and its seven hundred hectares of forest were fast becoming a small island, a Noah's ark desperately foundering at anchor in a storm called human population growth and human need.

We pushed on past Catemaco toward Veracruz. The kilometers zipped past, and soon we were out of the wet tropics. Near Veracruz we turned west toward Jalapa and began to climb toward the southeastern flank of the Sierra Madre Oriental through dry tropical forest, highly disturbed, of course, and mostly replaced by scrubby farmland. On rocky hillsides and in barrancas I saw a few bat plants: at least two species of columnar cacti, now barren of flowers; white-flowered morning glory trees; kapok trees bearing clusters of pale pink flowers; and calabash trees. This was *Leptonycteris* country. *Carollia* country was far south of us. By midafternoon we reached Jalapa, hilly and compact at an elevation of seventeen hundred meters. It was a beautiful sunny afternoon, but the weather was changing. A *norte* with its cool, strong winds and, sometimes, torrents of rain was approaching.

In Jalapa my thoughts turned to two earlier explorers of the Mexican fauna and flora, Edward Goldman and Edward Nelson. Six days earlier, on the drive from Mexico City to Jalapa, I had noted with considerable interest Cerro Malinche to the north of Highway 150 and the snow-capped volcanic peaks of Popocatépetl and Ixtacihuatl jutting majestically up from the Mexican plateau to the south. Malinche was the place where twenty-year-old Goldman was hunted by bandits in May 1893. Two months earlier, Goldman and Nelson had climbed both southern peaks in the incredibly short span of two weeks and ended up with severe cases of snow blindness. They worked around San Andres Tuxtla and Catemaco near Los Tuxtlas in the spring of 1894, moving no more than sixteen kilometers a day between collecting sites. On our drive that day in 1996 we covered in about four and a half hours a distance that would have taken the Biological Survey team six to ten days to cover on horseback if traveling at maximum speeds of thirty to fifty kilometers a day.

To me, the difference between those travel times symbolizes how much our physical world has changed in the last hundred years. At the end of the nineteenth century, large parts of the world, especially in the tropics, were accessible only with difficulty. That no longer is the case. Two of my recent graduate students, Andy Mack and Debra Wright, for example, studied cassowaries and the seeds they disperse at a remote site in the mountains of eastern Papua New Guinea. There they built a comfortable field station about a day's hike from the nearest village. But if they had run into an emergency, a helicopter could have reached them in less than an hour. In fact, in constructing their field station, they hired a helicopter to bring in most of the building materials, including a flush toilet.

Greater accessibility, of course, has made the natural world much more vulnerable to human modification. Such modification has been ongoing throughout human history and earlier, but its pace has accelerated enormously in the last century, in step with the exponential growth of our population. About 67 percent of Costa Rica, for example, was still covered in primary forest in 1940, thirty years before I began to work there. Now, less than 17 percent of its primary forest remains, and over fifty thousand hectares of forest are being cleared annually. At current rates of deforestation, the only primary forest left will soon be restricted to its widely scattered national parks and forest reserves.

Tremendous changes have also occurred in our intellectual world in the last hundred years. At the end of the nineteenth century, certain branches of biology, including ecology, animal behavior, genetics, and evolution, were just beginning to emerge as self-conscious disciplines and were very much in the shadow of more venerable disciplines such as anatomy and physiology. The first Ph.D.'s in mammalogy in the United States were not awarded until 1919, well after Goldman and Nelson had completed their historic biological surveys of Mexico. Now, we have sequenced the human genome, and ecology, behavior, and evolution are flourishing branches of biology (but still remain in the shadow of the medical sciences in terms of public recognition and research funding).

Symbolic of the burgeoning technology that we now use to ask and answer biological questions is the laptop computer that I routinely carry with me on trips and to my field sites. This marvelous machine, which weighs a couple of pounds, has much more computing power than ENIAC (shorthand for "electronic numerical integrator and computer"), the world's first electronic computer. Weighing about thirty tons and containing nineteen thousand vacuum tubes, ENIAC ushered in the "information age" in February 1946, when I was almost four years old and the year of Edward Goldman's death. During my visit to Jalapa's Instituto de Ecología, I was impressed by the number of computers and the amount of computer-related technology, including geographic information systems (GIS) and satellite imagery, that were being used by its graduate students and researchers. During my visit, the institute's buildings were being wired with many kilometers of snakelike gray cable for an Ethernet connection that will give most of their computers access to the Internet and the World Wide Web of information.

While computers have become essential tools for acquiring, analyzing, and synthesizing our data as well as a major means of communicating with colleagues down the hall or around the world, we biologists run the risk of losing contact with the real world and the organisms that initially attracted us to our profession. As someone once wryly noted to me, "Just add a couple more computers to the main laboratory building at La Selva and no one will have to go out into the forest anymore!" Somehow, I can't imagine field biologists spending most of their time staring at a computer screen, but the temptation, at least, will always be there.

My own career reflects in small part the intellectual and technological changes that have occurred since the time when Edward Goldman was an active field biologist. Like Goldman in the 1890s and beyond, I began my career collecting mammals for the Smithsonian Institution, but in Panama rather than in Mexico. The collecting protocols I used were essentially the same ones used by Goldman and Nelson in Mexico a hundred years ago. Since my Panama days, my research has progressed to more sophisticated studies using radiotelemetry and starch gel electrophoresis in the 1970s and 1980s and stable isotope analysis and the amplification of DNA by the polymerase chain reaction in the 1990s. I suspect that the kinds of biological questions we can ask and answer with today's technology would have amazed and fascinated Edward Goldman. We now have the intellectual and technological tools to probe some of life's deepest secrets, including, as my former colleague David Hillis likes to say, "the phylogeny of life." I personally look forward to the day when we will have a clear picture of the evolutionary history of columnar cacti and their nectar- and pollen-eating bats. When and where did these magnificent plants and animals evolve? For how long have these organisms been central players in the ecology of New World deserts and arid lands?

In the midst of this technological explosion and intellectual excitement, however, field biologists like me face a dark reality: the world is currently experiencing a biodiversity crisis. Biological diversity—that incomparably rich variety of life on earth and the web of interactions among species—is the bread and butter of field biology. Though he wasn't directly concerned with theoretical issues involving biodiversity, Edward Goldman certainly dealt with it on a daily basis, both in the field and in later life when he described and cataloged the mammal faunas of North America, Mexico, and Panama. He and his mentor Edward Nelson also were active in conservation of this diversity, most notably in the conservation of migrant birds, a problem of crucial importance today.

Today, those of us studying biological diversity are in a race—often called a desperate race—against time. Species are being lost at an accelerating rate as humans convert wildlands into domesticated and urbanized, biologically depauperate landscapes. A simple but poignant example of this loss comes from the observations of Luther Goldman, Edward's youngest son, who was born in 1908. In my first conversation with him in 1994, Luther told me that he originally was interested in mammals as a result of his father's influence but didn't want to follow in his dad's illustrious footsteps. So he specialized in ornithology and joined the Department of the Interior as a manager of wildlife refuges, initially in southern California and eventually on the Gulf coast of Texas.

Later in his career Luther worked in Washington, D.C., and he and his family moved into his wife's childhood home on an acre of land in nearby College Park, Maryland, in 1959. Shortly thereafter, he began to list every day on his kitchen calendar the birds he saw on his property. His list eventually included over 125 species. In recent years, however, he has seen far fewer species than he did in the early 1960s. Birds that have been missing from his property for years include yellow-billed cuckoos, red-headed woodpeckers, tree swallows, eastern bluebirds, indigo buntings, and chipping sparrows. He has witnessed sharp decreases in the abundance of many other species, including ruby-throated hummingbirds, eastern kingbirds, hermit thrushes, and yellow warblers. In contrast, birds that live comfortably in close contact with man, such as mourning doves, starlings, and house finches, have been increasing in abundance.

None of the birds that Luther Goldman no longer sees on his Maryland property have yet gone extinct. The distributions of some species have simply shifted as urbanization has crowded out forests and fields. But populations of a number of neotropical migrants have also decreased, for two major reasons. First, owing to the fragmentation of forests throughout the eastern United States, adults and young of many species that build open, cup-shaped nests have become much more vulnerable to predators such as raccoons, opossums, and cats as well as the brown-headed cowbird, an avian nest parasite. And second, the winter habitats of many tropical migrants have shrunk as a result of deforestation and habitat modification in tropical America.

How much longer can our tropical migrant birds sustain these two assaults on their well-being? According to John Terborgh, one of America's leading students of neotropical migrant birds, by the time tropical countries reach a population density of about one hundred per square kilometer, they have lost virtually all of their native forest. In 1986, four countries—the Dominican Republic, El Salvador, Haiti, and Trinidad and Tobago—had densities this high. Of these, El Salvador and Haiti are completely deforested. In less than thirty years, population densities in Cuba, Guatemala, Costa Rica, and Honduras—all important wintering grounds for tropical migrants—will exceed the threshold of one hundred humans per square kilometer. Terborgh writes, "When this happens, the remaining natural habitat is doomed, and along with it the essential requirement for overwinter survival of many songbird species" (1989, 166). From this, it seems likely that Luther Goldman's list of birds no longer seen will continue to grow.

And what about my study subjects, bats? How are they faring in the face of man's burgeoning global impact? Throughout the world, populations of many species of bats have been declining in recent decades. In the United States, for example, five species—the gray bat, Indiana bat, lesser and greater long-nosed bats, and Townsend's big-eared bat (two subspecies)—and the Hawaiian subspecies of the hoary bat are currently classified as federally endangered because of their low population numbers. As I noted in chapter 11, eleven species of bats have gone extinct, and sixty-two species are currently listed as endangered by the International Union for the Conservation of Nature and Natural Resources. Most of the extinct or threatened species live (or lived) on Pacific islands where native humans have also had a strong negative impact on birds.

Fortunately, neither of the bats that I have studied intensively—*Carollia perspicillata* and *Leptonycteris curasoae*—are likely to go extinct in the foreseeable future, although *Leptonycteris* is highly vulnerable to massive population losses because of its habit of living in large numbers in a few widely scattered caves. *Leptonycteris* shares this roosting behavior, and hence its vulnerability, with a number of other New World bats, including the Mexican free-tailed bat. *Tadarida* is even more gregarious than *Leptonycteris.* Female free-tails sometimes form maternity colonies containing millions of individuals. Bracken Cave near San Antonio, Texas, for example, houses an estimated twenty million adult Mexican free-tails from March to October each year.

In late September 1988, Merlin Tuttle took me to watch the sunset emergence of Mexican free-tails from Bracken Cave. Well before dark, a steady stream of bats began to swirl out of the wide cave mouth and gain altitude before flying east and then south. The numbers of departing bats quickly swelled, and soon the area in front of the cave was filled with circling bats. I stood in the middle of this roaring mass of flashing wings and plucked a single bat out of the air with my bare hand. Gentle in disposition, the velvety little gray-brown creature sat quietly in my cupped hand for a minute before I gently released it to join its roost mates in the exit flight. As the number of bats continued to increase, they frequently began to collide with me. Reluctantly, I left the vortex of bats and climbed to a safer location to watch the exodus. In the fading sunlight, I could see streams of bats, looking like wispy clouds of smoke, twisting and turning toward the horizon. A pair of red-tailed hawks slowly intercepted a column of bats and plucked a meal from the air in the darkening sky.

As Merlin and I drove back toward Austin, I wondered how many future generations will be lucky enough to see the thrilling exit flight of twenty million Mexican free-tailed bats? Was I witnessing a scene comparable to the mass movements of millions of passenger pigeons and plains bison that occurred in North America a little over a century ago—scenes that are gone forever? When, I wondered, would our species stop its wanton destruction of our world's biological heritage?

It is sunset, and I am sitting on a large rock, still warm from the day's heat, near the Sierra Kino cave. As the sun slips behind Isla Tiburón, far to my left, shadows lengthen in the barranca I have just scrambled up. Small caves of brick-red lava rock that dot the Sierra's flanks begin to fade from view. A canyon wren sings its haunting song of descending notes, evoking a memory of the soulful music of Miles Davis, my high school jazz hero. Columnar cacti—the cardóns, pithayas, and saguaros—cast thin shadows on creosote bushes and gray-green paloverde trees in the gravelly flatlands below me. As the sunlight continues to fade, cactus flowers slowly begin to open, revealing their waxy white tepals and creamy masses of pollen-covered anthers. Before it is totally dark, *Leptonycteris* bats start leaving the cave. Their long, narrow wings slice through the air, making their kite-rippling sound. Flying rapidly, they hurtle down the canyon toward the cactus-filled flatlands.

It is dark now, but I can still hear Leptos swooshing past me. In my imagination, I picture two rangy Americans dressed in worn field clothes, sitting silently by a campfire somewhere in Baja California, picking at a meal of quail, rabbit, beans, and tortillas. They have set their rat traps before sunset and will wait a few hours before picking them up. Then they and their Mexican *mozos* (assistants) will wearily mount their horses and continue their journey south among cardóns in the cool of the night, a waxing moon illuminating their way. As they ride, they will occasionally glance up at the towering cardóns but know nothing of the connection between these giants and *Leptonycteris* bats nor of cardón's odd breeding system. For the rest of their lives, their jobs will be to describe systematically the flora and fauna of poorly known regions of North America. My job, nearly a century later, is to build on their work and begin to describe how some of those plants and animals interact. Both jobs, I think, are satisfying. Both jobs are certainly full of adventures. El Duende practically guarantees it.

APPENDIX 1

A Brief Overview of Bat Diversity

Bats belong to order Chiroptera ("hand wing"), which is the second largest order of mammals (after rodents). Over nine hundred species are currently recognized, and these are classified in two suborders: Megachiroptera, the megabats (flying foxes and their allies); and Microchiroptera, the microbats (echolocating bats). Suborder Megachiroptera contains only one family; suborder Microchiroptera contains seventeen families.

SUBORDER MEGACHIROPTERA

Old World Fruit Bats or Flying Foxes (Family Pteropodidae). This family of fruit- and nectar-feeding bats contains about 175 species. Its distribution encompasses the Old World tropics and subtropics, including many islands in the South Pacific. Ranging in size from about eleven grams to nearly two kilograms, these bats are important seed dispersers and/or pollinators of many species of tropical plants, including commercially important tree species such as durian, eucalyptus, figs, guava, kapok, mango, and neem. Genera mentioned in this book include *Hypsignathus, Macroglossus, Nyctimene, Pteropus,* and *Syconycteris.*

SUBORDER MICROCHIROPTERA

Sac-winged or Sheath-tailed bats (Family Emballonuridae). This pantropical family contains about fifty-one species of relatively small insectivores. They have long, narrow wings, which contain scent-producing glands in shallow sacs in some species. Genera mentioned in this book include *Balantiopteryx* and *Saccopteryx.*

Bumble-bee Bats (Family Craseonycteridae). This family contains a single species, *Craseonycteris thonglongyai,* found only in Thailand. Weighing only 1.5 grams, this is perhaps the world's smallest mammal.

Mouse-tailed Bats (Family Rhinopomatidae). This family's three insectivorous species inhabit arid to semiarid regions of the Old World. Their long, mouselike tails extend far beyond the short tail membrane.

Old World False Vampire Bats (Family Megadermatidae). This family of five species is widely distributed in the paleotropics. Its species are medium to relatively large in size (up to about 150 grams). The large species are carnivores and eat a variety of small vertebrates, including fish, frogs, reptiles, birds, rodents, and other bats.

Old World Slit-faced Bats (Family Nycteridae). Mostly African insectivorous and carnivorous bats, this family contains twelve species. Its name comes from a split nose leaf that extends along the snout from the nostrils. These and other bats with nose leaves emit their echolocation calls through their nostrils rather than through their mouths.

Old World Leaf-nosed Bats (Families Hipposideridae and Rhinolophidae). These two closely related families, containing a total of about 130 species, are widely distributed in the Old World tropics and subtropics; a few species live in temperate regions. All species have a nose leaf of varying complexity. All are insectivorous and often hunt for their prey in densely vegetated areas, where they glean insects from foliage.

New Zealand Short-tailed Bats (Family Mystacinidae). This family of small, omnivorous bats contains two species, one of which is likely to be recently extinct. These short-legged bats are as adept at feeding on insects, fruit, and nectar on the ground as they are at pursuing insects on the wing. They occur only in New Zealand.

New World Fishing Bats (Family Noctilionidae). The larger of the two members of this neotropical family is morphologically well-adapted for gaffing small fish with its large hind feet and long, sharp claws. The smaller species is insectivorous. The single genus is *Noctilio.*

Mustached Bats (Family Mormoopidae). This neotropical family of insectivorous bats, which shares a common ancestor with the Noctilionidae and Phyllostomidae, contains eight species. All have either funnel-like lips surrounded by hairs or, in the case of the ghost-faced bat, a complex series of wrinkles covering the face. These bats live in very large colonies in "hot" caves. Both genera *(Pteronotus, Mormoops)* are mentioned in this book.

New World Leaf-nosed Bats (Family Phyllostomidae). This neotropical family of about 150 species exhibits the greatest range of feeding adaptations of all bats (and, indeed, of all mammals). Traditional subfamilies tend to reflect different feeding adaptations as follows: subfamily Phyllostominae—insectivory and carnivory; some of the smallest (a few grams) and the largest species (190 grams) occur in this subfamily; some species have long, spear-shaped nose leaves; subfamilies Glossophaginae and Lonchophyllinae—flower-visiting and fruit-eating bats with elongated snouts and reduced teeth and nose leaves; subfamily Phyllonycterinae—flower-visiting bats found only in the Greater Antilles; subfamily Brachyphyllinae—fruit-eating bats endemic to the West Indies; subfamily Carolliinae—fruit-eating bats feeding mostly on forest understory fruits; subfamily Stenodermatinae—fruit-eating bats with flattened faces and feeding mostly on fruits in the forest canopy; subfamily Desmodontinae—true vampire bats that are physiologically and morphologically specialized (e.g., reduced nose leaves and very modified dentition) for feeding on the blood of birds and mammals. Genera mentioned in this book include *Trachops, Phyllostomus, Chrotopterus,* and *Vampyrum* (Phyllostominae); *Glossophaga* and *Leptonycteris* (Glossophaginae); *Carollia* (Carolliinae); *Artibeus* and *Uroderma* (Stenodermatinae); and *Desmodus* (Desmodontinae).

Vesper Bats (Family Vespertilionidae). This large, cosmopolitan family (about 330 species) is perhaps most familiar to temperate-zone inhabitants. Nearly all species are relatively small and insectivorous, but different species employ a wide variety of hunting styles (e.g. aerial pursuit, flycatching, gleaning, fish gaffing). Roosting habits range from highly gregarious cave dwellers to solitary foliage roosters. Many temperate species hibernate in the winter, whereas a few are long-distance migrants. Genera mentioned in this book include *Antrozous, Eptesicus, Myotis,* and *Pipistrellus.*

Funnel-eared Bats (Family Natalidae). This small family (four species) of very small insectivorous bats occurs only in the New World tropics. These bats often live in large colonies in "hot" caves with mormoopid bats. The single genus is *Natalus.*

Smoky Bats (Family Furipteridae). This is another small family (two species) of small bats that occur only in South America. One species lives in the Atacama Desert of coastal Peru.

Disk-winged Bats (Family Thyropteridae). These small bats (two species) have suction cups on their wrists and feet, which they use to cling to the insides of rolled up banana-like leaves in New World tropical forests.

Old World Sucker-footed Bats (Family Myzopodidae). The single species of this family occurs in Madagascar. Like the previous family, this bat has suction cups on its wrists and ankles and roosts in rolled up banana-like leaves.

Free-tailed Bats (Family Molossidae). Broadly distributed in the world's tropics and subtropics, this family contains nearly ninety species. All species are insectivorous, and many are fast-flying aerial pursuers that forage well away from vegetation. One species, the Mexican free-tailed bat, is perhaps the world's most gregarious mammal. Some of its colonies contain an estimated ten million to twenty million bats. Genera mentioned in the text include *Eumops, Molossus,* and *Tadarida.*

APPENDIX 2

Some Common and Scientific Names Used in the Text

Higher Taxon	Common Name	Scientific Name
Amphibians		
Bufonidae	marine toad	*Bufo marinus*
Leptodactylidae	Túngara frog	*Physalaemus pustulosus*
Dendrobatidae	strawberry frog	*Dendrobates pumilio*
Hylidae	red-eyed tree frog	*Agalychnis callidryas*
Reptiles		
Crocodylidae	Central American caiman	*Caiman crocodilus*
Teiidae	western whiptail lizard	*Cneimodophorus tigris*
Helodermatidae	gila monster	*Heloderma suspectum*
Iguanidae	basilisk	*Basiliscus basiliscus*
	common iguana	*Iguana iguana*
	ctenosaur	*Ctenosaura similis*
	zebra-tailed lizard	*Callisaurus draconoides*
	desert iguana	*Dipsosaurus dorsalis*
Boidae	common boa	*Boa constrictor*
Colubridae	lyre snake	*Trimorphodon biscutatus*
	eastern garter snake	*Thamnophis sirtalis*

	northern brown snake	*Storeria dekayi*
	banded water snake	*Nerodia fasciata*
	western coachwhip	*Masticophis flagellum*
Viperidae	eastern massasauga	*Sistrurus catenatus*
	sidewinder	*Crotalus cerastes*
	western diamondback rattlesnake	*Crotalus atrox*
	South American rattlesnake	*Crotalus durissus*
	fer-de-lance	*Bothrops asper*
	bushmaster	*Lachesis stenophrys*
Elapidae	taipan	*Oxyuranus scutellatus*
Mammals		
Didephidae	water opossum	*Chironectes minimus*
	common opossum	*Didelphis marsupialis*
	gray four-eyed opossum	*Philander opossum*
	brown four-eyed opossum	*Metachirus nudicaudatus*
	Central American woolly opossum	*Caluromys derbianus*
	Robinson's mouse opossum	*Marmosa robinsoni*
Peramelidae	northern brown bandicoot	*Isoodon macrourus*
Myrmecophagidae	northern tamandua	*Tamandua mexicana*
	silky anteater	*Cyclopes didactylus*
Dasypodidae	nine-banded armadillo	*Dasypus novemcinctus*
Pteropodidae	hammer-headed bat	*Hypsignathus monstrosus*
	Queensland tube-nosed bat	*Nyctimene robinsoni*
	short-faced fruit bat	*Cynopterus sphinx*
	black flying fox	*Pteropus alecto*
	Indian giant flying fox	*Pteropus giganteus*
	grey-headed flying fox	*Pteropus poliocephalus*
	little red flying fox	*Pteropus scapulatus*
	Queensland blossom bat	*Syconycteris australis*
	northern blossom bat	*Macroglossus minimus*
	dawn bat	*Eonycteris spelea*
	Wahberg's epauletted bat	*Epomophorus wahlbergi*
Emballonuridae	greater white-lined bat	*Saccopteryx bilineata*
	gray sac-winged bat	*Balantiopteryx plicata*
	northern ghost bat	*Diclidurus albus*
Noctilionidae	greater fishing bat	*Noctilio leporinus*

Mormoopidae	ghost-faced bat	*Mormoops megalophylla*
	common mustached bat	*Pteronotus parnellii*
	Davy's naked-backed bat	*Pteronotus davyi*
Phyllostomidae	California leaf-nosed bat	*Macrotus californicus*
	Waterhouse's bat	*Macrotus waterhousii*
	tiny big-eared bat	*Micronycteris minuta*
	fringe-lipped bat	*Trachops cirrhosus*
	pale spear-nosed bat	*Phyllostomus discolor*
	greater spear-nosed bat	*Phyllostomus hastatus*
	false vampire bat	*Vampyrum spectrum*
	woolly false vampire bat	*Chrotopterus auritus*
	common long-tongued bat	*Glossophaga soricina*
	western long-tongued bat	*Glossophaga morenoi*
	lesser long-nosed bat	*Leptonycteris curasoae*
	greater long-nosed bat	*Leptonycteris nivalis*
	Mexican long-tongued bat	*Choeronycteris mexicana*
	orange nectar bat	*Lonchophylla robusta*
	short-tailed fruit bat	*Carollia perspicillata*
	little yellow-shouldered bat	*Sturnira lilium*
	Heller's broad-nosed bat	*Platyrrhinus helleri*
	Honduran white bat	*Ectophylla alba*
	pygmy fruit-eating bat	*Artibeus phaeotis*
	Jamaican fruit-eating bat	*Artibeus jamaicensis*
	great fruit-eating bat	*Artibeus lituratus*
	common tent-making bat	*Uroderma bilobatum*
	common vampire bat	*Desmodus rotundus*
Natalidae	Mexican funnel-eared bat	*Natalus stramineus*
Hipposideridae	diadem horseshoe bat	*Hipposideros diadema*
Megadermatidae	Australian ghost bat	*Macroderma gigas*
Vespertilionidae	greater broad-nosed bat	*Nycticeius ruppellii*
	evening bat	*Nycticieus humeralis*
	western pipistrelle	*Pipistrellus hesperus*
	black myotis	*Myotis nigricans*
	cave myotis	*Myotis velifer*
	little brown bat	*Myotis lucifugus*
	gray bat	*Myotis grisescens*
	Indiana bat	*Myotis sodalis*
	big brown bat	*Eptesicus fuscus*
	spotted bat	*Euderma maculatum*
	eastern red bat	*Lasiurus borealis*

	hoary bat	*Lasiurus cinereus*
	Townsend's big-eared bat	*Idionycteris townsendii*
	pallid bat	*Antrozous pallidus*
Molossidae	western mastiff bat	*Eumops perotis*
	Wagner's bonneted bat	*Eumops glaucinus*
	black mastiff bat	*Molossus ater*
	little mastiff bat	*Molossus molossus*
	Sinaloan mastiff bat	*Molossus sinaloae*
	Mexican free-tailed bat	*Tadarida brasiliensis*
Leporidae	forest rabbit	*Sylvilagus brasiliensis*
	desert cottontail rabbit	*Sylvilagus audubonii*
	antelope jackrabbit	*Lepus alleni*
	black-tailed jackrabbit	*Lepus californicus*
Callithricidae	Geoffroy's tamarin	*Saguinus geoffroyi*
Cebidae	western night monkey	*Aotus lemurinus*
	red-backed squirrel monkey	*Saimiri oerstedii*
	white-faced capuchin	*Cebus capucinus*
	mantled howler	*Alouatta palliata*
	Central American spider monkey	*Ateles geoffroyi*
Sciuridae	Indian giant tree squirrel	*Ratufa indica*
	Indian giant flying squirrel	*Petaurista petaurista*
	red-tailed squirrel	*Sciurus granatensis*
	antelope ground squirrel	*Ammospermophilus harrisii*
	round-tailed ground squirrel	*Spermophilus tereticaudus*
Heteromyidae	Panamanian spiny pocket mouse	*Liomys adspersus*
	Salvin's spiny pocket mouse	*Liomys salvini*
	forest spiny pocket mouse	*Heteromys desmarestianus*
	Merriam's kangaroo rat	*Dipodomys merriami*
Dasyproctidae	Central American agouti	*Dasyprocta punctata*
	paca	*Agouti paca*
Echimyidae	Tome's spiny rat	*Proechimys semispinosus*
Muridae	vesper rat	*Nyctomys sumichrasti*
	white-throated woodrat	*Neotoma albigula*
	northern pygmy rice rat	*Oryzomys fulvescens*
	Talamancan rice rat	*Oryzomys talamancae*
	dusky rice rat	*Melanomys caliginosus*

	hispid cotton rat	*Sigmodon hispidus*
	common cane rat	*Zygodontomys brevicauda*
	white-tailed rat	*Uromys caudimaculatus*
Procyonidae	northern raccoon	*Procyon lotor*
	white-nosed coati	*Nasua narica*
	kinkajou	*Potos flavus*
	olingo	*Bassaricyon gabbii*
Mustelidae	tayra	*Eira barbara*
	black-footed ferret	*Mustela nigripes*
	neotropical river otter	*Lutra longicaudus*
Canidae	coyote	*Canis latrans*
Felidae	bobcat	*Felis rufus*
	ocelot	*Leopardus pardalis*
	margay cat	*Leopardus wiedii*
	jaguarundi	*Herpailurus yaguaroundi*
	puma	*Puma concolor*
	jaguar	*Panthera onca*
Tapiridae	Baird's tapir	*Tapirus bairdii*
Tayassuidae	collared peccary	*Tayassu tajacu*
	white-lipped peccary	*Dicotyles pecari*
Cervidae	white-tailed deer	*Odocoileus virginianus*
Delphinidae	common dolphin	*Delphinus delphis*

Plants

	almendro tree	*Dipteryx panamensis*
	balsa tree	*Ochroma lagopus*
	banana	*Musa sapientum*
	baobab tree	*Adansonia digitata*
	blue gum tree	*Eucalyptus teriticornis*
	boojum tree	*Fouquieria columnaris*
	brittlebush	*Encelia farinosa*
	calabash tree	*Crescentia cujete*
	candelabra cactus	*Myrtillocactus cochal*
	cardón cactus	*Pachycereus pringlei*
	ceiba or kapok tree	*Ceiba pentandra*
	coastal agave	*Agave shawii*
	creosote bush	*Larrea tridentata*
	durian tree	*Durio zibethinus*

elephant grass	*Hyparrhenia rufa*
elephant tree	*Bursera microphyllum*
espavé tree	*Anacardium excelsum*
Guanacaste tree	*Enterolobium cyclocarpum*
guava	*Psidium guajava*
gumbo limbo tree	*Bursera simaruba*
hecho cactus	*Pachycereus pecten-aboriginum*
hog plum tree	*Spondias mombin*
ironwood tree	*Olneya testota*
jackfruit tree	*Artocarpus heterophyllus*
laurél tree	*Cordia alliodora*
lawyer palm	*Calamus muelleri*
mango tree	*Mangifera indica*
manzanita	*Arctostaphylos pungens*
mesquite tree	*Prosopis glandulosa*
morning glory tree	*Ipomoea arborescens*
neem tree	*Azadirachta indica*
ocotillo	*Fouquieria splendens*
organ pipe cactus (pithaya)	*Stenocereus thurberi*
paloblanco tree	*Lysiloma candida*
palo santo tree	*Triplaris melaeno-dendron*
paloverde tree	*Cercidium microphyllum*
papaya	*Carica papaya*
pejibaye palm	*Bactris gasipaes*
petai tree	*Parkia speciosa*
pithaya agria cactus	*Stenocereus gummosus*
pochote tree	*Bombacopsis quinatum*
poro-poro tree	*Cochlospermum vitafolium*
saguaro cactus	*Carnegiea gigantea*
soursop tree	*Annona reticulata*
sotol	*Dasylirion wheeleri*
stinging tree	*Dendrocnide* sp.

Sources of names include: Dobat 1985; Figgis 1985; Fujita and Tuttle 1991; McMahon 1985; Reid 1997; Roberts 1989; Scott, Savage, and Robinson 1983; Stebbins 1985; and Strahan 1983.

References

GENERAL BOOKS ABOUT BATS

Altringham, J. D.
1996 *Bats, biology and behaviour.* Oxford University Press, Oxford.
Fenton, M. B.
1992 *Bats.* Facts on File, New York.
Hall, L., and G. Richards.
2000 *Flying foxes, fruit and blossom bats of Australia.* Krieger Publishing Company, Malabar, Florida.
Hill, J. E., and J. D. Smith.
1984 *Bats, a natural history.* University of Texas Press, Austin.
Kunz, T. H., and M. B. Fenton (eds.).
2003 *Bat ecology.* University of Chicago Press, Chicago.
Nowak, R. M.
1994 *Walker's bats of the world.* Johns Hopkins University Press, Baltimore.
Wilson, D. E.
1997 *Bats in question.* Smithsonian Institution Press, Washington, D.C.

CHAPTER SOURCES

Aldrich, J. W., and B. P. Bole, Jr.
1937 The birds and mammals of the western slope of the Azuero Peninsula. *Scientific Publications of the Cleveland Museum of Natural History* 7: 1–196.

Allen, G. M.
1939 *Bats.* Harvard University Press, Cambridge, Massachusetts.
Allen, W.
2001 *Green phoenix: Restoring the tropical forests of Guanacaste, Costa Rica.* Oxford University Press, New York.
Bates, H.
1863 *The naturalist on the river Amazons.* Vol. 2. John Murray, London.
Bonaccorso, F. J., and T. J. Gush.
1987 An experimental study of the feeding behaviour and foraging strategies of phyllostomid fruit bats. *Journal of Animal Ecology* 56: 907–920.
Boza, M.
1986 *Parques nacionales: Costa Rica.* Incafo, S. A., Madrid.
Ceballos, G., T. H. Fleming, C. Chavez, and J. Nassar.
1997 Population dynamics of *Leptonycteris curasoae* (Chiroptera: Phyllostomidae) in Jalisco, Mexico. *Journal of Mammalogy* 78: 1220–1230.
Cockrum, E. L., and Y. Petryszyn.
1991 The long-nosed bat, *Leptonycteris:* An endangered species in the Southwest? *Occasional Papers of the Museum, Texas Tech University* 142: 1–32.
Cosson, J-F., J-M. Pons, and D. Masson.
1999 Effects of forest fragmentation on frugivorous and nectarivorous bats in French Guiana. *Journal of Tropical Ecology* 15: 515–534.
Cox, P. A., T. Elmqvist, E. D. Pierson, and W. E. Rainey.
1991 Flying foxes as strong interactors in South Pacific island ecosystems: A conservation hypothesis. *Conservation Biology* 5: 448–454.
Ditmars, R. L.
1936 A *Vampyrum spectrum* is born. *Bulletin of the New York Zoological Society* 39: 162–163.
Dobat, K.
1985 *Blüten und fledermäuse.* Dr. Waldemar Kramer, Frankfurt am Main.
Dobzhansky, T.
1950 Evolution in the tropics. *American Scientist* 38: 209–221.
Eby, P.
1991 Seasonal movements of grey-headed flying foxes, *Pteropus poliocephalus* (Chiroptera: Pteropodidae), from two maternity camps in northern New South Wales. *Australian Wildlife Research* 18: 547–559.
Ehrenfeld, D.
1981 *The arrogance of humanism.* Oxford University Press, Oxford.
Ehrlich, P., and P. H. Raven.
1964 Butterflies and plants: A study in coevolution. *Evolution* 18: 586–608.
Enders, R. K.
1935 Mammalian life histories from Barro Colorado Island, Panama. *Bulletin of the Museum of Comparative Zoology* 78: 387–501.
Figgis, P. (ed.).
1985 *Rainforests of Australia.* Weldons Pty. Ltd., McMahons Point, Australia.

Fleming, T. H.

1971a *Artibeus jamaicensis:* Delayed embryonic development in a neotropical bat. *Science* 171: 402–404.

1971b Population ecology of three species of neotropical rodents. *Miscellaneous Publications of the University of Michigan Museum of Zoology* 143: 1–77.

1974 Population ecology of two species of Costa Rican heteromyid rodents. *Ecology* 55: 493–510.

1981 Winter roosting and foraging behaviour of the Pied Wagtail. *Ibis* 123: 463–476.

1985 Coexistence of five sympatric *Piper* (Piperaceae) species in a Costa Rican dry forest. *Ecology* 66: 688–700.

1988 *The short-tailed fruit bat, a study in plant-animal interactions.* University of Chicago Press, Chicago.

1993 Plant-visiting bats. *American Scientist* 81: 460–467.

2000 Pollination of cacti in the Sonoran Desert. *American Scientist* 88: 432–439.

Fleming, T. H., and E. R. Heithaus.

1981 Frugivorous bats, seed shadows, and the structure of tropical forests. *Biotropica* 13: 45–53.

Fleming, T. H., and C. F. Williams.

1990 Phenology, seed dispersal, and recruitment in *Cecropia peltata* (Moraceae) in Costa Rican tropical dry forest. *Journal of Tropical Ecology* 6: 163–178.

Fleming, T. H., E. T. Hooper, and D. E. Wilson.

1972 Three Central American bat communities: Structure, reproductive cycles, and movement patterns. *Ecology* 53: 655–670.

Fleming, T. H., R. A. Nuñez, and L. da Silviera Lobo Sternberg.

1993 Seasonal changes in the diets of migrant and non-migrant nectarivorous bats as revealed by carbon stable isotope analysis. *Oecologia* 94: 72–75.

Fleming, T. H., M. D. Tuttle, and M. A. Horner.

1996 Pollination biology and the relative importance of nocturnal and diurnal pollinators in three species of Sonoran Desert columnar cacti. *Southwestern Naturalist* 41: 257–269.

Fleming, T. H., S. Maurice, S. Buchmann, and M. D. Tuttle.

1994 Reproductive biology and the relative fitness of males and females in a trioecious cactus, *Pachycereus pringlei. American Journal of Botany* 81: 858–867.

Fleming, T. H., C. T. Sahley, J. N. Holland, J. D. Nason, and J. L. Hamrick.

2001 Sonoran Desert columnar cacti and the evolution of generalized pollination systems. *Ecological Monographs* 71: 511–530.

Frankie, G. W., H. G. Baker, and P. A. Opler.

1974 Comparative phenological studies of trees in tropical wet and dry forests in the lowlands of Costa Rica. *Journal of Ecology* 62: 881–919.

Fujita, M. S., and M. D. Tuttle.
1991 Flying foxes (Chiroptera: Pteropodidae): Threatened animals of key ecological and economic importance. *Conservation Biology* 5: 455–463.

Goldman, E. A.
1920 Mammals of Panama. *Smithsonian Miscellaneous Collections* 69 (5): 1–309.
1951 Biological investigations in Mexico. *Smithsonian Miscellaneous Collections* 115: 1–476.
Fieldbooks, books 2 and 3. No. 1070201. Smithsonian Institution Archives, Washington, D.C.

Gomez, L. D., and J. M. Savage.
1983 Searchers on that rich coast: Costa Rican field biology, 1400–1980. Pp. 1–11 in D. H. Janzen (ed.), *Costa Rican natural history.* University of Chicago Press, Chicago.

Goodwin, G. G.
1946 Mammals of Costa Rica. *Bulletin of the American Museum of Natural History* 87: 271–474.

Goodwin, G. G., and A. M. Greenhall.
1961 A review of the bats of Trinidad and Tobago. *Bulletin of the American Museum of Natural History* 122: 187–302.

Greenhall, A. M.
1968 Notes on the behavior of the false vampire. *Journal of Mammalogy* 49: 337–340.

Griffin, D. R.
1958 *Listening in the dark.* Yale University Press, New Haven, Connecticut.

Guevara, S., and J. Laborde
1993 Monitoring seed dispersal at isolated standing trees in tropical pastures: Consequences for local species availability. Pp. 319–338 in T. H. Fleming and A. Estrada (eds.), *Frugivory and seed dispersal: Ecological and evolutionary aspects.* Kluwer Academic Publishers, Dordrecht, Netherlands.

Handley, C. O., Jr., D. E. Wilson, and A. L. Gardner.
1991 Demography and natural history of the common fruit bat, *Artibeus jamaicensis,* on Barro Colorado Island, Panama. *Smithsonian Contributions to Zoology* 511: 1–173.

Hartmann, W. K.
1989 *Desert heart, chronicles of the Sonoran Desert.* Fisher Books, Tucson, Arizona.

Heideman, P. D.
1989 Delayed development in Fischer's pygmy fruit bat, *Haplonycteris fischeri,* in the Philippines. *Journal of Reproductive Fertility* 85: 1–20.

Heithaus, E. R., and T. H. Fleming.
1978 Foraging movements of a frugivorous bat, *Carollia perspicillata* (Phyllostomidae). *Ecological Monographs* 48: 127–143.

Heithaus, E. R., T. H. Fleming, and P. A. Opler.
1975 Patterns of foraging and resource utilization in seven species of bats in a seasonal tropical community. *Ecology* 56: 841–856.
Herbst, L. H.
1985 The role of nitrogen from fruit pulp in the nutrition of a frugivorous bat, *Carollia perspicillata. Biotropica* 18: 39–44.
Herrera, L. G., T. H. Fleming, and J. S. Findley.
1993 Geographic variation in the carbon composition of the pallid bat, *Antrozous pallidus,* and its dietary implications. *Journal of Mammalogy* 74: 601–606.
Horner, M. A., T. H. Fleming, and C. T. Sahley.
1998 Foraging behaviour and energetics of a nectar-feeding bat, *Leptonycteris curasoae* (Chiroptera: Phyllostomidae). *Journal of Zoology* 244: 575–586.
Howell, D. J.
1976 Plant-loving bats, bat-loving plants. *Natural History* 85 (2): 52–59.
Howell, D. J., and B. S. Roth.
1981 Sexual reproduction in agaves: The benefits of bats: Cost of semelparous advertising. *Ecology* 62: 3–7.
Hutson, A. M., S. P. Mickleburgh, and P. A. Racey (comps.).
2001 *Microchiropteran bats, global status survey and conservation action plan.* International Union for the Conservation of Nature and Natural Resources, Gland, Switzerland.
Janzen, D. H.
1970 Herbivores and the number of tree species in tropical forests. *American Naturalist* 104: 501–528.
Kingsolver, B.
1988 *The bean trees.* Harper and Row, New York.
Koopman, K. F., and D. W. Steadman.
1995 Extinction and biogeography of bats on 'Eua, Kingdom of Tonga. *American Museum Novitates* 3125: 1–13.
Krutch, J. W.
1961 *The forgotten peninsula.* University of Arizona Press, Tucson.
Lack, D.
1947 *Darwin's finches.* Harper and Brothers, New York.
Lake, A.
1989 *Somoza falling.* Houghton Mifflin, Boston.
Law, B. S.
1995 The effect of energy supplementation on the local abundance of the common blossom bat, *Syconycteris australis,* in south-eastern Australia. *Oikos* 72: 42–50.
Law, B. S., and M. Lean.
1999 Common blossom bats *(Syconycteris australis)* as pollinators in fragmented Australian tropical rainforest. *Biological Conservation* 91: 201–212.

MacArthur, R. H., and E. R. Pianka.
1966 On the optimal use of a patchy environment. *American Naturalist* 100: 603–609.
MacArthur, R. H., and E. O. Wilson.
1967 *The theory of island biogeography.* Monographs in Population Biology No. 1. Princeton University Press, Princeton, New Jersey.
McCracken, G. F., and J. W. Bradbury.
1981 Social organization and kinship in the polygynous bat *Phyllostomus hastatus. Behavioral Ecology and Sociobiology* 8: 11–34.
McMahon, J. A.
1985 *Deserts.* Alfred Knopf, New York.
Mickleburgh, S. P., A. M. Hutson, and P. A. Racey (comps.).
1992 *Old World fruit bats, an action plan for their conservation.* International Union for the Conservation of Nature and Natural Resources, Gland, Switzerland.
Mitchell, A. W.
1986 *The enchanted canopy.* William Collins Sons and Company, Glasgow.
Morrison, D. W.
1978 Foraging ecology and energetics of the frugivorous bat *Artibeus jamaicensis. Ecology* 59: 716–723.
1979 Apparent male defense of tree hollows in the fruit bat, *Artibeus jamaicensis. Journal of Mammalogy* 60: 11–15.
Morton, P. A.
1989 *Murciélagos tropicales americanos.* World Wildlife Fund, U.S.A.
Navarro, L.
2000 *Flores para Lucía, la murciélaga.* University of Texas Press, Austin, Texas.
Nelson, E. W.
1916 Lower California and its natural resources. *Memoirs of the National Academy of Sciences* 16: 1–194.
Nowick, A.
1973 Bats aren't all bad. *National Geographic* 143: 615–637.
Packer, C.
1994 *Into Africa.* University of Chicago Press, Chicago.
Petit, S., and L. Pors.
1996 Survey of columnar cacti and carrying capacity for nectar-feeding bats on Curaçao. *Conservation Biology* 10: 769–775.
Pettigrew, J. D.
1986 Flying primates? Megabats have the advanced pathway from eye to midbrain. *Science* 231: 1304–1306.
Ratcliffe, F.
1932 Notes on the fruit bats (*Pteropus* spp.) of Australia. *Journal of Animal Ecology* 1: 32–57.
1947 *Flying fox and drifting sand.* Angus and Robertson, Sydney.

Reid, F. A.
1997 *A field guide to the mammals of Central America and southeast Mexico.* Oxford University Press, New York.

Richards, G. C., and L. S. Hall.
1998 Conservation biology of Australian bats: Are recent advances solving the problem? Pp. 271–281 in T. H. Kunz and P. A. Racey (eds.), *Bat biology and conservation.* Smithsonian Institution Press, Washington, D.C.

Roberts, N. C.
1989 *Baja California plant field guide.* Natural History Publishing Company, La Jolla, California.

Russell, R.
1985 *Daintree, where the rainforest meets the reef.* Kevin Weldon and Associates, McMahons Point, Australia.

Sader, S. A., and A. T. Joyce.
1988 Deforestation rates and trends in Costa Rica, 1940–1983. *Biotropica* 20: 11–19.

Schaller, G. B.
1964 *The year of the gorilla.* Ballantine Books, New York.

Schoener, T. W.
1971 Theory of foraging strategies. *Annual Review of Ecology and Systematics* 2: 369–404.

Schulze, M. D., N. E. Seavy, and D. F. Whitacre.
2000 A comparison of the phyllostomid bat assemblages in undisturbed neotropical forest and in forest fragments of a slash-and-burn farming mosaic in Petén, Guatemala. *Biotropica* 32: 174–184.

Scott, N. J., J. M. Savage, and D. C. Robinson.
1983 Checklist of reptiles and amphibians. Pp. 367–374 In D. H. Janzen (ed.), *Costa Rican natural history.* University of Chicago Press, Chicago.

Simmons, N. B.
1995 Bat relationships and the origin of flight. Pp. 27–43 in P. A. Racey and S. M. Swift (eds.), *Ecology, evolution and behaviour of bats.* Clarendon Press, Oxford.

Slud, P.
1960 The birds of finca "La Selva," a tropical wet forest locality. *Bulletin of the American Museum of Natural History* 121: 49–148.

Snow, D.
1971 Evolutionary aspects of fruit-eating by birds. *Ibis* 113: 194–202.

Spencer, H. J., and T. H. Fleming.
1989 Roosting and foraging behaviour of the Queensland tube-nosed bat, *Nyctimene robinsoni*: Preliminary radio-tracking observations. *Australian Wildlife Research* 16: 413–420.

Stebbins, R. C.
1985 *A field guide to western reptiles and amphibians.* Houghton Mifflin, Boston.

Sterling, K. B.
1977 *The last of the naturalists: The career of C. Hart Merriam.* Arno Press, New York.
Stone, D. E.
1988 The Organization for Tropical Studies (OTS): A success story in graduate training and research. Pp. 143–187 in F. Almed and C. M. Pringle (eds.), *Tropical rainforests: Diversity and conservation.* California Academy of Sciences, San Francisco.
Strahan, R. (ed.).
1983 *The Australian Museum complete book of Australian mammals.* Angus and Robertson Publishers, London.
Sykes, B.
2001 *The seven daughters of Eve.* W. W. Norton, New York.
Terborgh, J.
1989 *Where have all the birds gone?* Princeton University Press, Princeton, New Jersey.
Time.
1977 Nicaragua: Somoza's reign of terror. 109, no. 11 (14 March): 29–30.
Tuttle, M. D.
1991 Bats, the cactus connection. *National Geographic* 179 (6): 131–140.
Vehrencamp, S. L., F. G. Stiles, and J. W. Bradbury.
1977 Observations on the foraging behavior and avian prey of the neotropical carnivorous bat, *Vampyrum spectrum. Journal of Mammalogy* 58: 469–478.
von Helversen, O.
1993 Adaptations of flowers to the pollination by glossophagine bats. Pp. 41–59 in W. Barthlott et al. (eds.), *Plant-animal interactions in tropical environments.* Museum Alexander Koenig, Bonn.
Wafer, L.
1729 *A new voyage and description of the isthmus of America.* 3d ed. London.
Ward, P., and A. Zahavi.
1973 The importance of certain assemblages of birds as "information-centres" for food-finding. *Ibis* 115: 517–534.
Wetmore, A. Papers. Collection Division 9: Panama fieldwork files, box 153, folder 7. Smithsonian Institution Archives, Washington, D.C.
Whittaker, R. J., and S. H. Jones.
1994 The role of frugivorous bats and birds in the rebuilding of a tropical forest ecosystem, Krakatau, Indonesia. *Journal of Biogeography* 21: 245–258.
Wilkinson, G. S.
1984 Reciprocal food sharing in the vampire bat. *Nature, London* 308: 181–184.
1985a The social organization of the common vampire bat. 1. Pattern and cause of association. *Behavioral Ecology and Sociobiology* 17: 111–121.

1985b The social organization of the common vampire bat. 2. Mating system, genetic structure, and relatedness. *Behavioral Ecology and Sociobiology* 17: 123–134.

Wilkinson, G. S., and T. H. Fleming.

1996 Migration and evolution of lesser long-nosed bats, *Leptonycteris curasoae,* inferred from mitochondrial DNA. *Molecular Ecology* 5: 329–339.

Williams, C. F.

1986 Social organization of the bat *Carollia perspicillata* (Chiroptera: Phyllostomidae). *Ethology* 71: 265–282.

Subject Index

Name Index

Compositor: Integrated Composition Systems
Text: 10/13 Aldus
Display: Aldus
Printer and Binder: Edwards Brothers, Inc.